FUNDAMENTALS OF SURFACE AND THIN FILM ANALYSIS

FUNDAMENTALS OF SURFACE AND THIN FILM ANALYSIS

Leonard C. Feldman
AT&T Bell Laboratories

James W. Mayer
Department of Materials Science and Engineering
Cornell University

P T R Prentice Hall
Englewood Cliffs, New Jersey 07632

Library of Congress Cataloging-in-Publication Data

Feldman, Leonard C.
  Fundamentals of surface and thin film analysis.

  Includes bibliographies and index.
  1. Surfaces (Technology)—Analysis. 2. Thin
films—Analysis. I. Mayer, James W., 1930–
II. Title.
QD506.F39 1986   530.4′1   86-2479
ISBN 0-444-00989-2

Current printing (last digit):
10 9 8 7 6 5

Manufactured in the United States of America

© 1986 by P T R Prentice-Hall, Inc.
A Simon & Schuster Company
Englewood Cliffs, New Jersey 07632

Printed in the United States of America
10 9 8 7 6 5 4 3 2 1

ISBN 0-13-500570-1

Prentice-Hall International (UK) Limited, *London*
Prentice-Hall of Australia Pty. Limited, *Sydney*
Prentice-Hall Canada Inc., *Toronto*
Prentice-Hall Hispanoamericana, S. A., *Mexico*
Prentice-Hall of India Private Limited, *New Delhi*
Prentice-Hall of Japan, Inc., *Tokyo*
Simon & Schuster Asia Pte. Ltd., *Singapore*
Editora Prentice-Hall do Brasil, Ltda., *Rio de Janeiro*

To our wives and children,
Betty, Greg, and Dana,
and
Betty, Jim, John, Frank, Helen, and Bill

CONTENTS

PREFACE

One of the striking features of modern technology is the important role of the surface and near-surface regions of materials. Ion implantation and pulsed electron beams and lasers are used to modify composition and structure, thin films are deposited from a variety of sources, epitaxial layers are grown from molecular beams, and oxidation and catalytic reactions are studied under controlled conditions. The key to these methods has been the widespread availability of analytical techniques that are sensitive to the composition and structure of the outer surface of solids.

This book focuses on the physics underlying the techniques used to analyze the surface region of materials. New analytical techniques continue to emerge to meet technological requirements; yet all are based on a few processes that govern the interactions of particles and radiation with matter. The book addresses the fundamentals of these processes. From an understanding of processes that determine the energies and intensities of the emitted radiation, the application to materials analysis follows directly. The book is written for the materials scientist interested in the use of electron spectroscopies or ion beam analysis for sample characterization, for the materials analyst who needs information on techniques that are available outside his laboratory, and particularly for the seniors and graduate students who will use this new generation of analytical techniques in their research.

Modern materials analysis is based on the measurement of particles and radiation that emerge from a solid that is irradiated by photons, electrons, or heavy particles. Identification of elements is based on the energy of the emergent radiation; atomic concentration is based on the intensity of this emergent radiation. Coulomb scattering from the atom (Rutherford backscattering), formation of inner shell vacancies in the electronic structure (X-ray photoelectron spectroscopy), and transitions between levels (electron microprobe and Auger electron spectroscopy) are fundamental to materials analysis. They are discussed in detail in this book. Composition depth profiles are obtained with heavy-ion sputtering in combination with surface sensitive techniques (electron spectroscopies and secondary ion mass spectrometry). Depth profiles are also found from energy loss of light ions (Rutherford backscattering and prompt nuclear analysis). The structure of surface layers is discussed from diffraction (low energy electron diffraction) and particle (channeling) viewpoints. The book closes with a discussion of the use of nuclear techniques for composition analysis.

Because the book is concerned with the fundamentals of modern surface analysis, we have provided derivations of the basic parameters — energy and cross section or transition probability. The book is organized so that we start with the classical concepts of atomic collisions as applied to Rutherford scattering (Chapter 2), energy loss (Chapter 3), sputtering (Chapter 4), channeling (Chapter 5) and electron interactions (Chapter 6). An overview is given of diffraction techniques (Chapter 7) used in electron and X-ray structural analysis. Wave mechanics is required for an understanding of photoelectric cross sections and fluorescence yields; we review the wave equation and perturbation theory in Chapter 8. We use these relations to discuss photoelectron spectroscopy (Chapter 9), radiative transitions (Chapter 10), and nonradiative transitions (Chapter 11). Chapter 12 discusses the application of nuclear techniques to thin film analysis.

Both of us have been engaged heavily in research programs centered on materials analysis; we realize the need for a comprehensive treatment of the analytical techniques used in thin film and near surface analysis. We find that a basic understanding of the processes is important in a field that is rapidly changing. Instruments may change, but the fundamental processes will remain the same.

We have used the material in this book in a senior/graduate level course at Cornell University, as well as in short courses to scientists and engineers in industry. We wish to thank the Cornell undergraduate and graduate students as well as our colleagues for their comments and suggestions for revisions of the text. We acknowledge with gratitude the contribution of Dr. R. Pretorius of the National Accelerator Centre, Faure. He is largely responsible for the content of Chapter 12. We thank Jane Jorgensen and Ali Avcisoy for their drawings and artwork.

SYMBOLS USED IN THE TEXT

Symbol	Name	Equation, Table, or Figure	Page
a	Thomas–Fermi screening radius	Eq. (4.13)	75
A	mass number	—	28
a_0	Bohr radius, $a_0 = \hbar^2/me^2 = 0.529$ Å	Eq. (1.12)	9
b	impact parameter	Figure 2.6	21
C^b	bulk atomic concentration	—	89
C^S	surface atomic concentration	—	89
d	distance of closest approach	Eq. (2.19)	24
e	electronic charge, $e^2 = 14.4$ eV Å	Eq. (1.5)	5
E_B	electron binding energy	—	131
E_0	energy of incident particle	Eq. (2.5)	15
E_1	energy of detected particle	Eq. (2.5)	15
E_2	energy transferred to the target atom	Eq. (2.7)	16
f_{AB}	sputtering factor	Eq. (4.44)	89
F_D	density of deposited energy	Eq. (4.2)	73
I	average excitation energy of an electron	Figure 3.4	44
K	kinematic factor	Eq. (2.5)	15
m	rest mass of the electron	Eq. (1.9)	7
M_1	mass of the incident particle	Figure 2.1	15
M_2	mass of the target atom	Figure 2.1	15
N	atomic density (atoms/cm³)	—	6
N_A	Avogadro's number	Eq. (1.6)	6

Symbol	Name	Equation, Table, or Figure	Page
N_S	number of target atoms/cm^2	—	19
Q	number of incident particles	—	20
Q_D	number of detected particles	Eq. (2.10)	20
R	nuclear radius	Eq. (2.23)	28
R	charged particle range	Eq. (6.21)	142
S_n	nuclear stopping cross section	Eq. (4.29)	74
$[S]$	backscattering energy loss factor	Eq. (3.20)	48
t	thickness	—	19
T	energy transferred in a collision	Eq. (3.3)	43
u^2	mean square thermal vibration	—	158
u	unified mass unit$)$	Eq. (1.7)	6
v_0	Bohr velocity, $v_0 = e^2/\hbar = 2.2 \times 10^8$ cm/s	Eq. (1.13)	10
W	transition probability per unit time	Eq. (8.24)	90
Y	sputtering yield	Eq. (4.1)	71
Z	atomic number	—	13
Γ	energy width of an excited state	Eq. (10.9)	241
ϵ	stopping cross section	—	40
θ	scattering angle	Figure 2.1	15
θ_D	Debye temperature	Eq. (7.4)	158
λ	wavelength	Eq. (1.3)	5
λ	escape depth	Eq. (6.1)	129
μ	linear attenuation coefficient	—	196
μ/ρ	mass attenuation coefficient (cm^2/g)	—	197
ρ	mass density (g/cm^3)	—	6
σ_e	electron impact cross section	Eq. (6.11)	131
σ_{ph}	photoeffect cross section	Eq. (8.37)	194
σ_{sc}	screened scattering cross section	Eq. (2.22)	27
$\sigma(\theta)$	scattering cross section	Eq. (2.9)	19
τ	mean life of an excited state	Eq. (10.9)	239
ϕ	recoil angle	Figure 2.1	15
χ	minimum yield in channeling	Figure 5.3	103
ψ_c	critical angle in axial channeling	Eq. (5.9)	102
ψ_p	critical angle in planar channeling	Eq. (5.14)	103
Ω	detector solid angle	Eq. (2.7)	19
Ω_B^2	Bohr value of energy straggling	Eq. (3.28)	58
ω_p	plasmon frequency	Eq. (6.14)	133

FUNDAMENTALS OF SURFACE AND THIN FILM ANALYSIS

1

AN OVERVIEW: CONCEPTS, UNITS, AND THE BOHR ATOM

1.1 Introduction

Our understanding of the structure of atoms and atomic nuclei is based on scattering experiments. Such experiments determine the interaction of a beam of elementary particles — photons, electrons, neutrons, ions, etc. — with the atom or nucleus of a *known* element. (In this context we consider all incident radiation as "particles" including photons.) The classical example is Rutherford scattering in which the scattering of incident alpha particles from a thin foil confirmed the picture of an atom as composed of a small positively charged nucleus surrounded by electrons in circular-like orbits. As these fundamental interactions became understood, the scientific community recognized the importance of the inverse process — namely, measuring the interaction of radiation with targets of unknown elements to determine atomic composition. Such determinations are called "materials analysis." For example, alpha particles scatter from different nuclei in a distinct and well-understood manner. Measurements of the intensity and energy of the scattered particles provides a direct measure of elemental composition. The emphasis in this book is twofold: (1) to describe in a quantitative fashion those fundamental interactions that are used in modern materials analysis and (2) to illustrate the use of this understanding in practical materials analysis problems.

1

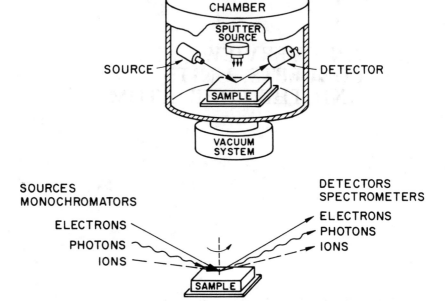

Figure 1.1 Schematic of radiation sources and detectors used in the materials analysis techniques described in this book. Analytical probes are represented by almost any combination of source and detected radiation, i.e., electrons in and electrons out or photons in and electrons out. Many chambers will also contain sample erosion facilities such as an ion sputtering as well as an evaporation apparatus for deposition of material onto a clean substrate under vacuum.

The emphasis in modern materials analysis is generally directed toward the structure and composition of the surface and outer microns of the material. The emphasis comes from the realization that the surface and near-surface regions control many of the mechanical and chemical properties of solids: corrosion, friction, wear, adhesion, and fracture. In addition, one can tailor the composition and structure of the outer layers by directed-energy processes utilizing lasers or electron and ion beams, as well as by more conventional techniques such as oxidation and diffusion.

In modern materials analysis one is concerned with the source of radiation, the beam of particles — photons, electrons, neutrons, or ions —, the interaction cross section, the emergent radiation, and the detection system. The primary interest of this book is the interaction of the beam with the material to be analyzed with emphasis on the energies and intensities of emitted radiation. As we will show, the energy of the emitted particle provides the signature or identification of the atom, the intensity tells the amount of atoms — i.e., sample composition. The radiation source and the detection system are important topics in their own right; however, the main emphasis in this book is on the ability to carry out quantitative materials analysis which depends upon interactions within the target.

There is an impressive array of experimental techniques available for the analysis of solids. Figure 1.1 gives the flavor of the possible combinations.

In some cases the same incident and emergent radiation is employed (we will use the general terms of radiation and particles for photons, electrons, ions, etc.). Listed below are examples with commonly used acronymns in parentheses.

Electron in, electron out: Auger electron spectroscopy (AES)

Ion in, ion out: Rutherford backscattering (RBS)

X-ray in, X-ray out: X-ray fluorescence spectroscopy (XRF)

In other cases the incident and emergent radiation differ as indicated below:

X-ray in, electron out: X-ray photoelectron spectroscopy (XPS)

Electron in, X-ray out: electron microprobe analysis (EMA)

Ion in, target ion out: secondary ion mass spectrometry (SIMS)

A beam of particles incident on a target either scatters elastically or causes an electronic transition in an atom. The scattered particle or the energy of the emergent radiation contains the signature of the atom. The energy levels in the transition are characteristic of a given atom; hence, measurement of the energy spectrum of the emergent radiation allows identification of the atom.

The number of atoms per cm^2 in a target is found from the relation between the number, I, of incident particles and the number of interactions. The term cross section is used as a quantitative measure of an interaction between an incident particle and an atom. The cross section σ per atom for a given process is defined through the probability, P,

$$P = \frac{\text{Number of interactions}}{\text{Number of incident particles}}. \tag{1.1}$$

For a target containing Nt atoms per unit area perpendicular to an incident beam of I particles, the number of interactions is $I\sigma Nt$. From a knowledge of detection efficiency for measuring the emergent radi-

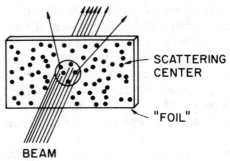

BEAM

Figure 1.2 Illustration of the concept of cross section and scattering. The central circle defines a unit area of a foil containing a random array of scattering centers. In this example there are five scattering centers/unit area. Each scattering center has an area (cross section for scattering) of $\sim \frac{1}{20}$ unit area; therefore, the probability of scattering is $\frac{5}{20}$ or 0.25. Then a fraction (0.25 in this example) of the incident beam will be scattered, i.e., 2 out of 8 trajectories in the drawing. A measure of the fraction of scattered beam is a measure of the probability [$P = Nt\sigma$, Eq.(1.1)]. If the foil thickness and density are known, Nt can be calculated, yielding a direct measure of the cross-section.

ation containing the signature of the transition, the number of atoms and ultimately the target composition can be found (Figure 1.2).

The information required from analytical techniques are species identification, concentration, depth distribution, and structure. The available analytical techniques have different capabilities to meet these requirements. The choice of analysis method depends upon the nature of the problem. For example, chemical bonding information can be obtained from techniques that rely upon transitions in the electronic structure around the atoms — the electron spectroscopies. Structural determination is found from diffraction or particle channeling techniques.

In the following chapters we are mostly concerned with materials analysis in the outer microns of the sample — the surface and near-surface regime. We emphasize the energy of the emergent radiation as an identification of the element and the intensity of the radiation as a measure of the amount of material. These are the basic principles that provide the foundation for the different analytical techniques.

1.2 Energies, Units, and Particles

With few exceptions it is the measurement of energy that is the hallmark of materials analysis. Although the SI (or MKS) system of units gives the joule (J) as the derived unit of energy, the electron volt (eV) is the traditional unit in materials analysis. The joule is so large that it is inconvenient as a unit in atomic interactions. The electron volt is defined as the kinetic energy gained by an electron accelerated from rest through a potential difference of 1 V. Since the charge on the electron is 1.602×10^{-19} Coulomb and a joule is a Coulomb-volt,

$$1 \text{ eV} = 1.602 \times 10^{-19} \text{ J}. \tag{1.2}$$

Commonly used multiples of the eV are the keV (10^3 eV) and MeV (10^6 eV).

In determination of crystal structure by X-ray diffraction, the diffraction conditions are determined by atomic spacing and hence the wavelength of the photon. The wavelength λ is the ratio of c/v, where c is the speed of light and v is the frequency, so that the energy E is

$$E = hv = \frac{hc}{\lambda} = \frac{12.4 \text{ keV}}{\lambda \text{ (Å)}}, \tag{1.3}$$

where Planck's constant $h = 4.136 \times 10^{-15}$ eV s, $c = 2.998 \times 10^8$ m/s, λ is in units of Å, and 1 angstrom (Å) is 10^{-10} m.

The energies of the emergent radiation provide the signature of the transition, the cross section determines the strength of the inter-action. Although the MKS unit for cross sectional area is m², the measured values are often given in cm². It is convenient to use cgs rather than SI units in relations involving the charge on the electron. The usefulness of cgs units is clear when considering the Coulomb force between two charged particles with Z_1 and Z_2 units of electronic charge separated by a distance r,

$$F = \frac{Z_1 Z_2 e^2 k_c}{r^2} \tag{1.4}$$

where the Coulomb law constant $k_c = (1/4\pi\epsilon_0) = 8.988 \times 10^9$ m/farad (F) in the SI system (where 1 F = 1 amp s/V) and $k_c = 1$ in the cgs system. In cgs units the value of $e = 4.803 \times 10^{-10}$ stat C, which leads to a quick conversion factor for Coulomb interactions of

$$e^2 \cong 1.44 \times 10^{-13} \text{ MeV cm} = 14.4 \text{ eV Å}. \tag{1.5}$$

In this book we use $k_c = 1$ and rely on Eq. (1.5) for e^2.

The masses of particles, given in kg in SI units, are generally expressed in unified mass units (u) which replaces the older atomic mass units, amu. The neutral carbon atom with 6 protons, 6 neutrons, and 6 electrons is the reference for the unified mass unit (u) which is defined as $\frac{1}{12}$ the mass of the neutral ^{12}C carbon atom (where the superscript indicates the mass number 12). Avogadro's number N_A is the number of atoms or molecules in a mole (mol) of a substance and is defined as the number of atoms of an element needed to equal its atomic mass in grams. Avogadro's number of ^{12}C atoms is equivalent to a mass of exactly 12 g and the mass of one ^{12}C atom is 12 mass units. The value of Avogadro's number, the number of atoms/mol, is

$$N_A = 6.0220 \times 10^{23}, \qquad (1.6)$$

and the unified mass unit u, the reciprocal of N_A, is

$$u = \frac{1}{N_A} = \frac{1\text{ g}}{6.022 \times 10^{23}} = 1.661 \times 10^{-24}\text{ g}. \qquad (1.7)$$

A large part of this book is devoted to the extraction of depth profiles — the atomic composition or impurity concentration as a function of depth below the surface. In terms of length measurement, the most natural unit is the angström, Å, where

$$1\text{ Å} = 0.1\text{ nm} = 10^{-10}\text{ m}.$$

For example the separation between atoms in a solid is about 3 Å.

The measurement techniques give depth scales in terms of areal density, the number Nt of atoms per cm², where t is the thickness and N the atomic density. For elemental solids the atomic density and the mass density ρ in g/cm³ are related by

$$N = N_A \rho / A, \qquad (1.8)$$

where A is the atomic mass number and N_A is Avogadro's number. Another unit of thickness is the mass absorption coefficient usually expressed as g/cm², the product of the mass density and linear thickness.

Each nucleus is characterized by a definite atomic number Z and mass number A. The atomic number Z is the number of protons and hence the number of electrons in the neutral atom; it reflects the atomic properties of the atom. The mass number gives the number of nucleons, protons and neutrons; isotopes are nuclei (often called nuclides) with the same Z and different A. The current practice is to represent each nucleus by the chemical name with the mass number as a superscript, i.e., ^{12}C. The chemical atomic weight (or atomic mass) of elements as listed in the periodic table gives the average mass, i.e., the average of the stable isotopes weighted by their abundance. Carbon, for example, has an atomic weight of 12.011 which reflects the 1.1% abundance of ^{13}C. Appendix 8 lists the elements and their relative abundance, atomic weight, atomic density, and specific gravity.

Table 1.1 Mass Energies of Particles and Light Nuclei

Particle	Symbol	Mass energy (MeV)
Electron	e or e^-	0.511
Proton	p	938.3
Neutron	n	939.6
Deuteron	d or $^2H^+$	1875.6
Alpha	α or $^4He^{++}$	3727.4

The masses of particles may be expressed in terms of energy through the Einstein relation

$$E = mc^2 \qquad (1.9)$$

which associates 1 J of energy with $1/c^2$ kg of mass. The mass of an electron is 9.11×10^{-31} kg, which is equivalent to an energy

$$E = (9.11 \times 10^{-31} \text{ kg})(2.998 \times 10^8 \text{ m/s})^2 = 8.188 \times 10^{-14} \text{ J} \\ = 0.511 \text{ MeV.} \qquad (1.10)$$

In materials analysis, the incident radiation is usually photons, electrons, neutrons, or low mass ions (neutral atoms stripped of one or more electrons). For example, the proton is an ionized hydrogen atom, the alpha particle is a helium atom with the two electrons removed ($^4He^+$ or $^4He^{++}$ is often used to denote a helium atom with one or two electrons removed) and the deuteron, $^2H^+$, is a neutron and proton bound together. The mass energies of some of these particles are given in Table 1.1. In analytical applications, the velocities of these particles are generally well below 10^7 m/s; hence relativistic effects do not enter, and the masses are independent of velocity.

1.3 Particle-Wave Duality and Lattice Spacings

In materials analysis, one tends to view the incident beam and emergent radiation as discrete particles — photons, electrons, ions. On the other hand, the interactions of radiation with matter and, in particular, the cross section for a transition is often based on the wave aspect of the radiation.

This wave–particle duality was of major concern in the early development of modern physics. The photon and the electron provide examples of the wave and particle nature of matter. For example, in the photoelectric effect, light behaves as if it were particlelike with each photon interacting with an atom to give up its energy, $E = h\nu$, to an electron which can escape from the solid. The diffraction of X-rays, on the other hand, from planes of atoms satisfies wave interference conditions.

Electrons and their diffraction from the crystal surfaces constitute a sensitive probe of surface structure. The classical, particle behavior of electrons on the other hand is illustrated in their deflection in electric and magnetic fields. One can associate both a wavelength λ and momentum p with the motion of an electron. Their connection is given by the de Broglie relation

$$\lambda = h/p, \tag{1.11}$$

where h is Planck's constant. Distances between lattice planes are of the order of an angström (0.1 nm). For diffraction, the wavelengths of electrons are of comparable magnitude. The electron velocity, $v = p/m$, corresponding to a wavelength of 1 Å is

$$v = \frac{h}{m\lambda} = \frac{6.6 \times 10^{-34}}{9.1 \times 10^{-31} \times 10^{-10}} = 7.25 \times 10^6 \text{ ms}^{-1},$$

where MKS units are used with $h = 6.6 \times 10^{-34}$ J s. The energy is

$$E = \frac{1}{2} mv^2 = \frac{9.1 \times 10^{-31}(7.25 \times 10^6)^2}{2} = 2.39 \times 10^{-17} \text{ J} = 149.5 \text{ eV}.$$

Electron diffraction studies of surfaces use electrons with low energies, between 40 eV and 150 eV, giving rise to the acronym LEED — low energy electron diffraction.

Energies of 1.0–2.0 MeV He$^+$ are commonly used in materials analysis; here the wavelengths are orders of magnitude smaller than the lattice spacing and the interactions of helium ions with solids are described on the basis of particle rather than wave behavior. For helium atoms, an energy of 2 MeV corresponds to a wavelength of 10^{-4} Å whereas distances between nearest neighbor atoms in a solid are the order of 2–5 Å.

The distances between atoms and atomic planes can be calculated from a known lattice constant and crystal structure. Aluminum, for example, contains $\sim 6 \times 10^{22}$ atoms/cm³, has a lattice constant of 4.04 Å, and has a face-centered cubic (fcc) crystal structure. One monolayer of atoms on the (100) surface then contains an areal atom density of 2 atoms/(4.04 Å)² or 1.2×10^{15} atoms/cm². Almost all solids have monolayer density values of 5×10^{14}/cm² to 2×10^{15}/ cm² on major crystallographic surfaces. In a loose way a monolayer is usually thought of as 10^{15} atoms/cm². The sensitivity of various surface spectroscopics is often measured in units of monolayers or atoms/cm²; bulk impurity determinations are usually given in atoms/cm³.

1.4 The Bohr Model

The identification of atomic species from the energies of emitted radiation was developed from the concepts of the Bohr model of the hydrogen atom. Particle scattering experiments established that the atom could be treated as a positively charged nucleus surrounded by a cloud of electrons. Bohr assumed that the electrons could move in stable circular orbits called stationary states and would emit radiation only in the transition from one stable orbit to another. The energies of the orbits were derived from the postulate that the angular momentum of the electron around the nucleus is an integral multiple of $h/2\pi$ ($h/2\pi$ is written as $\hbar$). In this section we give a brief review of the Bohr atom, which provides useful relations for simple estimates of atomic parameters.

For a single electron of mass m in a circular orbit of radius r about a fixed nucleus of charge Ze, the balance between the Coulomb and centripetal forces leads to

$$\frac{Ze^2}{r^2} = m \frac{v^2}{r}. \tag{1.12}$$

Bohr assumed that the angular momentum, mvr, has values given by an integer n times $\hbar$,

$$mvr = n\hbar.$$

From the above equations we have

$$v^2 = \frac{n^2\hbar^2}{m^2r^2} = \frac{Ze^2}{mr},$$

which can be rewritten to give the radii r_n of allowed orbits

$$r_n = \frac{\hbar^2 n^2}{mZe^2}. \tag{1.13}$$

For hydrogen, $Z = 1$, the radius a_0 of the smallest orbit, $n = 1$, is known as the Bohr radius and is given by

$$a_0 = \frac{\hbar^2}{me^2} = 0.5292 \times 10^{-10} \text{ m} = 0.53 \text{ Å}, \tag{1.14}$$

and the Bohr velocity of the electron in this orbit is

$$v_0 = \frac{\hbar}{ma_0} = \frac{e^2}{\hbar} = 2.188 \times 10^8 \text{ cm/s}. \tag{1.15}$$

The ratio of v_0 to the speed of light is known as the fine-structure constant α given by

$$\alpha = \frac{v_0}{c} = \frac{1}{137.04}. \tag{1.16}$$

The energy of the electron is defined here as zero when it is at rest at infinity. The potential energy, PE, of an electron in the Coulomb force field has a negative value, $-Ze^2/r$, in this convention and the kinetic energy (KE) is $Ze^2/2r$ [Eq. (1.12)] so that the total energy E is

$$E = \text{KE} + \text{PE} = \frac{Ze^2}{2r} - \frac{Ze^2}{r} = \frac{-Ze^2}{2r}$$

or for the nth orbital is

$$E_n = \frac{-Ze^2}{2r_n} = \frac{-me^4Z^2}{2\hbar^2 n^2} = -\frac{E_0 Z^2}{n^2}. \tag{1.17}$$

The electron bound to a positively charged nucleus has a discrete set of allowed energies,

$$E_n = \frac{-13.58Z^2}{n^2} \text{ eV}. \tag{1.18}$$

The binding energy E_B of such an electron is the positive value, $13.58Z^2/n^2$. The numerical value of the $n = 1$ state represents the energy required to ionize the atom by complete removal of the electron; for hydrogen, the ionization energy is 13.58 eV.

The Bohr theory does lead to the correct values for energy levels observed in H spectral lines. The nomenclature introduced by Bohr persists in the vocabulary of atomic physics: orbitals, Bohr radius, and Bohr velocity. The quantities v_0 and a_0 are used repeatedly in this book as they are the natural units to evaluate atomic processes.

Problems

1.1. Calculate the density of atoms in C (graphite), Si, Fe and Au. Express your answer in atoms/cm^3.

1.2. Calculate the number of atoms/cm² in one monolayer of Si(100) and Si(111) and in W(100).

1.3. Calculate the wavelength (in Å) of a 1 MeV He ion, a 150 eV electron and a 1 keV Ar ion.

1.4. Show that $e^2 = 14.4$ eV Å.

1.5. Find the ratio of velocity of a 1 MeV He ion to the Bohr velocity.

1.6. Use the literature and notes to state the incoming radiation (particles) and outgoing radiation (particles) in the following spectroscopies:

    AES —Auger electron spectroscopy
    RBS —Rutherford backscattering spectrometry
    SIMS—secondary ion mass spectroscopy
    XPS —X-ray photoelectron spectroscopy
    XRF —X-ray fluorescence
    SEM —scanning electron μ probe
    NRA —nuclear reaction analysis

In each case, state the nature of the atomic transition involved.

1.7. In this book we repeatedly make estimates using the Bohr model of the atom. Test the validity of this approximation by calculating the K shell binding energy, E_K ($n = 1$); the L shell binding energy, E_L ($n = 2$); the wavelength at the K shell absorption edge, ($\hbar\omega = E_K$); and the K X-ray energy ($E_K - E_L$) for Si, Ni and W. Compare with the accurate values given in the appendixes.

1.8. The Auger process, discussed in Chapter 11, corresponds to an electron transition involving the emission of an Auger electron with the energy ($E_K - E_L - E_L$) where K is for $n = 1$ and L is for $n = 2$. Show that in the Bohr model $ak = 1/\sqrt{2}$ where a is the K shell radius a_0/Z and $\hbar k$ is the momentum of the outgoing electron.

1.9. An incident photon of sufficient energy can eject an electron from an inner shell orbit. Such an excited atom may relax by rearranging the outer electrons to fill the vacancy. This is said to occur in a time equivalent to the orbital time. Calculate this characteristic atomic time for Ni. In later chapters we will show that the inverse of this time may be thought of as a rate for the Auger process.

References

1. R. D. Evans, *The Atomic Nucleus* (McGraw-Hill, New York, 1955).

2. J. D. McGervey, *Introduction to Modern Physics* (Academic Press, New York, 1971).

3. F. K. Richtmyer, E. H. Kennard, and J. N. Cooper, *Introduction to Modern Physics* (McGraw-Hill, New York, 1969), 6th ed.

4. R. L. Sproull and W. A. Phillips, *Modern Physics* 3rd Edition (John Wiley and Sons, New York 1980).

5. P. A. Tipler, *Modern Physics* (Worth Publishers, New York, 1978).

6. R. T. Weidner and R. L. Sells, *Elementary Modern Physics*, 3rd Edition (Allyn and Bacon, Boston, MA, 1980).

7. J. C. Willmott, *Atomic Physics* (John Wiley and Sons, New York, 1975).

2

ATOMIC COLLISIONS AND BACKSCATTERING SPECTROMETRY

2.1 Introduction

The model of the atom is that of a cloud of electrons surrounding a positively charged central core — the nucleus — that contains Z protons and $A - Z$ neutrons, where Z is the atomic number and A the mass number. Single-collision, large-angle, scattering of alpha particles by the positively charged nucleus not only established this model but also forms the basis for one modern analytical technique, Rutherford backscattering spectrometry. In this chapter we will develop the physical concepts underlying Coulomb scattering of a fast light ion by a more massive stationary atom.

Of all the analytical techniques, Rutherford backscattering spectrometry is perhaps the easiest to understand and to apply because it is based on classical scattering in a central-force field. Aside from the accelerator which provides a collimated beam of MeV particles (usually ^{4}He ions) the instrumentation is simple (Figure 2.1a). Semiconductor nuclear particle detectors are used which have an output voltage pulse proportional to the energy of the particles scattered from the sample into the detector. The technique is also the most quantitative as MeV He ions undergo close-impact scattering collisions which are governed by the well-known Coulomb repulsion between the positively charged nuclei of the projectile and target atom. The kinematics of the collision and the scattering cross section

13

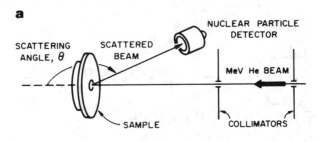

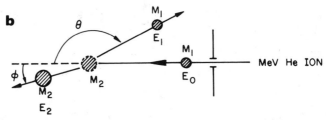

Figure 2.1 (a) Schematic of the experimental setup for Rutherford backscattering. A collimated beam of He ions is incident on a planar sample. Particles scattered to an angle θ are detected by a solid state nuclear particle detector. All this apparatus must be under vacuum. (b) Schematic representation of an elastic collision between a projectile of mass M_1, velocity v, and energy E_0 and a target mass M_2 which is initially at rest. After the collision, the projectile and the target mass have velocities and energies v_1, E_1 and v_2, E_2, respectively. The angles θ and ϕ are positive as shown. All quantities refer to a laboratory frame of reference.

are independent of chemical bonding, and hence backscattering measurements are insensitive to electronic configuration or chemical bonding within the target. To obtain information on the electronic configuration, one must employ analytical techniques such as photoelectron spectroscopy that rely on transitions in the electron shells.

In this chapter we treat scattering between two positively charged bodies of atomic number Z_1 and Z_2. The convention is to use the subscript 1 to denote the incident particle and the subscript 2 to denote the target atom. We first consider energy transfers during collisions as they provide the identity of the target atom. Then we calculate the scattering cross section which is the basis of the quantitative aspect of Rutherford backscattering. Here we are concerned with scattering from atoms on the sample surface or from thin layers. In a later chapter (Chapter 3) we discuss depth profiles.

2.2 Kinematics of Elastic Collisions

In Rutherford backscattering spectrometry, monoenergetic particles in the incident beam collide with target atoms and are scattered backwards into the detector–analysis system which measures the energies of the particles. In the collision, energy is transferred from the moving particle to the stationary target atom; the reduction in energy of the scattered particle depends on the masses of incident and target atoms and provides the signature of the target atoms.

The energy transfers or kinematics in elastic collisions between two isolated particles can be solved fully by applying the principles of conservation of energy and momentum. For an incident energetic particle of mass M_1, the values of the velocity and energy are v and E_0 ($E_0 = \frac{1}{2}M_1 v^2$) while the target atom of mass M_2 is at rest. After the collision, the values of the velocities v_1 and v_2 and energies E_1 and E_2 of the projectile and target atoms are determined by the scattering angle θ and recoil angle ϕ. The notation and geometry for the laboratory system of coordinate are given in Figure 2.1b.

Conservation of energy and conservation of momentum parallel and perpendicular to the direction of incidence are expressed by the equations

$$\tfrac{1}{2}M_1 v^2 = \tfrac{1}{2}M_1 v_1^2 + \tfrac{1}{2}M_2 v_2^2, \tag{2.1}$$

$$M_1 v = M_1 v_1 \cos\theta + M_2 v_2 \cos\phi, \tag{2.2}$$

$$0 = M_1 v_1 \sin\theta - M_2 v_2 \sin\phi. \tag{2.3}$$

Eliminating ϕ first and then v_2, one finds the ratio of particle velocities

$$\frac{v_1}{v} = [\pm (M_2^2 - M_1^2 \sin^2\theta)^{1/2} + M_1 \cos\theta]/(M_2 + M_1). \tag{2.4}$$

The ratio of the projectile energies for $M_1 < M_2$, where the plus sign holds, is

$$\frac{E_1}{E_0} = \left[\frac{(M_2^2 - M_1^2 \sin^2\theta)^{1/2} + M_1 \cos\theta}{M_2 + M_1} \right]^2. \tag{2.5}$$

The energy ratio, called the kinematic factor $K = E_1/E_0$, shows that the energy after scattering is determined only by the masses of the particle and target atom and the scattering angle. A subscript is usually added to K, i.e., K_{M_2}, to indicate the target atom mass. Tabulations of K values for different M_2 and θ values are given in Appendix 1 and are shown in Figure 2.2 for $\theta = 170°$. Such tables and figures are used routinely in the design of backscattering experiments. A summary of scattering relations is given in Table 3.1.

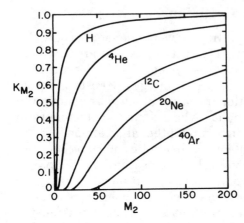

Figure 2.2 Graphic representation of the kinematic factor K_{M_2} [Eq. (2.5)] for a scattering angle $\theta = 170°$ as a function of the target mass M_2 for ^{1}H, ^{4}He, ^{12}C, ^{20}Ne, and ^{40}Ar.

For direct backscattering through 180°, the energy ratio has its lowest value given by

$$\frac{E_1}{E_0} = \left(\frac{M_2 - M_1}{M_2 + M_1}\right)^2,$$ (2.6a)

and at 90° is given by

$$\frac{E_1}{E_0} = \frac{M_2 - M_1}{M_2 + M_1}.$$ (2.6b)

In collisions where $M_1 = M_2$, the incident particle is at rest after the collision with all the energy transferred to the target atom, a feature well known in billiards. For $\theta = 180°$, the energy E_2 transferred to the target atom has its maximum value given by

$$\frac{E_2}{E_0} = \frac{4M_1M_2}{(M_1 + M_2)^2},$$ (2.7)

with the general relation given by

$$\frac{E_2}{E_0} = \frac{4M_1M_2}{(M_1 + M_2)^2} \cos^2 \phi.$$ (2.7′)

In practice, when a target contains two types of atoms that differ in their masses by a small amount ΔM_2, the experimental geometry is adjusted to produce as large a change ΔE_1 as possible in the measured energy E_1 of the projectile after the collision. A change of ΔM_2 (for fixed $M_1 < M_2$) gives the largest change of K when $\theta = 180°$. Thus $\theta = 180°$ is the preferred location for the detector ($\theta \cong 170°$ in practice because of detector size) — an experimental arrangement that has given the method its name of *backscattering* spectrometry.

The ability to distinguish between two types of target atoms that differ in their masses by a small amount ΔM_2 is determined by the ability of the experimental energy measurement system to resolve small differences ΔE_1 in the energies of backscattered particles. Most MeV ^{4}He backscattering apparatuses use a surface-barrier, solid state nuclear particle detector for measurement of the energy spectrum of the backscattered particles. As shown in Figure 2.3, the nuclear particle detector operates by the collection of the hole–electron pairs created by the incident particle in the depletion region of the reverse-biased Schottky barrier diode. The statistical fluctuations in

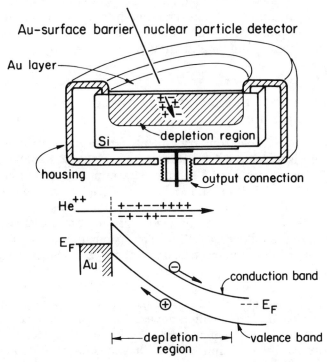

Figure 2.3 Schematic diagram of the operation of a gold surface barrier nuclear particle detector. The upper portion of the figure shows a cutaway sketch of the silicon disc with gold film mounted in the detector housing. The lower portion shows an alpha particle, He^{++} ion, forming holes and electrons over its penetration path. The energy band diagram of a reverse biased detector (positive polarity on n-type silicon) shows the electrons and holes swept apart by the high electric field within the depletion region.

the number of electron–hole pairs produce a spread in the output signal resulting in a finite resolution. Energy resolution values of 10–20 keV, full width at half maximum (FWHM), for MeV ^{4}He ions can be obtained with conventional electronic systems. For example, backscattering analysis with 2.0 MeV ^{4}He particles can resolve isotopes up to about mass 40 (the chlorine isotopes, for example). Around target masses close to 200, the mass resolution is about 20, which means that one cannot distinguish among atoms between ^{181}Ta and ^{201}Hg.

In backscattering measurements the signals from the semiconductor detector electronic system are in the form of voltage pulses. The heights of the pulses are proportional to the incident energy of the particles. The pulse height analyzer stores pulses of a given height in a given voltage bin or channel (hence the alternate description, multichannel analyzer). The channel numbers are calibrated in terms of the pulse height and hence there is a direct relationship between channel number and energy.

2.3 Rutherford Backscattering Spectrometry — An Example

In backscattering spectrometry, the mass differences of different elements and isotopes can be distinguished. Figure 2.4 shows a backscattering spectrum from a sample with approximately one monolayer of ^{63,65}Cu, ^{107,109}Ag, and ^{197}Au. The various elements are well separated in the spectrum and easily identified. Absolute coverages can be determined from a knowledge of the absolute cross section discussed in the following section. The spectrum is an illustration of the fact that heavy elements on a light substrate can be investigated at coverages well below a monolayer.

The limits of the mass resolution are indicated by the peak separation of the various isotopes. In Figure 2.4, the different isotopic masses of ^{63}Cu and ^{65}Cu which have a natural abundance of 69 and 31% have values of the energy ratio, or kinematic factor K, of 0.777 and 0.783 for $\theta = 170°$ and incident ^{4}He ions ($M_1 = 4$). For incident energies of 2.5 MeV, the energy difference of particles from the two masses is 17 keV, an energy value close to the energy resolution (FWHM = 14.8 keV) of the semiconductor particle-detector system. Consequently, the signals from the two isotopes overlap to produce the peak and shoulder shown in the figure. Particles scattered from the two Ag isotopes, ^{107}Ag and ^{109}Ag, have too small an energy difference, 6 keV, and hence the signal from Ag appears as a single peak.

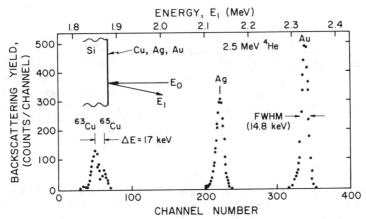

Figure 2.4 Backscattering spectrum ($\theta = 170°$) for 2.5 MeV He ions incident on a target with approximately one monolayer coverage of Cu, Ag, and Au. The spectrum is displayed as raw data from multichannel analyzer, i.e., in counts/channel and channel number.

2.4 Scattering Cross Section and Impact Parameter

The identity of target atoms is established by the energy of the scattered particle after an elastic collision. The number N_s of target atoms per unit area is determined by the probability of a collision between the incident particles and target atoms as measured by the total number Q_D of detected particles for a given number Q of particles incident on the target in the geometry shown in Figure 2.5. The connection between the number of target atoms N_s and detected particles is given by the scattering cross section. For a thin target of thickness t with N atoms/cm³, $N_s = Nt$.

The differential scattering cross section, $d\sigma/d\Omega$, of a target atom for scattering an incident particle through an angle θ into a differential solid angle $d\Omega$ centered about θ is given by

$$\frac{d\sigma(\theta)}{d\Omega} \cdot d\Omega \cdot N_s = \frac{\text{Number of particles scattered into } d\Omega}{\text{Total number of incident particles}}.$$

In backscattering spectrometry, the detector solid angle Ω is small (10^{-2} steradian or less) so that one defines an average differential scattering cross section $\sigma(\theta)$,

$$\sigma(\theta) = \frac{1}{\Omega} \int_{\Omega} \frac{d\sigma}{d\Omega} \cdot d\Omega, \qquad (2.8)$$

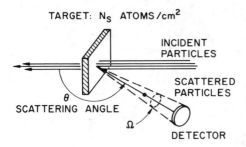

TARGET: N_S ATOMS/cm^2

INCIDENT PARTICLES

SCATTERED PARTICLES

θ

SCATTERING ANGLE

Ω

DETECTOR

Figure 2.5 Simplified layout of a scattering experiment to demonstrate the concept of the differential scattering cross section. Only primary particles that are scattered within the solid angle $d\Omega$ spanned by the detector are counted.

where $\sigma(\theta)$ is usually called the scattering cross section. For a small detector of area A a distance l from the target, the solid angle is given by A/l^2 in steradians.

For the geometry of Figure 2.5, the number N_s of target atoms/cm^2 is related to the yield Y or the number Q_D of detected particles (in an ideal, 100% efficient detector that subtends a solid angle Ω) by

$$Y = Q_D = \sigma(\theta) \cdot \Omega \cdot Q \cdot N_s, \qquad (2.9)$$

where Q is the total number of incident particles in the beam. The value of Q is determined by the time integration of the current of charged particles incident on the target. From Eq. (2.9) one can also note that the name "cross section" is appropriate in that $\sigma(\theta)$ has the dimensions of an area.

The scattering cross section can be calculated from the force that acts during the collision between the projectile and target atom. For most cases in backscattering spectrometry, the distance of closest approach during the collision is well within the electron orbit so that the force can be described as an unscreened Coulomb repulsion of two positively charged nuclei with charge given by the atomic numbers Z_1 and Z_2 of the projectile and target atoms. We derive this unscreened scattering cross section in Section 2.5 and treat the small correction due to electron screening in Section 2.7.

The deflection of the particles in a one-body formulation is treated as the scattering of particles by a center of force in which the kinetic energy of the particle is conserved. As shown in Figure 2.6, we can define the impact parameter b as the perpendicular distance between the incident particle path and the parallel line through the target nucleus. Particles incident with impact parameters between b and $b + db$ will be scattered through angles between θ and $\theta + d\theta$. With central forces, there must be complete symmetry around the axis of the beam so that

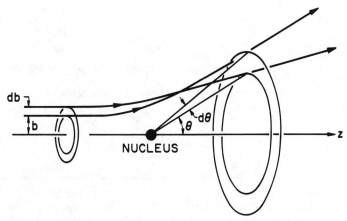

Figure 2.6 Schematic illustrating the number of particles between b and $b + db$ being deflected into an angular region $2\pi \sin \theta \, d\theta$. The cross section is, by definition, the proportionality constant; $2\pi b \, db = -\sigma(\theta)2\pi \sin \theta \, d\theta$.

$$2\pi b \, db = -\sigma(\theta) \cdot 2\pi \sin \theta \, d\theta. \tag{2.10}$$

In this case the scattering cross section $\sigma(\theta)$ relates the initial uniform distribution of impact parameters to the outgoing angular distribution. The minus sign indicates that an increase in the impact parameter results in less force on the particle so that there is a decrease in the scattering angle.

2.5 Central Force Scattering

The scattering cross section for central force scattering can be calculated for small deflections from the impulse imparted to the particle as it passes the target atom. As the particle with charge Z_1e approaches the target atom, charge Z_2e, it will experience a repulsive force that will cause its trajectory to deviate from the incident straight line path (Figure 2.7). The value of the Coulomb force F at a distance r is given by

$$F = \frac{Z_1 Z_2 e^2}{r^2}. \tag{2.11}$$

Let $\mathbf{p}_1$ and $\mathbf{p}_2$ be the initial and final momentum vectors of the particle. From Figure 2.8 it is evident that the total change in momen-

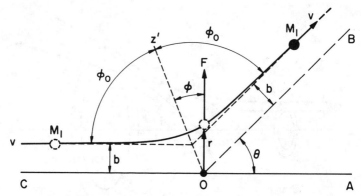

Figure 2.7 Rutherford scattering geometry. The nucleus is assumed to be a point charge at the origin O. At any distance r the particle experiences a repulsive force. The particle travels along a hyperbolic path that is initially parallel to line OA a distance b from it and finally parallel to line OB, which makes an angle θ with OA. The scattering angle θ can be related to the impact parameter b by classical mechanics.

tum $\Delta \mathbf{p} = \mathbf{p}_2 - \mathbf{p}_1$ is along the z′ axis. In this calculation the magnitude of the momentum does not change. From the isosceles triangle formed by $\mathbf{p}_1$, $\mathbf{p}_2$, and $\Delta \mathbf{p}$ shown in Figure 2.8 we have

$$\frac{\frac{1}{2}\Delta p}{M_1 v} = \sin \frac{\theta}{2}$$

or

$$\Delta p = 2M_1 v \sin \frac{\theta}{2}. \tag{2.12}$$

We now write Newton's law for the particle, $\mathbf{F} = d\mathbf{p}/dt$, or

$$d\mathbf{p} = \mathbf{F}\, dt.$$

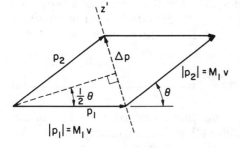

Figure 2.8 Momentum diagram for Rutherford scattering. Note that $|\mathbf{p}_1| = |\mathbf{p}_2|$, i.e., for elastic scattering the energy and speed of the projectile are the same before and after the collision.

The force F is given by Coulomb's law, and is in the radial direction. Taking components along the z' direction, and integrating to obtain Δp, we have

$$\Delta p = \int (dp)_{z'} = \int F \cos \phi \, dt = \int F \cos \phi \, \frac{dt}{d\phi} \, d\phi, \qquad (2.13)$$

where we have changed the variable of integration from t to the angle ϕ. We can relate $dt/d\phi$ to the angular momentum of the particle about the origin. Since the force is central (i.e., acts along the line joining the particle and the nucleus at the origin), there is no torque about the origin, and the angular momentum of the particle is conserved. Initially, the angular momentum has the magnitude $M_1 vb$. At a later time, it is $M_1 r^2 \, d\phi/dt$. Conservation of angular momentum thus gives

$$M_1 r^2 \frac{d\phi}{dt} = M_1 vb$$

or

$$\frac{dt}{d\phi} = \frac{r^2}{vb}.$$

Substituting this result and Eq. (2.11) for the force in Eq. (2.13), we obtain

$$\Delta p = \frac{Z_1 Z_2 e^2}{r^2} \int \cos \phi \, \frac{r^2}{vb} \, d\phi = \frac{Z_1 Z_2 e^2}{vb} \int \cos \phi \, d\phi$$

or

$$\Delta p = \frac{Z_1 Z_2 e^2}{vb} (\sin \phi_2 - \sin \phi_1). \qquad (2.14)$$

From Figure 2.7, $\phi_1 = -\phi_0$ and $\phi_2 = +\phi_0$, where $2\phi_0 + \theta = 180°$. Then $\sin \phi_2 - \sin \phi_1 = 2 \sin(90° - \tfrac{1}{2}\theta)$. Combining Eqs. (2.12) and (2.14) for Δp, we have

$$\Delta p = 2M_1 v \sin \frac{\theta}{2} = \frac{Z_1 Z_2 e^2}{bv} \, 2 \cos \frac{\theta}{2}. \qquad (2.15a)$$

This gives the relationship between the impact parameter b and scattering angle:

$$b = \frac{Z_1 Z_2 e^2}{M_1 v^2} \cot \frac{\theta}{2} = \frac{Z_1 Z_2 e^2}{2E} \cot \frac{\theta}{2}. \qquad (2.15b)$$

From Eq. (2.10), the scattering cross section can be expressed as

$$\sigma(\theta) = \frac{-b}{\sin\theta} \frac{db}{d\theta},\tag{2.16}$$

and from the geometrical relations $\sin\theta = 2\sin(\theta/2)\cos(\theta/2)$ and $d\cot(\theta/2) = -\frac{1}{2}d\theta/\sin^2(\theta/2)$,

$$\sigma(\theta) = \left(\frac{Z_1 Z_2 e^2}{4E}\right)^2 \frac{1}{\sin^4\theta/2}.\tag{2.17}$$

This is the scattering cross section originally derived by Rutherford. The experiments by Geiger and Marsden in 1911–1913 verified the predictions that the amount of scattering was proportional to $(\sin^4\theta/2)^{-1}$ and E^{-2}. In addition they found that the number of elementary charges in the center of the atom is equal to roughly half the atomic weight. This observation introduced the concept of the atomic number of an element which describes the positive charge carried by the nucleus of the atom. The very experiments that gave rise to the picture of an atom as a positively charged nucleus surrounded by orbiting electrons has now evolved into an important materials analysis technique.

For Coulomb scattering the distance of closest approach, d, of the projectile to the scattering atom is given by equating the incident kinetic energy E to the potential energy at d or

$$d = \frac{Z_1 Z_2 e^2}{E}.\tag{2.18}$$

The scattering cross section can be written as $\sigma(\theta) = (d/4)^2/\sin^4\theta/2$, which gives for 180° scattering $\sigma(180°) = (d/4)^2$. For 2 MeV He ions $(Z_1 = 2)$ incident on Ag $(Z_2 = 47)$.

$$d = \frac{(2)(47) \cdot (14.4 \text{ eV Å})}{(2 \times 10^6 \text{ eV})} = 6.8 \times 10^{-4} \text{ Å},$$

a value much smaller than the Bohr radius $a_0 = \hbar^2/m_e e^2 = 0.53$ Å, and the K shell radius of Ag, $a_0/47 \cong 10^{-2}$ Å. Thus the use of an unscreened cross section is justified. The cross section for scattering to 180° is

$$\sigma(\theta) = (6.8 \times 10^{-4} \text{ Å})^2/16 = 2.89 \times 10^{-8} \text{ Å}^2,$$

a value of 2.89×10^{-24} cm² or 2.89 barns, where the barn $= 10^{-24}$ cm².

2.6 Scattering Cross Section: Two-Body

In the previous section, we used central forces in which the energy of the incident particle was unchanged through its trajectory. From the kinematics (Section 2.2), we know that the target atom recoils from its initial position, and hence the incident particle loses energy in the collision. The scattering is elastic in that the total kinetic energy of the particles is conserved. Therefore, the change in energy of the scattered particle can be appreciable; for $\theta = 180°$ and ^{4}He ($M_1 = 4$) scattering from Si ($M_2 = 28$), the kinematic factor $K = (24/32)^2 = 0.56$ indicates that nearly one-half the energy is lost by the incident particle. In this section we evaluate the scattering cross section while including this recoil effect. The derivation of the center of mass to laboratory transformation is given in Section 2.10.

The scattering cross section [Eq. (2.17)] was based on the one-body problem of the scattering of a particle by a *fixed* center of force. However, the second particle is not fixed but recoils from its initial position as a result of the scattering. In general, the two-body central force problem can be reduced to a one-body problem by replacing M_1 by the reduced mass $\mu = M_1 M_2 / (M_1 + M_2)$. The matter is not quite that simple as indicated in Figure 2.9. The laboratory scattering angle θ differs from the angle θ_c calculated from the equivalent, reduced-mass, one-body problem. The two angles would only be the same if the second remains stationary during the scattering (i.e., $M_2 \gg M_1$).

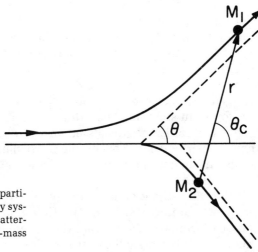

Figure 2.9 Scattering of two particles as viewed in the laboratory system showing the laboratory scattering angle θ and the center-of-mass scattering angle θ_c.

The relation between the scattering angles is

$$\tan \theta = \frac{\sin \theta_c}{\cos \theta_c + M_1/M_2},$$

derived in Eq. (2.24). The transformation gives

$$\sigma(\theta) = \left(\frac{Z_1 Z_2 e^2}{4E}\right)^2 \frac{4}{\sin^4 \theta} \frac{(\{1 - [(M_1/M_2) \sin \theta]^2\}^{1/2} + \cos \theta)^2}{\{1 - [(M_1/M_2) \sin \theta]^2\}^{1/2}}, \quad (2.19)$$

which can be expanded for $M_1 \ll M_2$ in a power series to give

$$\sigma(\theta) = \left(\frac{Z_1 Z_2 e^2}{4E}\right)^2 \left[\sin^{-4}\frac{\theta}{2} - 2\left(\frac{M_1}{M_2}\right)^2 + \cdots\right], \quad (2.20)$$

where the first term omitted is of the order of $(M_1/M_2)^4$. It is clear that the leading term gives the cross section of Eq. (2.17), and that the corrections are generally small. For He ($M_1 = 4$) incident on Si ($M_2 = 28$), $2 (M_1/M_2)^2 \cong 4\%$, even though appreciable energy is lost in the collision. For accurate quantitative analysis this correction should be included as the correction can be appreciable for scattering from light atoms such as carbon or oxygen. Cross section values given in Appendix 2 are based on Eq. (2.19). A summary of scattering relations and cross section formulae are given in Table 3.1.

2.7 Deviations from Rutherford Scattering at Low and High Energy

The derivation of the Rutherford scattering cross section is based on a Coulomb interaction potential $V(r)$, between the particle Z_1 and target atom Z_2. This assumes that the particle velocity is sufficiently large so that the particle penetrates well inside the orbitals of the atomic electrons. Then scattering is due to the repulsion of two positively charged nuclei of atomic number Z_1 and Z_2. At larger impact parameters found in small-angle scattering of MeV He ions or low-energy, heavy ion collisions (discussed in the chapter on sputtering), the incident particle does not completely penetrate through the electron shells and hence the innermost electrons screen the charge of the target atom.

We can estimate the energy where these electron screening effects become important. For the Coulomb potential to be valid for backscattering we require that the distance of closest approach d be smaller than the K shell electron radius, which can be estimated as a_0/Z_2, where $a_0 = 0.53$ Å, the Bohr radius. Using Eq. (2.18) for the

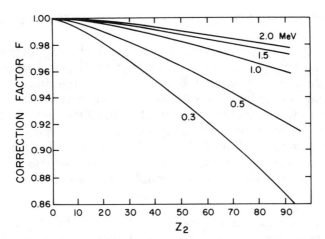

Figure 2.10 Correction factor F, which describes the deviation from pure Rutherford scattering due to electron screening for He^+ scattering from atoms, Z_2, at a variety of incident kinetic energies. [From J. L'Ecuyer, J. A. Davies, and N. Matsunami, *Rad. Eft.* **47**, 229 (1980).]

distance of closest approach d, the requirement for d less than the radius sets a lower limit on the energy of the analysis beam and requires that

$$E > Z_1 Z_2^2 e^2 / a_0.$$

This energy value corresponds to ~ 10 keV for He scattering from silicon and ~ 340 keV for He scattering from Au $(Z_2 = 79)$. However, deviations from the Rutherford scattering cross section occur at energies greater than the screening limit estimate given above as part of the trajectory is always outside of the electron cloud.

In Rutherford backscattering analysis of solids, the influence of screening can be treated to first order[1] by using a screened Coulomb cross section σ_{sc} obtained by multiplying the scattering cross section $\sigma(\theta)$ given in Eqs. (2.19) and (2.20) by a correction factor F,

$$\sigma_{sc} = \sigma(\theta)F, \tag{2.21}$$

where $F = (1 - 0.049\, Z_1 Z_2^{4/3}/E)$ and E is given in keV. Values of the correction factor are given in Figure 2.10. With 1 MeV ^{4}He ions incident on Au atoms, the correction factor corresponds to only 3%. Consequently, for analysis with 2 MeV ^{4}He ions, the screening correction can be neglected for most target elements. At lower analysis

[1] J. L'Ecuyer et al., *Nucl. Instr. and Methods* **160**, 337 (1979).

energies or with heavier incident ions, screening effects may be important.

At higher energies and small impact parameter values, there can be large departures from the Rutherford scattering cross section due to the interaction of the incident particle with the nucleus of the target atom. Deviations from Rutherford scattering due to nuclear interactions will become important when the distance of closest approach of the projectile-nucleus system becomes comparable to R, the nuclear radius. Although the size of the nucleus is not a uniquely defined quantity, early experiments with alpha-particle scattering indicated that the nuclear radius could be expressed as

$$R = R_0 A^{1/3}, \tag{2.22}$$

where A is the mass number and $R_0 \cong 1.4 \times 10^{-13}$ cm. The radius has values from a few times 10^{-13} cm in light nuclei to about 10^{-12} cm in heavy nuclei. When the distance of closest approach d becomes comparable to the nuclear radius, one should expect deviations from the Rutherford scattering. From equations (2.18) and (2.22), the energy where $R = d$ is

$$E = \frac{Z_1 Z_2 e^2}{R_0 A^{1/3}}.$$

For ^{4}He ions incident on silicon, this energy is about 9.6 MeV. Consequently, nuclear reactions and strong deviations from Rutherford scattering should not play a role in backscattering analyses at energies of a few MeV.

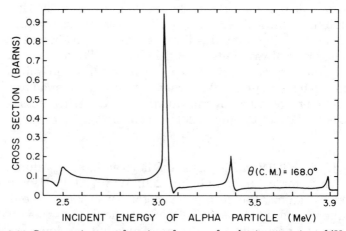

Figure 2.11 Cross section as a function of energy for elastic scattering of ^{4}He$^+$ from oxygen. The curve shows the anomalous cross section dependence near 3.0 MeV. For reference the Rutherford cross section at 3.0 MeV is ~0.037 barns.

One of the exceptions to the estimate given above is the strong increase (resonance) in the scattering cross section at 3.04 MeV for ^{4}He ions incident on ^{16}O, as shown in Figure 2.11. This reaction can be used to increase the sensitivity for the detection of oxygen. Indeed, many nuclear reactions are useful for element detection as described in Chapter 12.

2.8 Low Energy Ion Scattering (LEIS)

Whereas MeV ions can penetrate the order of microns into a solid, low energy ions ($\sim$keV) scatter almost predominantly from the surface layer and are of considerable use for "first monolayer" analysis. In low energy scattering incident ions are scattered, via binary events, from the atomic constituents at the surface and are detected by an electrostatic analyzer Figure 2.12. Such an analyzer detects only charged particles and in this energy range ($\simeq 1$ keV) particles that penetrate beyond a monolayer emerge nearly always as neutral atoms. Thus this experimental sensitivity to only charged particles further enhances the surface sensitivity of low energy ion scattering. The main reasons for the high surface sensitivity of low energy ion scattering is the charge selectivity of the electrostatic analyzer as well as the very large cross section for scattering.

The kinematic relations between energy and mass given in Eqs. (2.5)–(2.7) remain unchanged for the 1 keV regime. Mass resolution is determined as before by the energy resolution of the electrostatic detector. The shape of the energy spectrum, however, is considerably different than that with MeV scattering. The spectrum consists of a series of peaks corresponding to the atomic masses of the atoms in the surface layer.

Quantitative analysis in this regime is not straightforward for two primary reasons: (1) uncertainty in the absolute scattering cross section and (2) lack of knowledge of the probability of neutralization of the surface scattered particle. The latter factor is minimized by use of projectiles with a low neutralization probability and use of detection techniques that are insensitive to the charge state of the scattered ion.

Estimates of the scattering cross section are made using screened Coulomb potentials as discussed in the previous section. The importance of the screening correction is shown in Figure 2.13 which compares the pure Rutherford scattering cross section to two different forms of the screened Coulomb potential. As mentioned in the previous section, the screening correction for ~ 1 MeV He is only a few

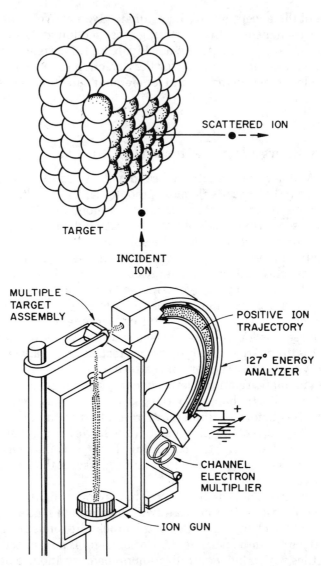

Figure 2.12 Schematic of self-contained electrostatic analyzer system used in low energy ion scattering. The ion source provides a beam of low energy ions which are scattered (to 90°) from samples held on a multiple target assembly and analyzed in a 127° electrostatic energy analyzer.

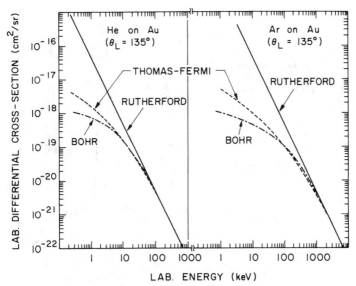

Figure 2.13 Energy dependence of the Rutherford, Thomas–Fermi, and Bohr cross sections for a laboratory scattering angle of 135°. The Thomas–Fermi and Bohr potentials are two common approximations to a screened Coulomb potential (a) He on Au and (b) Ar on Au. [From J. M. Poate and T. M. Buck, *Experimental Methods in Catalytic Research* Academic Press, New York, 1976, Vol. 3.]

percent (for He on Au) but is 2–3 orders of magnitude at ~1 keV. Quantitative analysis is possible if the scattering potential is known. The largest uncertainty in low energy ion scattering is not associated with the potential but with the neutralization probability, of relevance when charge sensitive detectors are used.

Low energy spectra for He^3 and He^4 scattered from an Fe–Re–Mo alloy are shown in Figure 2.14. The improved mass resolution associated with heavier mass projectiles is used to clearly distinguish the Mo from the Re. This technique is used in studies of surface segregation where relative changes in the surface composition can readily be obtained.

2.9 Forward Recoil Spectrometry

In elastic collisions particles are not scattered in a backward direction when the mass of the incident particle is equal to or greater than that of the target atom. The incident energy is transferred primarily to the lighter target atom in a recoil collision [Eq. (2.7)]. The energy of the

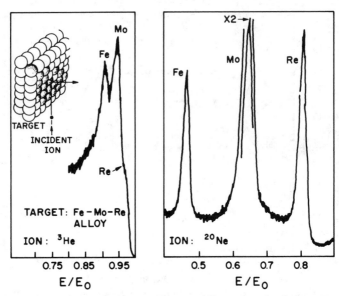

Figure 2.14 Energy spectra for ^{3}He scattering and ^{20}Ne scattering from an Fe–Mo–Re alloy. Incident energy was 1.5 keV. [From McKinney and Leys, 8th Natl. Conf. Electron Probe Anal., New Orleans, LA, 1973.]

recoils can be measured by placing the target at a glancing angle (typically 15°) with respect to the beam direction and by moving the detector to a forward angle ($\theta = 30°$) as shown in the inset of Figure 2.15. This scattering geometry allows detection of hydrogen and deuterium at concentration levels of 0.1 atomic percent and surface coverages of less than a monolayer.

The spectrum for ^{1}H and ^{2}H (deuteron) recoils from a thin polystyrene target are shown in Figure 2.15. The recoil energy from 3.0 MeV ^{4}He irradiation and recoil angle ϕ of 30° can be calculated from Eq. (2.7') to be 1.44 MeV and 2.00 MeV for ^{1}H and ^{2}H, respectively. Since ^{2}H nuclei recoiling from the surface receive a higher fraction ($\sim \frac{2}{3}$) of the incident energy E_0 than do ^{1}H nuclei ($\sim \frac{1}{2}$), the peaks in the spectrum are well separated in energy. The energies of the detected recoils are shifted to lower values than the calculated position due to the energy loss in the mylar film placed in front of the detector to block out He ions scattered from the substrate.

The application of forward recoil spectrometry to determine hydrogen and deuterium depth profiles is discussed in Chapter 3. The forward recoil geometry can also be used to detect other light mass species as long as heavy mass analysis particles are used.

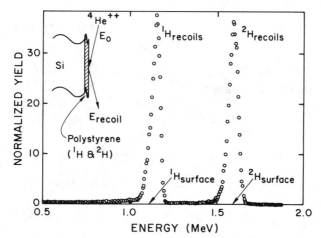

Figure 2.15 The forward recoil structure of ^{1}H and ^{2}H (deuterium) from 3.0 MeV ^{4}He ions incident on a thin (≈ 200 Å) polystyrene film on silicon. The detector is placed so that the recoil angle $\phi = 30°$ and a 10 micron (μm) thick mylar film is mounted in front of the detector.

2.10 Center of Mass to Laboratory Transformation

The derivation of the Rutherford cross section assumes a fixed center of force. In practice, the scattering involves two bodies, neither of which is fixed. In general, any two-body central force problem can be reduced to a one-body problem. However, since actual measurements are done in the laboratory, one must be aware of the appropriate transformation. The transformation equations yield finite and important "corrections" which must be incorporated in careful analytical work. These corrections are most important when the mass of the projectile, M_1, becomes comparable to the mass of the target M_2. Under these conditions, the recoil effects (nonfixed scattering center) become largest.

The relationship between the scattering angles in the laboratory system, θ, ϕ, and the angles in the center of mass (CM) system are illustrated in Figure 2.16a. The first step is to determine an analytical relation between the scattering angles in the two systems.

We use the following notation:

$\mathbf{r}_1$ and $\mathbf{v}_1$ are the position and velocity vectors of the incident particle in the laboratory system;

$\mathbf{r}_1'$ and $\mathbf{v}_1'$ are the position and velocity vectors of the incident particle in the center of mass system;

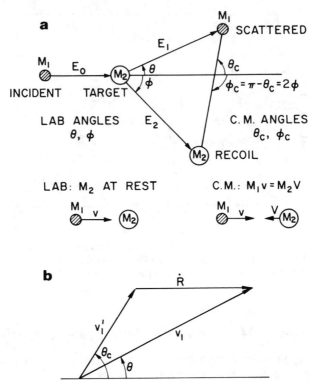

Figure 2.16 (a) The relationship between the scattering angles in the laboratory system, θ, ϕ, and the scattering angles in the center of the mass system, θ_c, ϕ_c. (b) Vector diagram illustrating the relationship between the velocity of particle 1 in the laboratory system, $\mathbf{v}_1$, and the velocity in the center of mass system, $\mathbf{v}_1'$.

$\mathbf{R}$ and $\dot{\mathbf{R}}$ are the position and velocity vectors of the center of mass in the laboratory system.

By definition

$$\mathbf{r}_1 = \mathbf{R} + \mathbf{r}_1'$$

so that

$$\mathbf{v}_1 = \dot{\mathbf{R}} + \mathbf{v}_1'.$$

The geometrical relationship between vectors and scattering angles shown in Figure 2.16b indicates the relation

$$\tan\theta = \frac{v_1' \sin\theta_c}{v_1' \cos\theta_c + |\dot{\mathbf{R}}|} \tag{2.23}$$

The definition of the center of mass vector, **R**, is

$$(M_1 + M_2)\mathbf{R} = M_1\mathbf{r}_1 + M_2\mathbf{r}_2$$

so that

$$(M_1 + M_2)\dot{\mathbf{R}} = M_1\dot{\mathbf{r}}_1 + M_2\dot{\mathbf{r}}_2,$$

where M_2, $\mathbf{r}_2$ refers to the target atom. From the vector diagram

$$\mathbf{v}_1' = \mathbf{v}_1 - \dot{\mathbf{R}}$$

or

$$\mathbf{v}_1' = \frac{M_2}{M_1 + M_2}(\dot{\mathbf{r}}_1 - \dot{\mathbf{r}}_2).$$

Since the system is conservative the relative velocity, $\dot{\mathbf{r}}_1 - \dot{\mathbf{r}}_2$, is the same before and after the collision. Initially $\dot{\mathbf{r}}_2 = 0$ so that

$$v_1' = \frac{M_2}{M_1 + M_2}v,$$

where v is the initial velocity of the particle.

The constant velocity of the CM can also be derived from the definition

$$(M_1 + M_2)\dot{\mathbf{R}} = M_1\mathbf{v}.$$

Substituting the relations for $\dot{\mathbf{R}}$ and v_1' in (2.23), we have

$$\tan\theta = \frac{\sin\theta_c}{\cos\theta_c + M_1/M_2}. \tag{2.24}$$

When $M_1 \ll M_2$, the angles in the two systems are approximately equal; the massive scatterer M_2 suffers little recoil.

A useful form of Eq. (2.24) is written as

$$\cot\theta = \cot\theta_c + x\csc\theta_c,$$

where $x = M_1/M_2$. This can be rearranged to yield

$$\cot\theta - \cot\theta_c = x\csc\theta_c$$

or

$$\sin\theta_c\cos\theta - \cos\theta_c\sin\theta = x\sin\theta$$

so that

$$\sin(\theta_c - \theta) = x\sin\theta. \tag{2.25}$$

For simplicity we let $\Delta = \theta_c - \theta$. From 2.25 we have

$$\cos \Delta \; d\Delta = x \cos \theta \; d\theta$$

and

$$\frac{d\Delta}{d\theta} = \frac{d\theta_c}{d\theta} - 1 = \frac{x \cos \theta}{\cos \Delta}$$

or

$$\frac{d\theta_c}{d\theta} = \frac{\sin \theta_c}{\sin \theta \cos \Delta}.$$

Then

$$\frac{d\sigma}{d\Omega} = \left(\frac{Z_1 Z_2 e^2}{2E}\right)^2 \left[\frac{(1 + x) \sin \theta_c}{2 \sin \theta \sin^2 \theta_c/2}\right]^2 \Big/ \cos \Delta,$$

where E, the energy in lab coordinates, is given by

$$E = E_c(1 + x).$$

It is useful to derive an expression for the cross section simply in terms of θ and x. We make use of the fact that

$$1 + x = (\sin \Delta + \sin \theta)/\sin \theta,$$

$$\sin \theta_c/2 \sin^2(\theta_c/2) = \cot(\theta_c/2),$$

$$\frac{\sin \theta + \sin \Delta}{\cos \theta + \cos \Delta} = \tan \frac{\theta_c}{2}$$

so that

$$\frac{(1 + x) \sin \Delta_c}{2 \sin^2(\theta_c/2)} = \frac{\cos \theta + \cos \Delta}{\sin \theta}$$

and

$$\frac{d\sigma}{d\Omega} = \left(\frac{Z_1 Z_2 e^2}{2E}\right)^2 \frac{(\cos \theta + \cos \Delta)^2}{\sin^4 \theta \cos \Delta}.$$

Noting that $\cos \Delta = (1 - \sin^2 \Delta)^{1/2} = (1 - x^2 \sin^2 \theta)^{1/2}$, we obtain

$$\frac{d\sigma}{d\Omega} = \left(\frac{Z_1 Z_2 e^2}{2E}\right)^2 \frac{[\cos \theta + (1 - x^2 \sin^2 \theta)^{1/2}]^2}{\sin^4 \theta (1 - x^2 \sin^2 \theta)^{1/2}}, \qquad (2.26)$$

which is the form given in Eq. (2.19).

Problems

2.1. $^4\text{He}^{++}$ particles are scattered from a thin foil of an elemental material with atomic number Z_1, mass density ρ_1, number A_1, thickness, t_1, and are observed at some fixed angle θ. The first foil is replaced with a second one (Z_2, ρ_2, A_2, t_2). What is the ratio of the number of particles observed at θ for the first and second foils?

2.2. A beam of 2 MeV helium ions is incident on a silver foil 10^{-6} cm thick and undergoes *Coulomb scattering* in accordance with the Rutherford formula.
(a) What is the distance of closest approach?
(b) Find the impact parameter for He ions scattered through 90°.
(c) What fraction of the incident 2 MeV He ions will be backscattered (i.e., $\theta > 90°$)?
The density of silver is 10.50 g/cm³, and its atomic weight is 107.88 g mol. [*Hint:* The integrated cross section for scattering through angles 0° to 90° is $\int_0^{\pi/2} d\sigma$.]

2.3. An α particle, $^4\text{He}^{++}$, makes a head-on collision with **(a)** a gold nucleus, **(b)** a carbon nucleus, **(c)** an α particle, and **(d)** an electron, each initially at rest. What fraction of the α particle's initial kinetic energy is transferred to the struck particle in each instance?

2.4. (a) Using the formula for the Rutherford scattering cross section in center of mass coordinates (CM), the relations for the recoil energy [Eq. (2.7)] and noting that $\phi = \pi/2 - \theta_c/2$ (Figure 2.16), calculate an expression for $d\sigma/dE_2$, the cross section for transferring an energy E_2 to a nucleus. *Hint:*

$$\frac{d\sigma}{d\theta_c} \cdot \frac{d\theta_c}{dE_2} = \frac{d\sigma}{dE_2}.$$

(b) Using the result of part (a), integrate $d\sigma/dE_2$ from E_{MIN} to E_{MAX} to find the *total* cross section for transferring an energy greater than E_{MIN}.
(c) Evaluate the result of part (b) in cm² for the case of 1.0 MeV He ions bombarding Si. Use $E_{\text{MIN}} = 14$ eV, the displacement energy of an Si atom bound in an Si lattice. Compare this cross section to σ_{Ruth} $(\theta = 180°)$.
(d) Use the result of (c) to calculate the fraction of atoms displaced (i.e., undergoing an energy transfer greater than 14 eV) for 1 μC of He ions incident on a target where the He beam diam $= 1$ mm. This result is only a lower limit to the displacements since we have ignored displacements due to recoiling Si atoms.

2.5. A carbon film is known to contain surface contaminants of Au, Ag, and Si. Sketch the backscattering spectrum indicating the energies of the various peaks and their relative heights.

2.6. An accelerator produces an He^+ ion current of 50 nA at 1.0 MeV. Using a 1 cm² detector 5 cm from the target at a scattering angle of 170°, determine the smallest amount of Au (atoms/cm²) that can be detected. Detectability is arbitrarily defined as 100 counts in 1 h. Under similar conditions, what is the detection limit for oxygen? Compare these limits to the number of atoms/cm² in a monolayer ($\cong 10^{15}$ atoms/cm²).

2.7. Derive the expression for E_2 (Eq. 2.7′), the energy transferred to the target atom using conservation of energy and momentum relations. Give the expression for E_2/E_0 for $\theta = 90°$.

2.8. Use the small angle approximation ($\sin \theta = \theta$) to show that the scattering cross section can be expressed as $\sigma(\theta) = (Z_1 Z_2 e^2/E)^2 (\theta)^{-4}$. Derive this expression using the impulse approximation in which the force of $Z_1 Z_2 e^2/b^2$ acts on the particle for an effective time $t = l/v$, where $l = 2b$. [Hint: An intermediate step in the derivation is to show that $b = Z_1 Z_2 e^2/E\theta$.]

References

1. W. K. Chu, J. W. Mayer, and M.-A. Nicolet, *Backscattering Spectrometry* (Academic Press, New York, 1978).
2. H. Goldstein, *Classical Mechanics* (Addison-Wesley, Reading, MA, 1959).
3. J. W. Mayer and E. Rimini, *Ion Beam Handbook for Material Analysis* (Academic Press, New York, 1977).
4. F. K. Richtmyer, E. H. Kennard, and T. N. Cooper, *Introduction to Modern Physics*, 6th Edition (McGraw-Hill, New York, 1969).
5. P. A. Tipler, *Modern Physics* (Worth Publishers, New York, 1978).

3

ENERGY LOSS
OF LIGHT IONS AND
BACKSCATTERING
DEPTH PROFILES

3.1 Introduction

In the previous chapter, it was tacitly assumed that the atoms to be identified were at the surface of the materials. In this chapter, we consider composition depth profiles which can be obtained from Rutherford backscattering spectrometry (RBS) and nuclear reaction analysis. In this case the depth scale is established by the energy loss dE/dx of light (H^+, d^+, and He^+) ions at high energies (0.5–5 MeV) during their passage through the solid (Figure 3.1). The energy lost in penetration is directly proportional to the thickness of material traversed so that a depth scale can be assigned directly and quantitatively to the energy spectra of detected particles. The yield of backscattered particles or reaction products is proportional to the scattering or reaction cross sections so the composition depth profile can be found from knowledge of energy loss and cross sections.

3.2 General Picture of Energy Loss and Units of Energy Loss

For light ions such as ^4He penetrating a solid, the energetic particles lose energy primarily through excitation and ionization in inelastic collisions with atomic electrons—termed "electronic-energy loss."

39

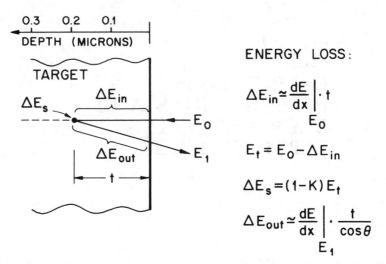

Figure 3.1 Energy loss components for a projectile that scatters from depth t. The sequence is: energy lost via electronic stopping on inward path, ΔE_{in}; energy lost in the elastic scattering process, ΔE_s; and energy lost to electronic stopping on the outward path, ΔE_{out}. Then $E_1 = E_0 - \Delta E_{in} - \Delta E_s - \Delta E_{out}$.

Microscopically, energy loss due to excitation and ionization is a discrete process. Macroscopically, however, it is a good assumption that the moving ions lose energy continuously. All we are concerned with here is the average energy loss during the penetration of ions into a given material.

To measure energy loss, we must determine two quantities: the distance Δt that the ions traverse in the target, and the energy loss ΔE in this distance. The mass density ρ or the atomic density N are frequently combined with the distance, in the form $\rho \, \Delta t$ or $N \, \Delta t$, to express the amount of material per unit area or the number of atoms per unit area that the projectiles have traversed in losing energy ΔE to the target material. Energy loss can be expressed in several different ways. Some frequently used units are

$$dE/dx: \text{eV}/\text{Å},$$

$$(1/\rho) \, dE/dx: \text{eV}/(\mu g/cm^2),$$

$$\epsilon = (1/N) \, dE/dx: \text{eV}/(atoms/cm^2), \text{eV cm}^2.$$

Recently, most authors have adopted $(1/N)dE/dx$ (eV cm^2) as the stopping cross section ϵ; we give the ^{4}He stopping cross section in these units in Appendix 3.

3.3 Energy Loss of MeV Light Ions in Solids
3.3.1 Applicable Energy Ranges

When an He or H ion moves through matter, it loses energy through interactions with electrons which are raised to excited states or ejected from atoms. The radii of atomic nuclei are so small compared with atomic dimensions that nuclear scattering is rare compared to interactions with electrons; therefore, in a first approximation, nuclear interactions may be neglected in the slowing down process.

Theoretical treatments of inelastic collisions of charged particles with target atoms or molecules are separated into fast collisions and slow collisions. The criterion is the velocity of the projectile relative to the mean orbital velocity of the atomic or molecular electrons in the shell or subshell of a given target atom. When the projectile velocity v is much greater than that of an orbital electron (fast-collision case), the influence of the incident particle on an atom may be regarded as a sudden, small external perturbation. This picture leads to Bohr's theory of stopping power.[1,2] The collision produces a sudden transfer of energy from the projectile to the target electron. The energy loss of a fast particle to a stationary nucleus or electron can be calculated from scattering in a central-force field. The stopping cross section decreases with increasing velocity because the particle spends less time in the vicinity of the atom. In the low energy (slow velocity) regime this argument does not hold, and it is found that the stopping power is proportional to velocity. The maximum in the stopping cross section is found at the energy separating these two regions. In backscattering spectrometry we are concerned with the region near and above the maximum.

One useful estimate of the lower energy limit of the fast collision case is to compare the particle velocity or energy with the Bohr velocity v_0 of an electron in the innermost orbit of a hydrogen atom.

$$v_0 = \frac{e^2}{\hbar} = \frac{c}{137} = 2.2 \times 10^8 \text{ cm/s}. \tag{3.1}$$

This velocity is equivalent to that of a 0.1 MeV He ion or 25 keV H ion. As shown in Figure 3.2, the values of the energy loss reach a maximum around 0.5 MeV for He in Al.

In this high energy, fast collision regime, values of dE/dx are proportional to Z_1^2 at the same velocity. The dashed curve and squares

---

[1] N. Bohr, *Phil. Mag.* **25**(6), 10 (1913).
[2] N. Bohr, *Mat. Fys. Medd. Dan. Vid. Selsk.* **18** (8), (1948).

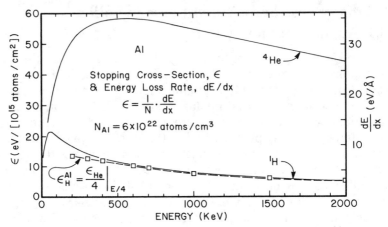

Figure 3.2 Stopping cross section ϵ and energy loss rate dE/dx for ^{4}He and ^{1}H in Al. The open squares for the hydrogen data are scaled from the He data by evaluating the stopping powers at the same velocity $(E/4)$ and scaling by 4 for the Z_1 dependence. The atomic density of Al is 6×10^{22} atoms/cm^3.

in Figure 3.2 are the values for the energy loss of ^{4}He ions in Al reduced by a factor of four $([Z_1^2(H)/Z_1^2(He)] = 1/4)$ and plotted at an energy of one-quarter of that for the He ion $([M(H)/M(He)] = 1/4)$, i.e., at the same velocity. The incident He or H particle is considered as fully ionized (He^{++} or α particle, H$^+$ or proton) in its passage through matter — the velocity of the particle is sufficiently great so that it is stripped of its electrons. At lower velocities the average charge of the projectile becomes lower and the number of electrons available for excitation decreases; therefore, the energy loss decreases.

3.3.2 Derivation of dE/dx

In 1913 Bohr derived an expression for the rate of energy loss of a charged particle on the basis of classical considerations. He considered a heavy particle, such as an α particle or a proton, of charge Z_1e, mass M, and velocity v passing an atom electron of mass m at a distance b (Figure 3.3). As the heavy particle passes, the Coulomb force acting on the electron changes direction continuously. If the electron moves negligibly during the passage of the heavy particle, the impulse, $\int F\,dt$, parallel to the path is zero by symmetry, since for each position of the incident particle in the $-x$ direction there is a corresponding position in the $+x$ direction which makes an equal and opposite contribution to the x component of the momentum. However, throughout the passage, there is a force in the y direction, and momentum Δp is transferred to the electron. This problem of energy transfer is similar to the Coulomb force scattering used to derive the Rutherford scattering law in Chapter 2 if we think of the electron moving towards the "stationary" projectile at the same velocity v. The momentum transferred to the electron during the full passage is therefore

$$\Delta p = \frac{2Z_1 e^2}{bv}, \tag{3.2}$$

where we have used Eq. (2.15a) with $\theta \cong 0$, a small angle approximation. If the electron has not achieved a relativistic velocity, its kinetic energy is given by

$$\frac{\Delta p^2}{2m} = \frac{2Z_1^2 e^4}{b^2 m v^2} = T, \tag{3.3}$$

where T is the energy transfer in the collision.

The differential cross section, $d\sigma(T)$, for an energy transfer between T and $T + dT$ is

$$d\sigma(T) = -2\pi b db, \tag{3.4}$$

and the energy loss per unit path length, dE/dx, is

$$-\frac{dE}{dx} = n \int_{T_{min}}^{T_{max}} T \, d\sigma, \tag{3.5}$$

where n is the number of electrons per unit volume. In terms of impact parameter b,

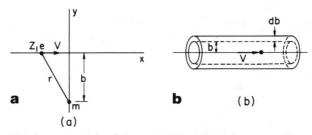

Figure 3.3 (a) A heavy particle of charge $Z_1 e$ passing an electron at distance b. (b) A shell of radius b and thickness db with its axis the path of the heavy charged particle.

$$-\frac{dE}{dx} = n \int_{b_{min}}^{b_{max}} T 2\pi b \; db, \qquad (3.6)$$

which reduces to

$$-\frac{dE}{dx} = \frac{4\pi Z_1^2 e^4 n}{mv^2} \ln \frac{b_{max}}{b_{min}}. \qquad (3.7)$$

To choose a meaningful value for b_{min}, we observe that if the heavy particle collided head on with the electron, the maximum velocity transferred to a stationary electron is $2v$. The corresponding maximum kinetic energy (for a nonrelativistic v) is $T_{max} = \frac{1}{2}m(2v)^2 = 2mv^2$. If this value of T_{max} is inserted in Eq. (3.3), the corresponding b_{min} becomes

$$b_{min} = \frac{Z_1 e^2}{mv^2}. \qquad (3.8)$$

If b_{max} is allowed to become infinite, $-dE/dx$ goes to infinity because of the contribution of an unlimited number of small energy transfers given to distant electrons. But the smallest energy an atomic electron can accept must be sufficient to raise it to an allowed excited state. If I represents the average excitation energy of an electron, we choose $T_{min} = I$, and find

$$b_{max} = \frac{2Z_1 e^2}{\sqrt{2mv^2 I}}. \qquad (3.9)$$

When Eqs. (3.8) and (3.9) are substituted in Eq. (3.7), we obtain

$$-\frac{dE}{dx} = \frac{2\pi Z_1^2 e^4 n}{mv^2} \ln \frac{2mv^2}{I}.$$

This calculation is based on direct collisions with electrons in the solid. There is another term of comparable magnitude due to distant resonant energy transfer. The derivation is outside the scope of this book but leads in its simplest form, to a total stopping power twice that shown above, i.e.,

$$-\frac{dE}{dx} = \frac{4\pi Z_1^2 e^4 n}{mv^2} \ln \frac{2mv^2}{I} \qquad (3.10)$$

or

$$-\frac{dE}{dx} = \frac{2\pi Z_1^2 e^4}{E} \cdot NZ_2 \cdot \left(\frac{M_1}{m}\right) \ln \frac{2mv^2}{I},$$

where $E = M_1 v^2 / 2$ and $n = N Z_2$ with N given by the atomic density in the stopping medium.

Thus we can regard the electronic interactions as composed of two contributions: (1) close collisions with large momentum transfers where the particle approaches within the electronic orbits and (2) distant collisions with small momentum transfers where the particle is outside the orbits. The two contributions are nearly equal (equipartition rule) for particle velocities used in Rutherford backscattering.

The average excitation energy I for most elements is roughly $10Z_2$ in eV, where Z_2 is the atomic number of the stopping atoms. Experimental and calculated values of I are given in Figure 3.4. The description of stopping power so far ignores the shell structure of the atoms and variations in electron binding. Experimentally these effects show up as small deviations (except for the very light elements) from the "$10Z_2$" approximations as shown in Figure 3.4.

The complete energy loss formula (often referred to as the Bethe formula) contains corrections which include relativistic terms at high velocities and corrections for the nonparticipation of the strongly

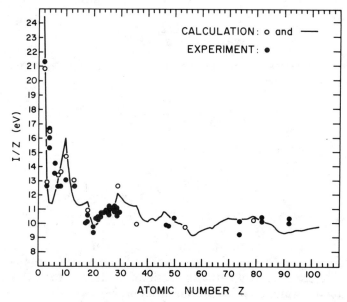

Figure 3.4 Calculation of mean excitation energy by Lindhard and Scharff's theory with a Hartree–Fock–Slater charge distribution. The calculation I/Z vs. atomic number Z reveals structure, as was observed in many experimental measurements. [From W. K. Chu and D. Powers, *Phys. Lett.* **40A**, 23 (1972).]

bound inner shell electrons. For helium ions in the energy regime of a few MeV, relativistic effects are negligible and nearly all the target electrons participate ($n = NZ_2$) in the stopping process. Consequently, Eq. (3.10) can be used to estimate values of dE/dx. In analysis, it is preferable to use tabulated or numerical values such as those listed in Appendix 3.

For example, the electronic energy loss of 2 MeV ^{4}He ions in Al has a value [calculated from Eq. (3.10)] of 31.5 eV/Å using values of $n = NZ_2 = 0.78/Å^3$ and $I = 10Z_2 = 130$ eV. The value given in Appendix 3 is $\epsilon = 44.25$ eV/(10^{15} atoms/cm^2) or a value of $dE/dx = 26.6$ eV/Å ($dE/dx = \epsilon N$). Thus the first-order treatment gives values to within 20% of the experimental values.

3.3.3 Comparison of Energy Loss to Electrons and to Nuclei

A penetrating He ion can also transfer energy to the nuclei of the solid through small angle scattering events. This component to the total energy loss of an energetic ion is termed *nuclear energy loss*. Nuclear energy loss is much smaller than the electronic loss. If the derivation of Eq. (3.7) is repeated for collisions with target atoms, we would find

$$-\frac{dE}{dx}\bigg|_n = \frac{4\pi Z_2^2 Z_1^2 e^4 N}{M_2 v^2} \ln \frac{b_{max}}{b_{min}}, \tag{3.11}$$

where b_{min} represents the maximum energy transfer to a target atom in a head-on collision,

$$T_{max} = \frac{4M_1 M_2}{(M_1 + M_2)^2} E \tag{3.12}$$

so that b_{min} becomes

$$b_{min} = \frac{2Z_1 Z_2 e^2}{vp} \tag{3.13}$$

and b_{max} can be approximated by Eq. (3.9) with I now representing a displacement energy.

In comparison of Eqs. (3.7) and (3.11), the major differences are the mass (m or M_2) in the denominator and the charge Z_2 of the target atom. For protons, neglecting the ratio of log terms, the ratio of nuclear to electronic energy loss per atom is

$$\frac{dE/dx|_n}{dE/dx|_e} \cong N \frac{Z_2^2}{M_2} \times \frac{m}{n} = \frac{Z_2}{M_2} m \cong \frac{1}{3600}, \tag{3.14}$$

where the number of electrons per volume $n \simeq Z_2 N$, $M_2 \simeq 2 Z_2 m_p$, and $m_p \simeq 1836$ m is the mass of the proton.

3.4 Energy Loss in Compounds — Bragg's Rule

The process by which a particle loses energy when it moves swiftly through a medium consists of a random sequence of independent encounters between the moving projectile and an electron attached to an atom of the solid. For a target that contains more than one element the energy loss is the sum of the losses of the constituent elements weighted by the abundance of the elements. This postulate is known as Bragg's rule and states that the stopping cross section $\epsilon^{A_m B_n}$ of a solid of composition $A_m B_n$ is given by

$$\epsilon^{A_m B_n} = m\epsilon^A + n\epsilon^B, \tag{3.15}$$

where ϵ^A and ϵ^B are the stopping cross sections of the atomic constituents A and B.

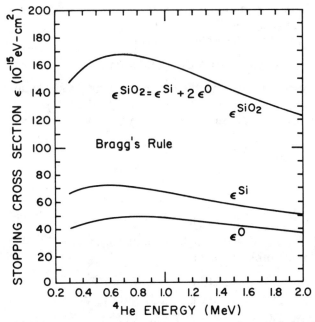

Figure 3.5 Stopping cross sections for ^{4}He ions in Si, O, and SiO_2. The oxide stopping cross section was determined on a molecular basis ϵ^{SiO_2}, on the assumption that Bragg's rule of linear additivity holds with 2.3×10^{22} SiO_2 molecules/cm^3.

To take a specific example of SiO_2 on a molecular basis,

$$\epsilon^{SiO_2} = \epsilon^{Si} + 2\epsilon^0, \tag{3.16}$$

where ϵ^{SiO_2} is now the stopping power/molecule so that $dE/dx = N\epsilon^{SiO_2}$, where N is the number of molecules/volume. Figure 3.5 shows the stopping cross section for SiO_2 on a molecular basis. The energy loss value, dE/dx, for 2.0 MeV He is 28.3 eV/Å, close to the value of elemental Si — 24.6 eV/Å.

3.5 The Energy Width in Backscattering

As MeV He ions traverse the solid, they lose energy along their incident path at a rate dE/dx between 30 and 60 eV/Å. In thin film analysis, to a good approximation, the total energy loss ΔE into a depth t is proportional to t. That is,

$$\Delta E_{in} = \int^t \frac{dE}{dx} \, dx \cong \frac{dE}{dx}\bigg|_{in} \cdot t, \tag{3.17}$$

where $dE/dx|_{in}$ is evaluated at some average energy between the incident energy E_0 and $E_0 - t(dE/dx)$.

The energy of a particle at depth t is

$$E(t) = E_0 - t \cdot dE/dx|_{in}. \tag{3.18}$$

After large angle scattering, the particle energy is $KE(t)$, where K is the kinematic factor defined in Eq. (2.5). The particle loses energy along the outward path and emerges with an energy

$$E_1(t) = KE(t) - \frac{t}{|\cos \theta|} \frac{dE}{dx}\bigg|_{out}$$

$$= -t \left(K \frac{dE}{dx}\bigg|_{in} + \frac{1}{|\cos \theta|} \frac{dE}{dx}\bigg|_{out} \right) + KE_0, \tag{3.19}$$

where θ is the scattering angle. The energy width ΔE of the signal from a film of thickness Δt is

$$\Delta E = \Delta t \left(K \frac{dE}{dx}\bigg|_{in} + \frac{1}{|\cos \theta|} \frac{dE}{dx}\bigg|_{out} \right) = \Delta t[S]. \tag{3.20a}$$

The subscripts "in" and "out" refer to the energies at which dE/dx is evaluated, and $[S]$ is often referred to as the backscattering energy loss factor.

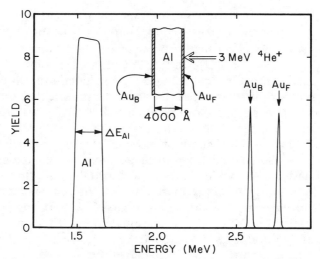

Figure 3.6 The backscattering spectrum ($\theta = 170°$) for 3.0 MeV He ions incident on a 4000 Å Al film with thin Au markers on the front and back surfaces.

The backscattering spectrum at $\theta = 170°$ for 3 MeV ^{4}He incident on a 4000 Å Al film with thin Au markers ($\simeq 3$ monolayers of Au) on the front and back surfaces is shown in Figure 3.6. The energy loss rate dE/dx along the inward path in Al is $\simeq 22$ eV/Å at energies of 3 MeV and is $\simeq 29$ eV/Å on the outward path at energies of about 1.5 MeV ($K_{Al} \simeq 0.55$). Inserting these values into Eq. (3.20), we obtain an energy width ΔE_{Al} of 165 keV. The energy separation between the two Au peaks is slightly larger, 175 keV, because one uses K_{Au} in Eq. (3.20) along with dE/dx(Al).

The assumption of constant values for dE/dx or ϵ along the inward and outward tracks leads to a linear relation between ΔE and the depth t at which scattering occurs. For thin films, $\Delta t \lesssim 1000$ Å, the relative change in energy along the paths is small. In evaluating dE/dx, one can use the "surface energy approximation" in which $(dE/dx)_{in}$ is evaluated at E_0 and $(dE/dx)_{out}$ is evaluated at KE_0. In this approximation the energy width ΔE_0 from a film of thickness Δt is

$$\Delta E_0 = \Delta t[S_0] = \Delta t \left[K \frac{dE}{dx} \bigg|_{E_0} + \frac{1}{|\cos \theta|} \frac{dE}{dx} \bigg|_{KE_0} \right], \quad (3.20b)$$

where the subscripts denote the surface energy approximation.

When the film thickness or the path length becomes appreciable, a better approximation can be made by selecting a constant value of dE/dx at a mean energy $\bar{E}$ intermediate between that at the end points of each track. For the inward track, the incident particle enters at energy E_0 and has an energy $E(\Delta t)$ before scattering at Δt so that $\bar{E}_{in} = \frac{1}{2}[E(\Delta t) + E_0]$. After scattering, the particle has an energy $KE(\Delta t)$ so that $\bar{E}_{out} = \frac{1}{2}[E_1 + KE(\Delta t)]$. In this "mean energy approximation", the energy $E(\Delta t)$ before scattering can be calculated from values of dE/dx or can be further approximated by assuming that the energy difference ΔE is measured or known and that this loss is subdivided equally between the incident and the outward path so that E is approximately $E_0 - \frac{1}{2}\Delta E$. Then $\bar{E}_{in} = E_0 - \frac{1}{4}\Delta E$ and $\bar{E}_{out} = E_1 + \frac{1}{4}\Delta E$.

A comparison between the surface energy and the mean energy approximation is shown in Figure 3.7 for 2.0 MeV He scattering from a Pt film. In the surface energy approximation the conversion between energy width ΔE and thickness Δt is 148.5 eV/Å. In the mean energy approximation the ΔE versus Δt relation deviates from a straight line relation and the value of ΔE for a 5000 Å film exceeds by about 3% the

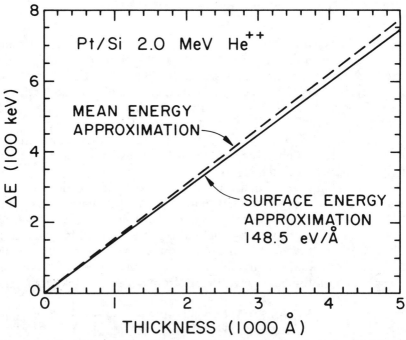

Figure 3.7 Comparison of the calculated relationship between the energy width, ΔE, and thickness for a Pt film. In the surface energy approximation the values of dE/dx are evaluated at the incident energy, E_0 on the inward path and KE_0 on the outward path. In the mean energy approximation dE/dx is evaluated at appropriate average energies as discussed in the text.

value from the surface energy approximation. The comparison between the mean energy and surface approximation serves as a quick
estimate of the probable error introduced by using the surface energy
approximation. The main point here is that a backscattering spectrum can be thought of as a *linear* depth profile of the elements within
the sample.

3.6 The Shape of the Backscattering Spectrum

The energy spectrum from an infinitely thick target has a characteristic slope (Figure 3.8) which can be understood from these relations
between depth and energy loss and the energy dependence of the
Rutherford cross section.

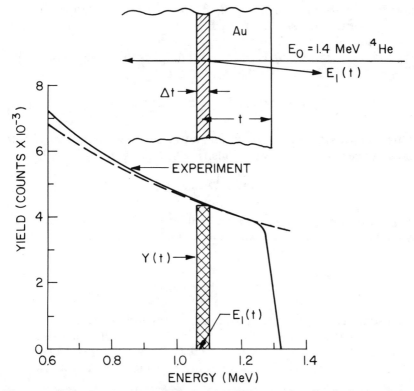

Figure 3.8 Backscattering spectrum for 1.4 MeV He ions incident on a thick Au sample.
The dashed line is calculated based on Eq. (3.25) and normalized to the experimental
curve at 1.3 MeV.

In backscattering measurements, the detector subtends a solid angle Ω so that the total number of detected particles Q_D or yield Y from a thin layer of atoms, Δt, is

$$Y = \sigma(\theta) \cdot \Omega \cdot Q \cdot N\Delta t, \tag{3.21}$$

where Q is the measured number of incident particles and $(N\Delta t)$ is the number of target atoms/cm^2 in the layer.

For thicker layers or bulk targets projectiles can scatter from any depth t, resulting in a continuous energy spectrum to low energy. The yield from a slice of width Δt at depth t is given by (for $\theta = 180°$)

$$Y(t) = \left(\frac{Z_1 Z_2 e^2}{4\ E(t)}\right)^2 NQ\Omega\Delta t, \tag{3.22}$$

where $E(t)$ is the energy of the particle at depth t [Eq. (3.18)] and N is the atomic density. In backscattering one measures the spectrum of particles emerging with energy E_1. To convert Eq. (3.22) to a spectrum $Y(E_1)dE_1$ of the measured energy E_1, we note that $E(t)$ is an intermediate energy between E_0 and E_1. If we denote ΔE_{in} as the energy lost on the inward path, $\Delta E_{in} = E_0 - E(t)$, and ΔE_{out} as the energy lost on the outgoing path, $\Delta E_{out} = KE(t) - E_1$, the ratio

$$A = \frac{\Delta E_{out}}{\Delta E_{in}} = \frac{KE(t) - E_1}{E_0 - E(t)} \approx \frac{dE/dx|_{out}}{dE/dx|_{in}}. \tag{3.23}$$

This is approximately constant for slowly varying energy loss values as is the case for 2.0 MeV He ions. Then the energy E at depth t is

$$E(t) = \frac{E_1 + AE_0}{K + A}. \tag{3.24}$$

The value of A can be determined explicitly; but for medium to heavy mass targets where $K \approx 1$ and $A \approx 1$, the value of $E(t) \approx (E_0 + E_1)/2$ and

$$Y(E_1) \propto \frac{1}{(E_0 + E_1)^2}. \tag{3.25}$$

This spectral shape for $E_0 = 1.4$ MeV is indicated in Figure 3.7.

The shape of backscattering spectra and depth profiles can be obtained from computer programs[3] which are used in both simulation and analysis of RBS data. Simulation of energy widths and signal

---

[3] L. R. Doolittle, *Nucl. Intr. and Meth.* **B9**, 344 (1985).

heights, for example, are a guide in design of sample configuration and scattering geometry.

3.7 Depth Profiles with Rutherford Scattering

The energy loss of light ions follows a well-behaved pattern in the MeV energy range. The values of dE/dx or ϵ can be used to obtain composition depth profiles from the energy spectra of backscattered particles or particles emitted in nuclear reactions. We illustrate the technique with backscattering spectra from an implanted Si substrate and a thin film on Si.

For dilute concentrations of an impurity, $\lesssim 1$ atomic %, the stopping power is simply determined by the host. Figure 3.9 shows a spectrum of As implanted into Si. The conversion of the energy scale to a depth scale is given by Eq. (3.20) using $K = K_{As}$ and dE/dx for

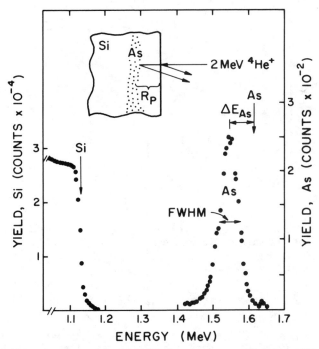

Figure 3.9 Energy spectrum of 2 MeV ⁴He ions backscattered from a silicon crystal implanted with a nominal dose of 1.2×10^{15} As ions/cm² at 250 keV. The vertical arrows indicate the energies of particles scattered from surface atoms of ²⁸Si and ⁷⁵As.

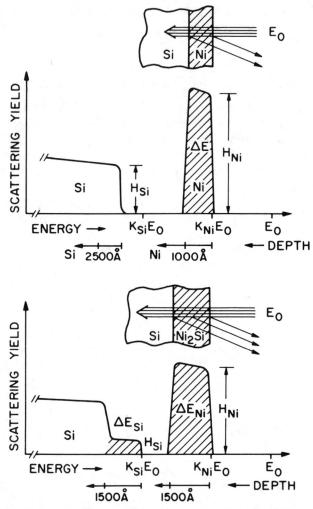

Figure 3.10 Schematic backscattering spectra for MeV ^{4}He ions incident on 1000 Å Ni film on Si *(top)* and after reaction to form Ni_2Si *(bottom)*. Depth scales are indicated below the energy axes.

silicon. The shift, ΔE_{As} indicates that the As is implanted below the surface of the Si.

The upper section in Figure 3.10 shows a 1000 Å Ni film on Si. Nearly all the incident ^{4}He beam penetrates microns into the target before it is stopped. Particles scattered from the front surface of the Ni have an energy given by the kinematic equation, $E_1 = E_0 K$, where the kinematic factor K for ^{4}He backscattered at a laboratory angle of 170° is 0.76 for Ni and 0.57 for Si.

As particles traverse the solid, they lose energy along their incident path at a rate of about 64 eV/Å (assuming a bulk density for Ni of 8.9 g cm^{-3}). In thin film analysis, to a good approximation, energy loss is linear with thickness. Thus, a 2 MeV particle will lose 64 keV penetrating to the Ni–Si interface. Immediately after scattering from the interface, particles scattered from Ni will have an energy of 1477 keV derived from $K_{Ni} \times (E_0 - 64)$. On their outward path, particles will have slightly different energy loss due to the energy dependence of the energy loss processes, in this case 69 eV Å^{-1}. On emerging from the surface, the ^{4}He ions scattered from Ni at the interface will have an energy of 1408 keV. The total energy difference ΔE between particles scattered at the surface and near the interface is 118 keV, a value which can be derived from Eq. (3.20).

In general, one is interested in reaction products or interdiffusion profiles, and the lower portion of Figure 3.10 shows schematically a Ni film reacted to form Ni$_2$Si. After reaction, the Ni signal ΔE_{Ni} has spread slightly, owing to the presence of Si atoms contributing to the energy loss. The Si signal exhibits a step corresponding to Si in the Ni$_2$Si. It should be noted that the ratio of the heights H_{Ni}/H_{Si} of Ni to Si in the silicide layer gives the composition of the layer. To a first approximation the expression of the concentration ratio is given by

$$\frac{N_{Ni}}{N_{Si}} \cong \frac{H_{Ni}}{H_{Si}} \frac{\sigma_{Si}}{\sigma_{Ni}} \cong \frac{H_{Ni}}{H_{Si}} \cdot \left(\frac{Z_{Si}}{Z_{Ni}}\right)^2, \qquad (3.26)$$

where we have ignored the difference in stopping cross sections along the outward path for particles scattered from Ni and Si atoms. The yield from the Ni or Si in the silicide is given closely by the product of signal height and energy with ΔE. Therefore a better approximation to the concentration ratio of two elements A and B uniformly distributed within a film is

$$\frac{N_A}{N_B} = \frac{H_A \Delta E_A \sigma_B}{H_B \Delta E_B \sigma_A}. \qquad (3.27)$$

In this case of Ni$_2$Si the difference between application of Eqs. (3.26) and (3.27) corresponds to a 5% difference in the determination of the stoichiometry of the silicide.

3.8 Depth Resolution and Energy Loss Straggling

With backscattering spectrometry one can determine composition changes with depth. In this section we consider the limits to depth resolution δt, in backscattering spectrometry. The relation between the energy resolution δE_1 and depth resolution is given by Eq. (3.20) as

$$\delta t = \delta E_1 / [S]. \tag{3.28}$$

3.8.1 Grazing Angle Techniques

Equations (3.20a) and (3.28) give the following formula for the depth resolution:

$$\delta t = \frac{\delta E_1}{K(dE/dx)_{in} + \dfrac{(dE/dx)_{out}}{|\cos \theta|}}. \tag{3.29}$$

This equation corresponds to the case of a flat planar sample with the beam incident normal to the surface and a scattering angle θ. For a given detector energy resolution depth sensitivity is optimum by maximizing the energy loss associated with scattering from a depth into the sample. In general, this is done by grazing angle techniques so that the path length and hence the energy loss is a maximum. Using the formula (3.29), the depth resolution is improved by observing the scattering at angles close to 90°, so that the cos θ term approaches zero. With this geometry depth resolutions as small as 20 Å have been observed using standard solid state detectors. There are three factors to consider which influence grazing angle methods of improving depth resolution.

1. *Finite detector acceptance angle.* In any useful system the detector possesses a finite detector angle which constitutes a broadening to the scattering angle set by the geometry. A common grazing exit angle configuration consists of a slit aperture to define the detector angle with a width of 1 mm in the scattering plane and a height of $\sim$ 1 cm perpendicular to the plane, yielding a total area of 0.1 cm². If such a detector is 6 cm from the target, it adds an additional uncertainty of approximately 1° to the scattering angle. This constitutes an appreciable broadening factor as the grazing angle becomes substantially less than 5°. Smaller acceptance angles become impractical when considering the total charge (time) required for an experiment.

2. *Surface roughness.* One of the most difficult parameters for the analyst to control is the surface roughness of the unknown sample. It is clear, however, that surface roughness will set a limit to any grazing angle technique. We note that polished semiconductor grade material is usually extraordinarily flat on the scale of a few degrees and lends itself to these types of analysis. In general, it is surface roughness that will set a final limit to depth resolution, even if acceptance angles are made extremely small.

3. *Straggling.* This contribution is discussed in the following section.

3.8.2 Straggling

The energy resolution is normally composed of two contributions: detector resolution δE_d and energy straggling δE_s. Assuming the two contributions are independent and satisfy Poisson's statistics, the total resolution, δE_1 is given by

$$(\delta E_1)^2 = (\delta E_d)^2 + (\delta E_s)^2. \tag{3.30}$$

An energetic particle that moves through a medium loses energy via many individual encounters. Such a discrete process is subject to statistical fluctuations. As a result, identical energetic particles, which have the same initial velocity, do not have exactly the same energy after passing through a thickness Δt of a homogeneous medium. The energy loss ΔE is subject to fluctuations. The phenomenon is called energy straggling. Energy straggling places a finite limit on the precision with which energy losses, and hence depths can be resolved.

Light particles such as ^1H or ^4He in the MeV energy range lose energy primarily by encounters with the electrons in the target and the dominant contribution to energy straggling is the statistical fluctuations in these electronic interactions. The distribution of energy loss ΔE for many particles passing through a foil gives a distribution that is approximately Gaussian, when ΔE is small compared to the incident energy E_0. In the Gaussian region the probability of finding an energy loss between ΔE and $d\Delta E$ is

$$P(\Delta E)d\Delta E = \frac{\exp\left[-\Delta E^2/2\Omega_B^2\right]}{(2\Omega_B^2\pi)^{1/2}} \, d \, \Delta E,$$

where Ω_B^2 is the mean square derivation. If we consider $d\sigma$ as the cross section for an energy transfer T, in a foil of thickness t containing n electrons/cm^3, then similar to Eq. (3.5),

$$\Delta E = nt \int T d\sigma$$

and

$$\Omega_B^2 = nt \int T^2 d\sigma. \tag{3.31}$$

From Eqs. (3.3) and (3.4) we have

$$d\sigma = \frac{2\pi Z_1^2 e^4}{mv^2 T^2} dT$$

so that

$$\Omega_B^2 = \frac{2\pi Z_1^2 e^4 nt}{mv^2} (T_{max} - T_{min}),$$

where $T_{max} = 2mv^2$, and $T_{min} = I$. For these swift particles, $T_{max} \gg I$ and $n = NZ_2$, giving

$$\Omega_B^2 = 4\pi Z_1^2 e^4 NZ_2 t, \tag{3.32}$$

an expression often referred to as the Bohr value of energy straggling. To determine the energy resolution in Eq. (3.30), we note that the full width at half maximum is $2(2\ln 2)^{1/2}$ times the standard deviation, $\delta E_s = 2.35\Omega_B$.

Bohr's theory predicts that energy straggling does not depend on the energy of the projectile and that the value of the energy variation increases with the square root of the electron density per unit area $NZ_2 t$ in the target. The quantity Ω_B^2/Nt for He ions is numerically equal to Z_2 within 4% when expressed in units of 10^{-12} (eV cm)^2. This rule of thumb allows one to construct a simple estimate of the thickness of target material (atoms/cm^2) that produces an energy straggling of 15 keV for 2 MeV ^{4}He. Figure 3.11 shows a calculation of straggling for elements throughout the periodic chart. In Si, for example ($Z_2 = 14$ and $N = 5 \times 10^{22}$ at. cm^{-3}), films approximately 5000 Å thick can be analyzed before straggling becomes comparable to normal detector resolution of 15–20 keV.

Straggling sets a fundamental limit to the depth resolution, possible with ion beam energy loss techniques. Since Ω_B^2 is proportional to t, the straggling is a function of the depth of penetration. For He ions

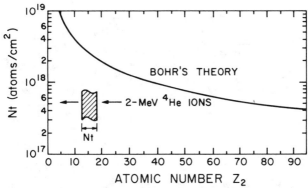

Figure 3.11 The amount of target material Nt (number of atoms/cm²) required to produce 15 keV (FWHM) of energy straggling in a transmission experiment or 21 keV (FWHM) in a backscattering experiment. The projectiles were ^{4}He at 2 MeV. [From W. K. Chu et al., *Backscattering Spectrometry* (Academic Press, New York, 1978).]

incident on layers, $\lesssim 1000$ Å the straggling is small compared to the resolution of a solid state detector and hence plays no role in the obtainable depth resolution. For depths greater than 2000 Å energy loss straggling sets the limit in depth resolution. Using a scattering angle of 95° (grazing exit angle of 5°) a particle scattering from a depth t in the sample corresponds to an energy loss pathlength of length $10t$. Thus straggling, which contributes to the energy resolution at depths the order of 2000 Å in conventional backscattering, now becomes important at depths of approximately 200 Å. The point is that improvement in depth resolution by grazing angle methods only represents a genuine improvement in the near surface region.

3.9 Hydrogen and Deuterium Depth Profiles

Forward recoil spectrometry (Section 2.9) is a method for nondestructively obtaining depth profiles of light elements in solids. For the geometry shown in Figure 3.12a, the technique can be used to determine hydrogen and deuterium concentration profiles in solid materials to depths of a few microns by using ^{4}He ions at energies of a few MeV. The forward recoil technique is similar to backscattering analysis but instead of measuring the energy of the scattered helium ion, the energies of the recoiling ^{1}H or ^{2}H nuclei are measured. Hydrogen is ligher than helium, and both particles are emitted in the forward direction. A mylar foil ($\approx 10\ \mu$m) is placed in front of the detector to block the penetration of the abundantly scattered helium ions while

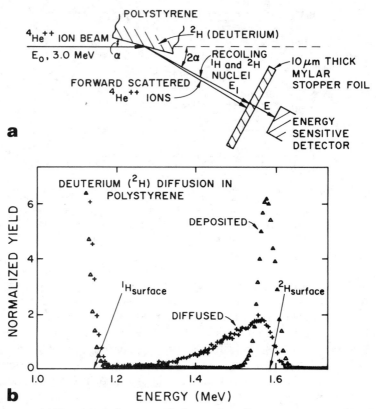

Figure 3.12 (a) Experimental geometry for forward recoil spectrometry experiments to determine depth profiles of ^{1}H and ^{2}H in solids. (b) Recoil spectrum of ^{2}H diffused in a sample of polystyrene for 1 h at 170°C. The sample consisted of a bilayer film consisting of 120 Å of deuterated polystyrene on a large molecular weight ($M_w = 2 \times 10^7$) film of polystyrene. [From Mills et al., *Appl. Phys. Lett.* **45**, 957 (1984).]

permitting the passage of the H ions. The stopping power of ^{1}H ions is sufficiently low compared to that of He ions (Figure 3.2) so that a 1.6 MeV ^{1}H ion only loses 300 keV in penetrating a film that completely stops 3 MeV ^{4}He ions. The mylar absorber does introduce energy straggling which combined with the energy resolution of the detector results in an energy resolution at the sample surface of about 40 keV.

Depth profiles are determined by the energy loss of the incident He ion along the inward path and the energy loss of the recoil ^{1}H or ^{2}H ion along the outward path. The diffusion of deuterium (^{2}H) in polystyrene can be determined from spectra such as shown in Figure 3.12b. In that case a ^{2}H ion detected at an energy of 1.4 MeV corresponds to a collision that originated a ^{2}H recoil from a depth about 4000 Å below the surface. The use of forward recoil spectrometry allows the determination of hydrogen and deuterium diffusion coefficients in the range of 10^{-12}–10^{-14} cm^2/s, a range that is difficult to determine by conventional techniques.

As an example of depth measurement, consider a layer of hydrocarbon on both sides of a self-supported 4000 Å Al film (instead of the Au markers shown in Figure 3.6). We will use a symmetrical scattering geometry with the sample inclined at an angle α to the beam and the detector at an angle 2α so that the path length to the back surface (sample thickness t) is the same, $t/\sin \alpha$, for both the inward He ion and the outward proton. Hydrogen recoils originating from the front surface will have an energy $E_2 = K'E_0$, where $K' = 0.480$ for $2\alpha = 30°$ [Eq. (2.7′), where K' denotes the recoil kinematic factor]. Hydrogen recoils originating from the back surface at t will have an energy $E_2(t)$ given by

$$E_2(t) = K'E_0 - K'\Delta E_{He} - \Delta E_H, \tag{3.33}$$

where ΔE_{He} is the energy loss of the He on the inward path and ΔE_H is the energy loss of the hydrogen on the outward path;

$$\Delta E_{He} = \frac{dE}{dx}\bigg|_{He} \cdot \frac{t}{\sin \alpha}$$

and

$$\Delta E_H = \frac{dE}{dx}\bigg|_H \cdot \frac{t}{\sin \alpha},$$

where $dE/dx|_{He}$ is evaluated at E_0 and $dE/dx|_H$ at E_2 or for simplicity at $K'E_0$.

The energy width ΔE between recoils from the front E_1 and back $E_1(t)$ surfaces is

$$\Delta E = \frac{t}{\sin \alpha} \cdot \frac{dE}{dx}\bigg|_{He} \left\{K' + \frac{dE}{dx}\bigg|_H \bigg/ \frac{dE}{dx}\bigg|_{He}\right\}, \tag{3.34}$$

where the stopping power ratio is about one-sixth. For a 4000 Å Al film, the energy width ΔE from H recoils at the front and back side is about 250 keV for 2 MeV He ions. This is sufficiently large to make forward recoil spectrometry a useful technique for hydrogen analysis. Hydrogen depth profiles can also be determined from nuclear reaction analysis (Chapter 12) or by use of secondary ion mass spectroscopy discussed in the following chapter (Chapter 4).

3.10 Ranges of H and He Ions

Hydrogen depth profile analysis by forward recoil spectrometry (Section 3.9) requires the use of mylar foils to stop the scattered He ions while permitting the hydrogen ions to penetrate the foil into the detector (Figure 3.12a). Mylar foils are also used in prompt radiation analysis using nuclear reactions (Chapter 12) in order to block the elastically scattered particles from the detector (Figure 12.11). In this section, we will discuss ranges of H and He ions in solids.

The penetration of alpha particles and protons in matter was a subject of great interest in the 1930s. The energies of these charged particles could be determined by measurements of their absorption in matter — in particular, air. These measurements were made by placing a collimated α-emitting source in air on a movable slide whose distance from a particle detector could be varied.[4] The number of detected α particles stayed practically constant with increased separation between source and detector up to a distance R, and then the number of detected particles dropped to zero. The distance R is the range of the particles. Values of R can be correlated with the initial energy of the particle; for example, the mean range of 5.3 MeV α particles (^{210}Po) is about 3.8 cm in air while that of 8.78 MeV particles (^{212}Po) is about 6.9 cm. More detailed discussions of range – energy relations in air are found in standard texts in atomic (Evans, 1955) or nuclear physics (Kaplan, 1962). Since the density of air is approximately 10^{-4} of a solid, we expect centimeter ranges in air to correspond to microns in solids.

Here, we are interested in the stopping of 1 – 5 MeV charged particles (protons, deuterons, and α particles) in foils and can use many of the approximations used in evaluating alpha particle ranges in air. The range R of a particle is given by

$$R = \int\limits_{0}^{E_0} \left(\frac{dE}{dx}\right)^{-1} dE, \qquad (3.35)$$

where E_0 is the initial kinetic energy and dE/dx is the energy lost per unit path length. For these swift nonrelativistic particles where electronic energy loss dominates, the rate of energy loss is given by Eq. (3.10).

---

[4] M. G. Holloway and M. S. Livingston, *Phys. Rev.* **54**, 18 (1938).

For our purposes, the ranges of particles can be found in North-cliffe and Schilling (1970) or Ziegler (1977). We can use the scaling parameter that the range is given, using velocity as a parameter, as

$$R = \frac{M_1}{Z_1^2} F(v), \qquad (3.36)$$

where $F(v)$ is a function of velocity. This expression is not exact for the neutralization phenomena at the end of the range, and other corrections are neglected; but it is sufficiently accurate for most cases, excluding very low energies. From Eq. (3.36) we conclude for the same incident velocities $R(^4He) = R(^1H)$ (see problem 3.1).

The range–energy relations for α particles (^4He) and protons (^1H) are given for mylar and silicon in Figure 3.13a. The data shows that for ^4He energies above 4 MeV, the ranges are equal to those of protons at the same velocity [E(He)/4]. The scaling only holds at energies above the maximum in the dE/dx curves shown in Figure 3.13b. In these curves dE/dx values at the same velocity scale as Z_1^2 so that the ^1H values are $\frac{1}{4}$ of the ^4He values.

A 10-μm mylar film is sufficient to stop a 2.5 MeV He ion. The energy loss of a 2.5 MeV proton in the film can be estimated from the range–energy relation. The range of a 2.5 MeV proton in mylar is 80 μm and a 70-μm film has a thickness equivalent to the range of a 2.3 MeV proton; hence the energy lost by a 2.5 MeV proton in traversing

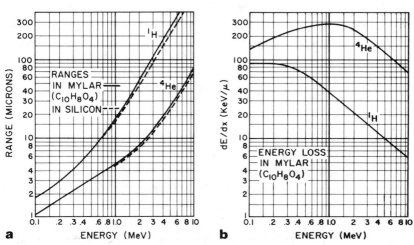

Figure 3.13 (a) The range, in microns, for protons and ^4He ions of energies of 0.1 to 10 MeV in mylar, $C_{10}H_8O_4$, and silicon. (b) The energy loss, in keV/μ, for protons and ^4He ions in mylar.

the film is about 0.2 MeV. This could also be estimated from the dE/dx curve in Figure 3.13b by noting that a 2.5 MeV proton loses 18 keV/μm in mylar.

The close agreement between the range–energy curves in mylar and silicon is fortuitous. In general, the energy loss in compounds is found from the weighted average of the stopping cross sections (Section 3.4).

3.11 Sputtering and Limits to Sensitivity

The Z^2 dependence of the Rutherford scattering cross section clearly indicates a large sensitivity to heavy elements. It is of interest to ask for the ultimate sensitivity, i.e., the smallest amount of material detectable by this technique.

The limit for the ion scattering technique is set by sputtering. Sputtering is a process in which an energetic ion impinging on a solid creates a collision cascade due to small angle nuclear events. Some fraction of the secondary ions acquire the correct momentum to escape the solid, giving rise to an erosion process. This erosion technique is an important process in surface analysis and is described completely in Chapter 4. In the case of Rutherford scattering this erosion process is an undesirable phenomena and sets the limit on sensitivity. The basic question is under what conditions will the erosion of the material occur before a measurement is complete. Sputtering is defined in terms of a sputtering yield Y, which is the number of atoms ejected from the solid per incident ion. In the following we calculate the limit to sensitivity set by the sputtering process.

We consider a thin layer of material (possibly submonolayer) containing N_s atoms/cm^2. The yield of scattered ions, Q_D, is given by the usual equation in a Rutherford backscattering measurement:

$$Q_D = \sigma(\theta)\Omega Q N_s,$$

where $\sigma(\theta)$ is the differential cross section, Ω is the solid angle of the detector, and Q is the number of incident ions.

For the same number of incident ions the loss of atoms from the layer, ΔN_s, due to sputtering by the incident ions is

$$\Delta N_s = YQ/a, \tag{3.37}$$

where a is the area of the probing beam spot. We require that the amount of erosion be less than the original film thickness, i.e.,

$$\Delta N_s < N_s, \tag{3.38}$$

Table 3.1 Summary of Scattering Relations

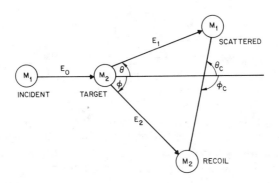

Lab Energy of Scattered Particle; Kinematic Factor, K_{M_2}	$K_{M_2}(\theta) = E_1/E_0 = \left[\dfrac{(M_2^2 - M_1^2 \sin^2\theta)^{1/2} + M_1\cos\theta}{M_2 + M_1} \right]^2;\ M_2 > M_1$ $E_1/E_0 = 1 - \dfrac{2M_1 M_2}{(M_1 + M_2)^2}(1 - \cos\theta_c)$
Lab Energy of Recoil Nucleus	$E_2/E_0 = 1 - E_1/E_0 = \dfrac{4M_1 M_2}{(M_1 + M_2)^2}\cos^2\phi;\quad \phi < \pi/2$
Lab Angle of Recoil Nucleus	$\phi = \frac{1}{2}(\pi - \theta_c);\quad \sin\phi = \left[\dfrac{M_1 E_1}{M_2 E_2}\right]^{1/2}\sin\theta;\quad \tan\theta = \dfrac{\sin 2\phi}{M_1/M_2 - \cos 2\phi}$
C. M. Angle of Scattered Particle	$\cos\theta_c = 1 - 2\cos^2\phi$
Lab Energy of Scattered Particle	$E_1/E_0 = \dfrac{M_2 - M_1}{M_2 + M_1},\quad \theta = 90°$ $E_1/E_0 = \left[\dfrac{M_2 - M_1}{M_2 + M_1}\right]^2,\quad \theta = 180°$
Rutherford Scattering Cross Section	$d\sigma/d\Omega = \left[\dfrac{Z_1 Z_2 e^2}{4E_c}\right]^2 \dfrac{1}{\sin^4\theta_c/2}$ $d\sigma/d\Omega = \left[\dfrac{Z_1 Z_2 e^2}{4E_0}\right]^2 \left[\sin^{-4}\dfrac{\theta}{2} - 2\left(\dfrac{M_1}{M_2}\right)^2 + \cdots\right];\ M_1 < M_2$
Backscattering Factor, $[S]$; Normal incidence	$\Delta E_1 = [S]\,\Delta t$ $[S] = K_{M_2}\dfrac{dE}{dx}\Big|_{in} + \dfrac{1}{|\cos\theta|}\dfrac{dE}{dx}\Big|_{out}$

M_1, Z_1 = mass, charge of incident particle
M_2, Z_2 = mass, charge of target atom
θ, ϕ = angle of scattered particle and recoil in laboratory system
θ_c, ϕ_c = angle of scattered particle and recoil in center-of-mass system
E_1, E_2 = the energy after scattering of the projectile, target atom
E_0 = the energy of the incident projectile in the laboratory system
$E_c = (M_2/M_2 + M_1)E_0$ = incident energy in center-of-mass-system.
$\dfrac{dE}{dx}\Big|_{in}, \dfrac{dE}{dx}\Big|_{out}$ = the rate of energy loss on the inward, outward path.

which can be expressed as a limit on the value of Q given by

$$Q < \left(\frac{Q_D a}{Y\sigma(\theta)\Omega}\right)^{1/2}, \tag{3.39}$$

and a corresponding minimum value of N_s given by

$$N_s > \left(\frac{Q_D Y}{\sigma(\theta)a\Omega}\right)^{1/2}. \tag{3.40}$$

In evaluating this quantity, we use the following standard numbers for the case of a layer of gold ($Z = 79$): $\sigma(\theta)$, the cross section for He scattering at 2 MeV to 170°, is 10^{-23} cm²/steradian; Ω, the solid angle of the detector corresponding to a 1 cm² detector 5 cm from the target, is 4×10^{-2} steradians; Y, the sputtering yield, is 10^{-3} (see Chapter 4); a, the area of the probing beam, is 10^{-2} cm²; and Q_D is arbitrarily taken as 10^2, the minimum number of counts required for a statistically significant measurement. With these values we find a minimum layer thickness of 5×10^{12} Au atoms/cm² or close to 1/1000 of a monolayer. Using these expressions, we see that the scattering parameters and geometry could be optimized in a number of ways to further increase the sensitivity. Experience suggests that the absolute best sensitivity realizable for this favorable case of a heavy scatterer is 5×10^{11} atoms/cm². Note that the incident charge required is not prohibitive. Equation (3.39) corresponds to an incident particle dose of 5×10^{13} ions or about 10 μCoulombs.

3.12 Summary of Scattering Relations

Throughout the book, and particularly in Chapters 2 and 3, we have developed and used a number of kinematic relations and cross sections and the backscattering factor. For the reader's convenience we have collected these scattering relations in Table 3.1.

Problems

3.1. Derive

$$R = \frac{M_1}{Z_1^2} F(v)$$

[Eq. (3.36)] using $dE = Mv\,dv$ and $dx = mv^2 dE\,(4\pi Z_1^2 e^4 n \ln(2mv^2/I))^{-1}$ and estimate the range of 1.0 MeV deuterons in Si using Eq. (3.36) and Figure 3.13.

3.2. Using Eq. (3.10):
(a) Evaluate dE/dx in eV/Å for 1 MeV He in Si.
(b) Convert to units of eV/(10^{15} atoms/cm^2) and compare your answer with Appendix 3.
(c) Determine dE/dx in eV/Å for 4 MeV C ions in silicon.

3.3. Show that the maximum in the stopping power is given by

$$E = \frac{MZ_2 I_0}{4m} \cdot e,$$

where e is 2.718 and $I_0 = 10$ eV. Evaluate for He in Si.

3.4. Draw the backscattering spectrum ($\theta = 180°$) for 2.0 MeV He$^+$ expected for
(a) A sample of Si containing a uniform 1% Au impurity;
(b) A sample of Au containing a uniform 1% Si impurity;
(c) A sample of Si covered with a 1000 Å Pt film;
(d) A sample of Si with a 1000 Å PtSi film on Si.
In all spectra show specific energies for the leading edge (and trailing edge in thin film cases) of the spectral features and indicate the relative heights.

3.5. For the scattering geometry in Figure 3.12a, what is the energy width of the recoiling hydrogen atoms for a 2000 Å thick mylar target? Use the energy loss values in mylar given in Figure 3.13 and ignore the influence of the stopper foil in front of the detector.

3.6. Determine the backscattering energy loss factor (S) [as in Eq. (3.20)] for the case of nonnormal incidence: incident beam at angle θ_1 with respect to the sample normal and scattering angle θ_2.

3.7. Consider an experiment involving 2.0 MeV ions backscattered to 180° from a 2000 Å SiO$_2$ target.
(a) Sketch the backscattering spectrum indicating the energies of the features in the spectrum.
(b) Assuming that the stopping power is energy independent show that the ratio of the energy widths of scattering from oxygen and Si can be written as

$$(K_{Si} + 1)/(K_0 + 1),$$

where K denotes the kinematic factor.
(c) Recognizing that energy loss is not energy independent, derive a precise form for the ratio of the energy widths from the two elements. Evaluate the ratio and compare this to the expression given in part (b).
(d) Using the stopping powers, the stoichiometry, and the cross sec-

tion, derive an equation for the heights of the Si and oxygen peaks and evaluate.

(e) Assume the oxide is of unknown stoichiometry, Si_xO_y. Derive an expression for the ratio of the heights in terms of the scattering cross section, the elemental stopping powers and x/y. [*Hint:* Express the stopping power in terms of x and y using Bragg's rule.]

3.8. Assume in backscattering measurements that energy straggling is given by $\Omega_{tot}^2 = (K\Omega_B|_{in})^2 + (\Omega_B|_{out})^2$. Calculate the amount of energy straggling in an RBS signal from a thin Cr layer underneath an Al film 4000 Å thick with an analysis beam of 2 MeV ^{4}He ions ($\theta = 180°$). What is the total signal width (FWHM) if the detector resolution is 15 keV? What thickness of Al does this correspond to?

References

1. H. H. Anderson and J. F. Ziegler, *Hydrogen Stopping Power and Ranges in all Elements* (Pergamon Press, New York, 1977).

2. W. K. Chu, J. W. Mayer, and M. A. Nicolet, *Backscattering Spectrometry* (Academic Press, New York, 1978).

3. R. D. Evans, *The Atomic Nucleus* (McGraw-Hill Book Co., New York 1955).

4. I. Kaplan, *Nuclear Physics* (Addison-Wesley, Reading, MA, 1964).

5. R. B. Leighton, *Principles of Modern Physics* (McGraw-Hill Book Co., New York, 1959).

6. L. C. Northcliffe and R. F. Schilling, Range and Stopping Power Tables for Heavy Ions, *Nucl. Data Tables* **7**(3-4), 733 (1970).

7. F. K. Richtmyer, E. H. Kennard, and J. N. Cooper, *Introduction to Modern Physics*, 6th Edition (McGraw-Hill Book Co., New York, 1978).

8. J. F. Ziegler, *Helium Stopping Powers and Ranges in all Elements* (Pergamon Press, New York, 1977).

4

SPUTTER DEPTH PROFILES AND SECONDARY ION MASS SPECTROMETRY

4.1 Introduction

This chapter deals with the erosion of the sample by energetic particle bombardment. In this process, called sputtering, surface atoms are removed by collisions between the incoming particles and the atoms in the near surface layers of a solid. Sputtering provides the basis for composition depth profiling with surface analysis techniques either by analysis of the remaining surface with electron spectroscopies or analysis of the sputtered material. Here we describe the most widely used of these latter techniques, secondary-ion-mass spectroscopy (SIMS).

In previous chapters we have been concerned with the energies and yields of particles scattered from the target material under analysis. With Rutherford backscattering using MeV He ions, the energy loss along the inward and outward paths provides the depth information (Figure 4.1). In other analytical techniques the atoms to be identified must lie at the surface of the materials. For example, the observation depth in X-ray photo-electron and Auger-electron spectroscopy (XPS and AES) can be as small as 10–20 Å. In order to use XPS and AES to determine depth profiles, it is necessary to remove controlled thicknesses of the surface layer. This surface layer removal is carried out in materials analysis by bombarding the surface with low energy (0.5–20 keV) heavy ions, such as O^+ or Ar^+, which eject or sputter target atoms from the surface. The yield of sputtered

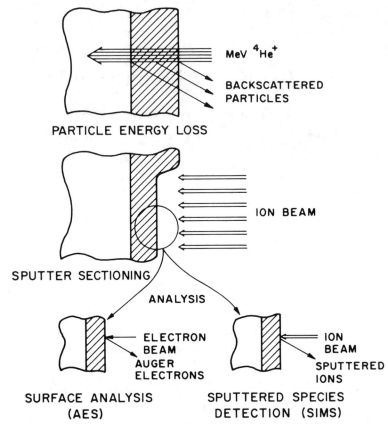

Figure 4.1 Schematic diagram of two approaches to obtain depth profiles in thin films. With particle energy loss techniques, the thickness of the layer is determined from the energy loss of the energetic particles. With sputter sectioning techniques the amount of material probed is determined by the sputtering yield. The surface composition can be directly analyzed either by electron spectroscopies or the amount of material removed by sputter species detection.

atoms, the number of sputtered atoms per incident ion, lies in the range of 0.5 – 20 depending upon ion species, ion energy, and target material. Surface sensitive techniques can then be used after each layer is removed to determine the composition of the new surface and hence deduce the depth profile of the atomic composition. It is also possible to analyze the sputtered atoms, generally the ionized species, to determine the composition of the sputter-removed materials. This technique of secondary-ion-mass spectroscopy or SIMS has been used extensively in depth profiling. One can also measure the characteristic radiation emitted from excited sputtered ions or atoms to determine the composition of the sputter removed material.

For sputtering, it is the energy lost in elastic collisions with the atomic cores — termed *nuclear energy loss* — that determines the energy transfer to and eventual ejection of surface atoms. For backscattering or nuclear analysis, the energetic particles lose energy primarily through electron excitation and ionization in inelastic collisions with atomic electrons — termed *electronic energy loss*. A good assumption is that electronic energy loss and nuclear energy loss can be treated separately and independently. In Chapter 3 we described electronic energy loss and showed that the amount of nuclear energy loss was small. In the sputtering regime the nuclear energy loss dominates.

4.2 Sputtering by Ion Bombardment — General Concepts

Surfaces of solids erode under ion bombardment. The erosion rates are characterized primarily by the sputtering yield Y, which is defined as

$$Y = \text{Sputtering yield} = \frac{\text{Mean number of emitted atoms}}{\text{Incident particle}}. \quad (4.1)$$

The sputtering yield depends on the structure and composition of the target material, the parameters of the incident ion beam, and the experimental geometry. Measured values of Y cover a range of over seven decades; however, for the medium mass ion species and keV energies of general interest in depth profiles, the values of Y lie between 0.5 and 20. The sputtering yields of MeV light ions for most materials are of the order of 10^{-3}. Consequently, Rutherford backscattering analysis will cause the sputtering of only a small fraction of a monolayer during a typical analysis (see Section 3.11).

Sputtering yields can be accurately predicted by theory for single element materials. Figure 4.2 shows the energy and incident particle dependence of the sputtering yield of Si. The experimental values, in good agreement with calculations (solid line) by Sigmund (1981), are based on nuclear energy loss mechanisms and the sharing of this energy loss among the large number of atoms which define the collision cascade. For any given ion–target combination it is desirable to refer to tabulated values or to determine the yield experimentally.

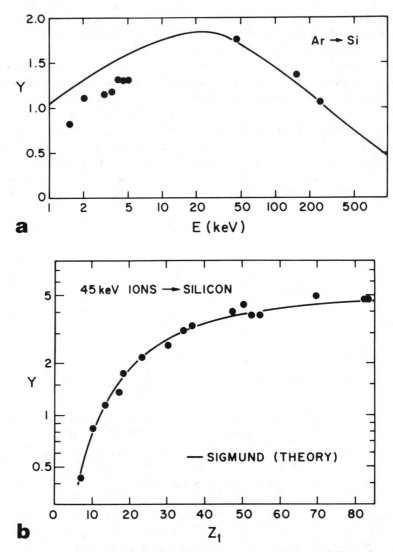

Figure 4.2 (a) Energy dependence of the Ar sputtering yield of Si and (b) incident ion dependence of the Si sputtering yield. The solid line represents the calculations of Sigmund and the data is from Andersen and Bay (1981).

There are a number of review articles and books on the topic of sputtering that are listed in the references at the end of this chapter.

In the sputtering process, atoms are ejected from the outer surface layers. The bombarding ion transfers energy in collisions to target atoms which recoil with sufficient energy to generate other recoils (Figure 4.3). Some of these backward recoils (about 1–2 atoms for a 20 keV Ar ion incident on Si) will approach the surface with enough energy to escape from the solid. It is these secondary recoils which make up most of the sputtering yield. For example, for the case of Ar on Si, target recoils in the backward direction toward the surface are kinematically forbidden as is Ar backscattering (see Chapter 2). The sputtering process involves a complex series of collisions (the collision cascade) involving a series of angular deflections and energy transfers between many atoms in the solid. It is possible to simulate the sputtering process on a computer via a *series* of binary events, but such simulations do not readily yield the dependencies of the sputtering process on various experimental parameters. The problem has been approached based on transport theory, which considers the dynamics of the collision cascade and derives the total energy flux in the backward direction. Such a derivation is beyond the scope of this book. However, we do extract the important parameters based on nuclear energy loss concepts. Clearly the most important parameter in the process is the energy deposited at the surface.

The sputtering yield should be proportional to the number of displaced or recoil atoms. In the linear cascade regime that is applicable for medium mass ions (such as Ar), the number of recoils is proportional to the energy deposited per unit depth in nuclear energy loss. We can then express the sputtering yield Y for particles incident normal to the surface as

$$Y = \Lambda F_D(E_0), \tag{4.2}$$

where Λ contains all the material properties such as surface binding energies and $F_D(E_0)$ is the density of deposited energy at the surface

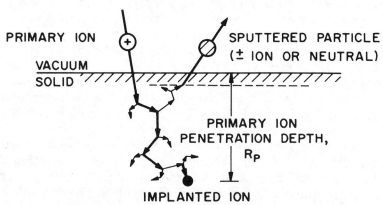

Figure 4.3 Schematic of the ion–solid interactions and the sputtering process.

and depends on the type, energy, and direction of the incident ion and the target parameters Z_2, M_2, and N.

The deposited energy at the surface can be expressed as

$$F_D(E_0) = \alpha NS_n(E_0), \tag{4.3}$$

where N is the atomic density of target atoms, $S_n(E)$ is the nuclear stopping cross section, and $NS_n(E) = dE/dx|_n$ the nuclear energy loss. In this equation α is a correction factor which takes into account the angle of incidence of the beam to the surface and contributions due to large angle scattering events which are not included in the development. The sputtering yield is calculated in Section 4.4.

The evaluation of $S_n(E)$ rests on the collision cross section for energy transfer to a substrate atom. In the keV sputtering regime where the particle velocity is much less than the Bohr velocity, screening of the nuclear charge by the electrons must be included in the description of the collisions. The procedure to obtain the sputtering yield is to first treat a screened potential from a description of the Thomas–Fermi approximations (Section 4.9), and then derive the collision cross section based on a screened potential to obtain the nuclear stopping cross section (Section 4.3).

4.3 Nuclear Energy Loss

A charged particle penetrating a solid loses energy through two processes: (1) energy transfer to electrons, electronic energy loss; and (2) energy transfer to the atoms of the solid, nuclear energy loss. In both cases the interaction is basically of a Coulomb type; for the electronic case it is pure Coulomb (see Chapter 3) while in the nuclear case it is a form of screened Coulomb potential. The two mechanisms have different energy dependencies — in the electronic case there is a peak in the cross section at projectile energies of the order of 0.1 – 1.0 MeV for light projectiles; in the nuclear case the peak in the loss cross section is at much lower energy, of the order of 0.1 – 10 keV. In penetration theory the electronic and nuclear energy losses are treated as uncorrelated and simply summed. In many cases, one or the other contribution is negligible and is simply ignored. Sputtering is governed by the energy deposited via nuclear energy loss at the surface of a solid. This is the mechanism which transfers momentum and energy to the atoms of the solid resulting in energetic secondaries and sputtering. In this section we give a simple description of nuclear energy loss and compare it to the more sophisticated treatments. As in other sections of this text our aim is to provide a simple mathematical description of the process in order to provide some insight into the quantitative understanding of the physics.

The derivation of nuclear energy loss uses two main assumptions: (1) a simple screened Coulomb potential and (2) the impulse approximation.

The interaction potential between two atoms Z_1 and Z_2 can be written in the form of a screened Coulomb potential (Section 4.9) using χ as the screening function,

$$V(r) = \frac{Z_1 Z_2 e^2}{r} \cdot \chi\left(\frac{r}{a}\right), \qquad (4.4)$$

where a is the Thomas–Fermi screening radius for the collision

$$a = \frac{0.885 a_0}{(Z_1^{1/2} + Z_2^{1/2})^{2/3}}. \qquad (4.5)$$

The values of a lie between 0.1 and 0.2 Å for most interactions. We take as a screening function

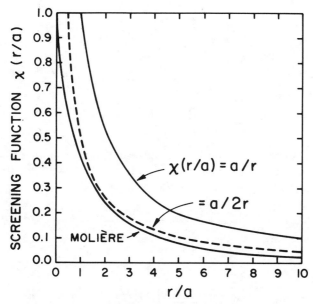

Figure 4.4 The screening function χ in the Molière approximation. At $r = a$, the value of the screening function has dropped to about 0.4 in support of the choice of a as an approximation for the atomic size. Shown also are the screening functions $\chi = a/r$ and $a/2r$.

$$\chi(r/a) = a/2r \tag{4.6}$$

leading to a potential of the form

$$V(r) = \frac{Z_1 Z_2 e^2 a}{2r^2}. \tag{4.7}$$

The screening functions a/r and $a/2r$ are illustrated in Figure 4.4 and the latter reasonably follows the shape of the more accurate Molière potential for $r/a > 1$. The $1/r^2$ potential is not a good approximation for $r/a < 1$ since it does not go into the pure Coulomb form as $r \to 0$. For low energy particles as used in sputtering, this is not a severe problem since the distance of closest approach is large thus preventing interactions at small r.

The impulse approximation is appropriate for the small-angle large-impact parameter collisions that dominate the sequence of scatterings which determine the charged particle trajectory. We have used this approximation in the derivation of electronic stopping given in Chapter 3. In the impulse approximation the change in momentum is given by

$$\Delta p = \int_{-\infty}^{\infty} F_\perp \, dt \tag{4.8}$$

or

$$\Delta p = \frac{1}{v} \int_{-\infty}^{\infty} F_\perp \, dx, \tag{4.9}$$

where $F_\perp$ is the component of the force acting on the ion perpendicular to its incident direction. By using the geometry of Figure 3.3, the force may be written with $r = \sqrt{x^2 + b^2}$ as

$$F_\perp = -\frac{\partial V(r)}{\partial y} = -\frac{\partial V(\sqrt{x^2 + b^2})}{\partial b}. \tag{4.10}$$

Then

$$\Delta p = -\frac{1}{v} \frac{\partial}{\partial b} \int_{-\infty}^{\infty} V(\sqrt{x^2 + b^2}) \, dx \tag{4.11}$$

or, using Eq. (4.7),

$$\Delta p = -\frac{1}{v}\frac{\partial}{\partial b}\int_0^\infty \frac{Z_1 Z_2 e^2 a}{(x^2 + b^2)}\,dx, \tag{4.12}$$

which reduces to

$$\Delta p = \frac{\pi Z_1 Z_2 e^2 a}{2vb^2}. \tag{4.13}$$

The energy transferred, T, to the recoiling nucleus is

$$T = \frac{\Delta p^2}{2M_2},$$

$$T = \frac{\pi^2 Z_1^2 Z_2^2 e^4 a^2}{8M_2 v^2 b^4}. \tag{4.14}$$

The cross section $d\sigma(T)$ for transfer of energy between T and $T + dT$ is

$$d\sigma = -2\pi b\,db$$

or

$$d\sigma = -\frac{\pi^2 Z_1 Z_2 e^2 a}{8\sqrt{(M_2/M_1)E}}\,T^{-3/2}\,dT, \tag{4.15}$$

where $E = M_1 v^2/2$.

It is convenient to express this result in terms of the maximum energy transfer T_m, where

$$T_m = \frac{4M_1 M_2}{(M_1 + M_2)^2}\,E,$$

then

$$d\sigma = -\frac{\pi^2 Z_1 Z_2 e^2 a}{4\,T_m^{1/2}}\frac{M_1}{(M_1 + M_2)}\,T^{-3/2}\,dT. \tag{4.16}$$

The nuclear stopping cross section S_n is given by

$$S_n = -\int T\,d\sigma$$

or

$$S_n = \frac{\pi^2 Z_1 Z_2 e^2 a}{2T_m^{1/2}}\left(\frac{M_1}{M_1 + M_2}\right)T^{1/2}\Big|_0^{T_m}$$

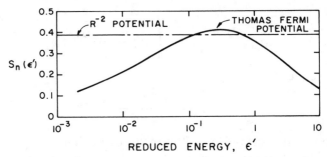

Figure 4.5 Reduced nuclear stopping cross section $S_n(\epsilon')$ (or $d\epsilon'/d\rho$) as a function of ϵ'. The Thomas–Fermi curve represents the most accurate value of S_n for the Thomas–Fermi potential; the horizontal line is the result for an R^{-2} potential [Eq. (4.17)].

$$S_n = \frac{\pi^2 Z_1 Z_2 e^2 a M_1}{2 (M_1 + M_2)}. \tag{4.17}$$

The nuclear energy loss is given by

$$\frac{dE}{dx}\bigg|_n = NS_n, \tag{4.18}$$

where N is the number of atoms/volume in the solid. Note that in this approximation $dE/dx|_n$ is independent of energy, i.e.,

$$\frac{dE}{dx}\bigg|_n = N \frac{\pi^2}{2} Z_1 Z_2 e^2 a \frac{M_1}{M_1 + M_2}. \tag{4.19}$$

Figure 4.5 compares this energy independent value to the values of the nuclear energy loss using the Thomas–Fermi potential.

Figure 4.5 follows the description of Lindhard in which the nuclear energy loss is expressed in terms of a reduced energy ϵ' given by the ratio of the Thomas screening distance to the distance of closest approach,

$$\epsilon' = \frac{M_2}{M_1 + M_2} E \cdot \frac{a}{Z_1 Z_2 e^2} \tag{4.20a}$$

and a reduced length ρ based on a cross section πa^2 and an energy ratio T_m/E,

$$\rho = xNM_2 4\pi a^2 M_1/(M_1 + M_2)^2. \tag{4.20b}$$

This form is then generally useful in that the stopping power for any combination of projectile and target at any energy can be found. In this formalism, $d\epsilon'/d\rho = S_n(\epsilon')$ so that using Eqs. (4.20a) and (4.20b)

$$\left.\frac{dE}{dx}\right|_n = 4\pi a N Z_1 Z_2 e^2 \frac{M_1}{M_1 + M_2} S_n(\epsilon'), \qquad (4.21)$$

where $S_n(\epsilon')$ depends on the form of V(r). In terms of the value of $(dE/dx)_n$ derived in Eq. (4.19), the energy-independent value of the nuclear energy loss in reduced units of energy and length is 0.393. This is slightly different than the value given by Lindhard et al. (1963) since we have used an impulse approximation while these authors evaluate the scattering integral more completely.

Note that our approximation, Eq. (4.19), gives the correct order of magnitude for the stopping power but deviates considerably in the energy dependence. Most significantly, it does not display the $1/E$ dependence at high energy. This is a result of using a $1/r^2$ potential rather than a $1/r$ potential. Clearly, the $1/r^2$ approximation is worst at high energies where close collisions are important. Accurate values of the nuclear energy loss $(dE/dx)_n$ can be derived from Eq. (4.21) and Figure 4.5, which gives $S_n(\epsilon')$ for the more accurate Thomas–Fermi potential (Section 4.9).

For 1 keV Ar ions incident on a medium mass target, Cu, $a = 0.103$ Å and $\epsilon' = 0.008$, and, for 10 keV O ions on Cu, $a = 0.115$ and $\epsilon' = 0.27$. (Ar and oxygen are generally used in sputter profiling.) Thus for ion energies from 1 to 10 keV, the values of ϵ' are in the range of 0.01–0.3; this is a range just below the plateau of dE/dx. As an approximation to estimate the magnitude of $dE/dx|_n$, an energy independent value (a rough average) of $S_n(\epsilon') \cong 0.39$ can be used. For Ar in Cu, $dE/dx|_n \cong 124$ eV/Å, and, for oxygen in Cu, $dE/dx|_n \cong 32$ eV/Å.

4.4 Sputtering Yield

The yield Y of sputtered particles from single element amorphous targets was expressed in Eq. (4.2) as the product of two terms: one, Λ, containing material parameters and the other, F_D, the deposited energy. The derivation of Λ involves a description of the number of recoil atoms that can overcome the surface barrier and escape from the solid. The details of the derivation for the linear cascade regime are given by Sigmund. The result is

$$\Lambda \cong \frac{0.042}{N U_0} \text{ (Å/eV)}, \qquad (4.22)$$

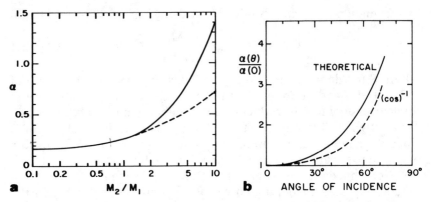

Figure 4.6 Factor α occurring in the backward-sputtering yield formula (4.3). (a) Dependence on mass ratio M_2/M_1: (——) theoretical, evaluated for elastic scattering only, no surface correction applied; (- - -) interpolated from experimental sputtering yields for 45 keV ions on Si, Cu, Ag, and Au. (The difference is mainly due to the neglect of the surface correction at large mass ratios.) (b) Dependence on angle of incidence: (——) theoretical for Ar ions on Cu; (- - -) $(\cos \theta)^{-1}$ dependence, valid mainly in the high-velocity limit. [From Sigmund, 1981.]

where N is the atomic density (in $\mathring{A}^{-3}$) and U_0 (in eV), the surface binding energy. The value of U_0 can be estimated from the heat of sublimation ($\cong$ heat of vaporization) and typically has values between 2 and 4 eV. For the deposited energy, Eq.(4.3),

$$F_D = \alpha \cdot N \cdot S_n,$$

the value of α is a function of the mass ratio and ranges between 0.2 and 0.4 as shown in Figure 4.6. The value of α increases with the angle of incidence because of increased energy deposition near the surface. A reasonable average value for normal incidence sputtering with medium mass ions is $\alpha = 0.25$.

For Ar incident on Cu, the value of $NS_n = 124$ eV/$\mathring{A}$. The surface binding energy, U_0 is ~ 3 eV based on a heat of vaporization of ≈ 3 eV. The sputtering yield with $N = 8.5 \times 10^{-2}$ atoms/$\mathring{A}^3$ is

$$Y = \frac{0.042 \times 0.25 \times 124 \text{ eV}/\mathring{A}}{8.45 \times 10^{-2}/\mathring{A}^3 \times 3 \text{ eV}} = 5.1,$$

which is in reasonable agreement with measured values of about 6.

These estimates hold for the ideal case of an amorphous single-element target. The sputtering yields from single crystal, polycrystalline, or alloy targets may deviate significantly from the simple estimates above. With polyatomic targets preferential sputtering of one of the elements can lead to changes in composition of the surface layer. These changes will be reflected in the Auger yields which give the composition of the altered, not original, layer. Another complication is ion beam mixing (redistribution within the collision cascade), which can lead to broadening of the interface when profiling layered targets. In many of these cases it is possible to use Rutherford back-scattering to establish layer thicknesses and the concentration of the major constituents. This will then provide a calibration for the sputter profile.

4.5 Secondary Ion Mass Spectrometry (SIMS)

Surface layers are eroded by the sputtering process and hence the relative abundance of the sputtered species provide a direct measure of the composition of the layer that has been removed. Sputtered

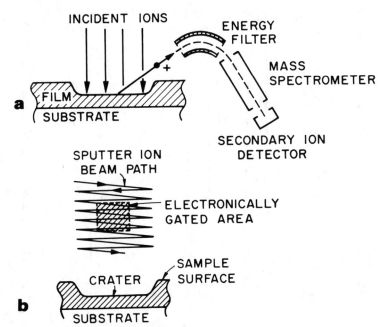

Figure 4.7 (a) Schematic of the SIMS apparatus. An incident ion beam results in sputtered ionic species which are passed through an electrostatic energy filter and a mass spectrometer and finally detected by an ion detector. (b) The beam is usually swept across a large area of the sample and the signal detected from the central portion of the sweep. This avoids crater edge effects.

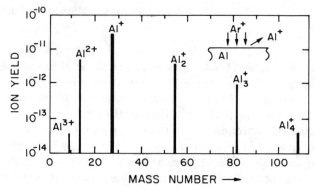

Figure 4.8 Secondary ion cluster spectrum from Ar bombardment of Al. Note that the ordinate is a log scale. The predominant species is Al^+ but Al_2^+ and Al_3^+ are also in abundance. [From Werner, 1978.]

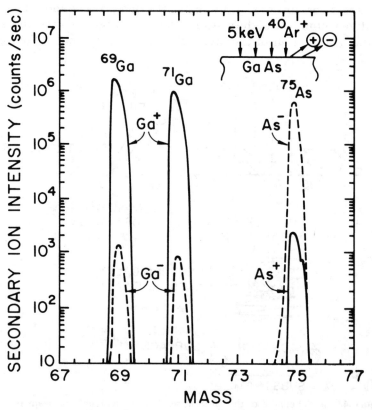

Figure 4.9 Positive (——) and negative (---) ion yields from GaAs bombarded with 5 keV Ar ions. The SIMS spectrum shows the large difference in sensitivity between ionized species for nearly identical sputtering yields of Ga and As from GaAs. [From Magee, *Nucl. Instr. Meth.* **191**, 297 (1981).]

species are emitted as neutrals in various excited states, as ions both positive and negative, singly and multiply charged, and as clusters of particles. The ratio of ionized to neutral species from the same sample can vary by orders of magnitude depending on the condition of the surface. Analysis of sputtered species is the most sensitive of the surface analysis techniques. The common use is the detection and measurement of low concentrations of foreign atoms in solids.

One of the most commonly used sputtering techniques is the collection and analysis of the ionized species — the secondary ions. As shown in Figure 4.7, the secondary ions enter an energy filter, usually an electrostatic analyzer, and then are collected in a mass spectrometer. This gives rise to the acronym SIMS for secondary ion mass spectrometry. All SIMS instruments possess a capability for surface and elemental depth concentration analysis. In one mode of operation the sputter ion beam is rastered across the sample where it

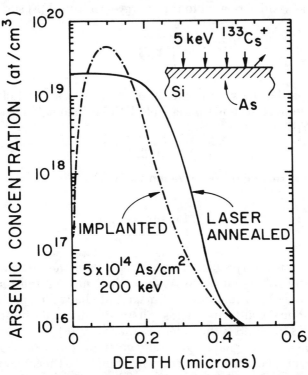

Figure 4.10 SIMS concentration profile of As implanted in Si and redistributed by pulsed laser melting of the outer Si layer. The measured concentration profile extends below levels of $10^{16}/cm^3$. [From C. Magee, RCA Laboratories, private communication.]

erodes a crater in the surface. To insure that ions from the crater walls are not monitored, the detection system is electronically gated for ions from the central portion of the crater. There are also direct imaging instruments — ion microscopes — in which the secondary ions from a defined micro-area of the sample are detected so that an image of the surface composition can be displayed.

The spectra of both positive and negative secondary ions are complex exhibiting not only singly and multiply charged atomic ions but all ionized clusters of target atoms. As shown in Figure 4.8, the mass spectrum from Ar bombarded Al shows not only singly ionized atoms but also doubly and triply ionized atoms and two, three, and four atom clusters. In most cases the yield of singly ionized atoms predominates.

Sputtered particles emerge from the solid with a distribution of energies corresponding to the fluctuations in the many individual events which make up the sputtering process. The sputtered particles have a total yield Y related to the energy spectrum $Y(E)$ such that

$$Y = \int_0^{E_m} Y(E) \, dE, \tag{4.23}$$

where E_m is the maximum energy of the sputtered particles. The positively ionized secondary ion yield $Y^+(E)$ is related to the sputtering yield $Y(E)$ by

$$Y^+(E) = \alpha^+(E)Y(E) \tag{4.24}$$

and the total secondary positive ion yield is

$$Y^+ = \int_0^{E_m} \alpha^+(E)Y(E) \, dE, \tag{4.25}$$

where the ionization probability $\alpha^+(E)$ depends on the particle energy and the nature of the substrate. As shown in Figure 4.9, the ionization yield can vary by three orders of magnitude between species with nearly identical sputtering yields. The major difficulty in quantitative analysis by SIMS is the determination of $\alpha^+(E)$.

The measured signal I^+, generally given in counts/s, of a mono-isotopic element of mass A at a concentration C_A in the target is given by

$$I_A^+ = C_A i_p \beta T \alpha^+(E,\theta) Y(E,\theta) \, \Delta\Omega \, \Delta E, \tag{4.26}$$

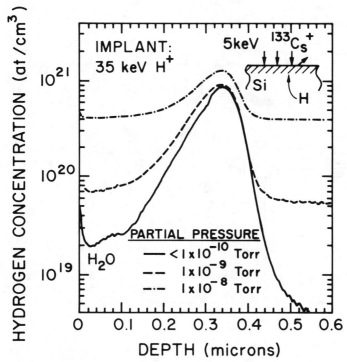

Figure 4.11 SIMS measurements of hydrogen depth profiles for 5 keV Cs sputtering of a silicon sample implanted at 35 keV with a dose of 1×10^{16} H ions/cm². The effect of the H_2O partial pressure in the analysis chamber upon the H dynamic range is evident. [From Magee and Botnick, 1981.]

where i_p is the primary beam current (ions/s), θ and E represent the angle and pass energy of the detector system, $\Delta\Omega$ and ΔE are the solid angle and "width" of the energy filter, and β and T are the detector sensitivity and the transmission of the system for the ion species measured. Both α^+ and Y are dependent on the sample composition. The composition dependence can frequently be neglected if concentration profiles of a low level constituent in a matrix of constant composition are to be determined. A good example of this application is measurement of the depth profile of ion implanted impurities in semiconductors (Figure 4.10). The maximum impurity concentration is less than 10^{-3} and hence the presence of the As has minimal effect on α^+. A strong feature of SIMS is the ability to analyze hydrogen over a wide range of concentrations as shown in Figure 4.11. In this case surface contamination by water vapor can influence the dynamic range.

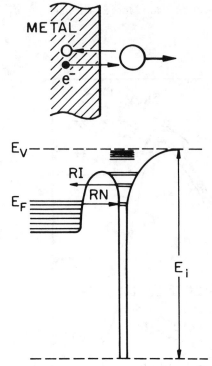

Figure 4.12 Model of the electronic structure of an ion or atom close to a metallic surface. E_F = Fermi energy; E_i = ionization energy; RI = resonance ionization and RN = resonance neutralization.

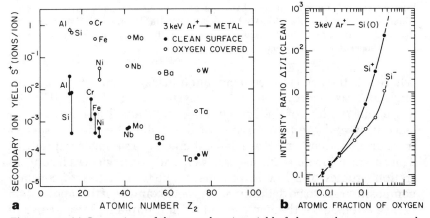

a ATOMIC NUMBER Z_2

b ATOMIC FRACTION OF OXYGEN

Figure 4.13 (a) Comparison of the secondary ion yield of clean and oxygen covered metals sputtered by 3 keV argon ions. (b) Intensity ratios $\Delta I/I_c$ for Si$^+$ and Si$^-$ ion yields from oxygen implanted Si versus oxygen concentration for 3 keV Ar sputtered silicon. The oxygen-induced intensity ΔI is given by $\Delta I = I - I_c$, where I is the measured intensity from oxygen-doped Si and I_c is the ion emission from clean Si. (Both (a) and (b) from Wittmaack, *Surface Sci.* **112**, 168 (1981).]

Secondary ion yields are very sensitive to the presence of either electropositive or electronegative ions at the target surface. The picture of neutralization of a positive ion leaving a surface involves the atomic energy levels of the emitted species and the availability of electrons at the solid surface to fill the ionized level. In one view, this process is most efficient when there are electrons in the solid at precisely the same binding energy as the unoccupied level. Under this condition, a resonance tunneling can occur which neutralizes the outgoing species (Figure 4.12). Thus the probability of neutralization depends on the band structure of the solid and the atomic levels of the sputtered ion. For high yields of ionized particles, one desires to reduce the neutralization probability. This could be accomplished by forming a thin oxide layer, which results in a large forbidden gap and a decrease in number of available electrons for neutralization. For example oxygen adsorption causes an enhancement of secondary ion yield. Figure 4.13 shows secondary ion yields for 3 keV Ar bombardment of clean and oxygen-covered metals. The enhancement in yield covers a wide range of two to three orders of magnitude. The enhancement for Si is shown in Figure 4.13b as a function of oxygen concentration in the Si. The sensitivity to an oxidized surface can be an advantage; for this reason, SIMS analysis is often carried out with the surface "flooded" with oxygen or bombarded with an oxygen beam.

4.6 Secondary Neutral Mass Spectrometry (SNMS)

As shown in Figure 4.13 the secondary ion yield from Si could vary over 3 orders of magnitude depending on the oxygen concentration. These matrix effects can be avoided when the sputtered neutral particles are used for composition analysis (Oechsner, 1984). The mass analysis system still requires ions for detection, and, in SNMS, the emission (sputtering) and ionization (charge transfer) processes can be decoupled by ionizing the sputtered neutral atoms after (postionization) emission from the sample surface.

An example of an SNMS system is shown in Figure 4.14, where the major difference from a conventional SIMS system (Figure 4.7) is the insertion of an ionizing plasma chamber in front of the mass spectrometer. The grids act as an electrical diaphragm between the sample and the chamber which prevents ions, of both signs, from

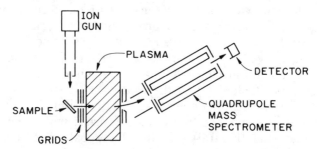

Figure 4.14 Apparatus for carrying out Secondary Neutral Mass Spectrometry (SNMS) in which neutral sputtered species enter a plasma environment for postionization. The ions are then extracted and detected in the quadruple mass spectrometer system.

entering or leaving the chamber. Thus only neutral species enter the ionizing chamber, and species ionized within the chamber cannot reach the sample.

Ionization of the neutral species can be achieved in the chamber by use of a low-pressure high-frequency plasma excited by electron cyclotron wave resonance (Oechsner, 1984). The postionization factor α_A^0 for the sputtered species A depends on the plasma parameters, electron impact ionization of A, and the travel time of A through the ionizer. Values of α_A^0 close to 10^{-1} are achieved for near noble ions and 10^{-2} for transition metal ions like Ta. The postionization factor α^0 is determined by the experimental conditions of the system and for a particular species A, the factor α_A^0 can be treated as a constant for the apparatus.

The measured signal I_A^0 of the neutral species A can be written as

$$I_A^0 = i_p Y_A \alpha_A^0 (1 - \alpha_A^+ - \alpha_A^-)\eta_A^0, \qquad (4.27)$$

where i_p is the primary beam current, Y_A is the sputtering yield of A, α_A^+ and α_A^- are ionization yields for the formation of secondary ions, and η_A is the instrumental factor. The ionization probabilities α^+ and α^- are usually well below unity so the factor $(1 - \alpha^+ - \alpha^-)$ can be treated as unity. Since sample matrix effects are small in the postionization factor α^0, calibration can be achieved readily by use of standards. The sensitivity of SNMS to low concentrations of impurities is comparable to that of SIMS with a detection value of about 1 ppm. In SNMS, however, one does not expect large variations in yield with variation in the properties of the substrate. Instead of plasmas, high powered lasers can be used to ionize the neutral species.

4.7 Preferential Sputtering and Depth Profiles

In a description of sputtering from a multicomponent system, the influence of preferential sputtering and surface segregation must be included. For a homogeneous sample with two atomic components A and B, the surface concentrations C^s are equal to those in the bulk, C^b, in the absence of segregation to the surface which might occur due to thermal processes. Then at the start of sputtering

$$C_A^s / C_B^s = C_A^b / C_B^b. \tag{4.28}$$

The partial yield of atomic species A and B is defined as in Eq. (4.1) by

$$Y_{A,B} = \frac{\text{Number of ejected atoms A,B}}{\text{Incident particle}}. \tag{4.29}$$

The partial sputtering yield Y_A of species A is proportional to the surface concentration C_A^s and similarly Y_B is proportional to C_B^s. The ratio of partial yields is given by

$$\frac{Y_A}{Y_B} = f_{AB} \cdot \frac{C_A^s}{C_B^s}, \tag{4.30}$$

where the sputtering factor f_{AB} takes into account differences in surface binding energies, sputter escape depths, and energy transfers within the cascade. Measured values of f_{AB} generally are in the range between 0.5 and 2.

In the case where f_{AB} is unity, $Y_A / Y_B = C_A^s / C_B^s$ and the yield of sputtered particles is a direct measure of the bulk concentration ratio. In the case where $f_{AB} \neq 1$, the surface concentrations and yields will change from their initial values, $C_A^s(0)$ and $Y_A(0)$, to their values, $C_A^s(\infty)$ and $Y_A(\infty)$, at long times when steady state is achieved.

At the start of sputtering, $t = 0$,

$$\frac{Y_A(0)}{Y_B(0)} = f_{AB} \cdot \frac{C_A^s(0)}{C_B^s(0)} = f_{AB} \cdot \frac{C_A^b}{C_B^b}. \tag{4.31}$$

At long times, when steady state conditions have been achieved, conservation of mass requires that the ratios of partial yields equal the bulk concentration ratio,

$$\frac{Y_A(\infty)}{Y_B(\infty)} = \frac{C_A^b}{C_B^b}. \tag{4.32}$$

For example, if there is preferential sputtering where $f_{AB} > 1$, the sputtering yield of A is greater than that of B, and the surface will be enriched in B. This enrichment of the surface produces an increase in the sputtering yield of B (more B atoms) and a decrease in the sputtering yield of A (less A atoms). As the process continues with macroscopic amounts (greater than 100 Å) of material removed, the increased concentration of B just balances out the preferential sputtering of A. Therefore, at steady state the surface concentration ratio will differ from that of the bulk when $f_{AB} \neq 1$:

$$\frac{C_A^s(\infty)}{C_B^s(\infty)} = \frac{1}{f_{AB}} \cdot \frac{C_A^b}{C_B^b}. \tag{4.33}$$

That is, the surface composition is rearranged so that the total sputtering yield gives the bulk composition in spite of differences in yields of the individual atomic species. Analysis of the composition of the remaining surface layer at this point would show a difference from that of the bulk composition.

An example of the change in composition of a silicide layer is shown in Figure 4.15 for PtSi that was sputtered with 20 keV argon ions and then analyzed with 2 MeV ^{4}He ions. The Rutherford backscattering spectrum shows an enrichment of the Pt concentration in the surface region. The ratio of Pt/Si increased from the value of unity associated with that of the bulk values to a value near two in the surface region. The increase in the Pt concentration is due to the fact

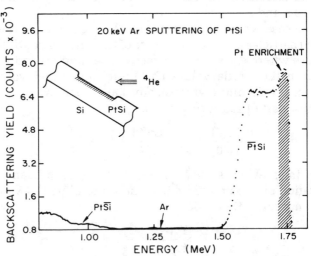

Figure 4.15 RBS spectrum of a PtSi film after sputtering with 20 keV Ar ions. The shaded portion in this Pt signal indicates an increase in the concentration of Pt in the near surface region as a result of the enhanced Si sputtering. [From Liau et al., *J. Appl. Phys.* **49**, 5295 (1978).]

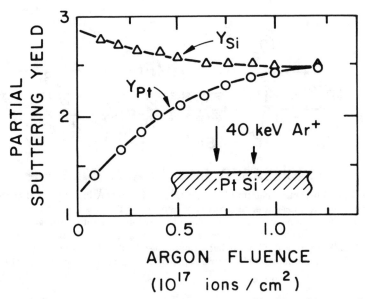

Figure 4.16 Dose dependence of the partial sputtering yields of Si and Pt emitted from PtSi for 40 keV Ar bombardment. [From Liau and Mayer, 1980.]

that the partial sputtering yield of Si is greater than that of Pt, $Y_{Si} >$ Y_{Pt}. Figure 4.16 shows the partial yields as a function of argon ion dose. As one would expect, at low bombardment doses the sputtering yield of Si is significantly greater than that of Pt. At the onset of sputtering the yield ratio $Y_{Si}(0)/Y_{Pt}(0) = 2.4$. As the bombardment proceeds, the partial sputtering yields merge into the same value. The equality of the Si and Pt yields merely reflects the fact that the yield ratio after steady state has been reached is equal to the bulk concentration ratio which for PtSi is unity.

4.8 Interface Broadening and Ion Mixing

One of the applications of sputtering is the removal of deposited or grown layers in thin film structures in order to analyze the composition at the interface between the film and substrate. In these applications, the penetration of the ions used in the sputtering beam can induce an intermixing between the film and substrate due to the strong atomic displacements and diffusion that occurs within the collision cascade around the track of the ion used in sputtering. This

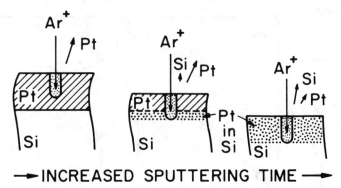

→ INCREASED SPUTTERING TIME →

Figure 4.17 Schematic diagram of Ar sputtering of a 1000 Å Pt layer on Si at three different times in the sputtering process. When the Ar range is less than the Pt film thickness, only Pt ions are sputtered. When the Ar ion penetrates through the Pt/Si interface, ion-induced intermixing occurs and a Si signal is found in the sputtering yield. After the initial Pt film has been removed, a Pt signal is still observed due to mixing of Pt into the substrate Si. [From Liau et al., *J. Vac. Sci. Technol.* **16**, 121 (1979).]

intermixing leads to an artificial broadening of the concentration depth profiles at the interface.

Sputtering requires bombardment of the surface with primary ions of appreciable energy (typically 1–20 keV) whose range far exceeds the escape depth of the sputtered ions and often exceeds the observation depth in electron spectroscopies. Therefore, due to the ion-induced intermixing in the collision cascade, a zone of altered material precedes the "analytical" zone during layer removal. Ion mixing is illustrated in Figure 4.17 for SIMS analysis of a Pt film deposited on Si and sputtered by argon ions. When the argon ions penetrate through the Pt/Si interface some of the Si atoms in the substrate will be transported to the top surface of the Pt film where they can be sputtered. Thus a silicon signal will appear before the Pt film is sputtered away. Platinum is also intermixed with the Si, and, consequently, a Pt signal will persist in the SIMS spectrum at depths well beyond the thickness of the original deposited layer of Pt.

An estimate of the interfacial broadening in such systems can be made by setting the range R of the sputtering ion equal to the half-width of the broadened signal. The ion range is given by

$$R = \int_{E_0}^{0} \frac{1}{dE/dx}\, dE, \tag{4.34}$$

which can be approximated for medium mass ions in the keV energy range by assuming that nuclear energy loss dominates and has an energy independent value,

$$R = E_0/(dE/dx)_n, \tag{4.35}$$

where $(dE/dx)_n$ is given by Eq. (4.19). The value of $(dE/dx)_n$ for Ar ions in Cu is about 100 eV/Å, which is the basis for the rule of thumb that the altered layer extends 10 Å/keV.

The amount of interface broadening can be minimized by proper choice of ion energies and incident angles during sputter profiling. In many cases, sputter depth profiles can have a better depth resolution than that obtained with backscattering spectrometry.

When possible it is advantageous to use two or more analytical techniques that provide complementary data. Figure 4.18 shows the analysis of a tungsten silicide film on polycrystalline silicon (polySi). Secondary ion mass spectroscopy (Figure 4.18a) is used to determine the phosphorus concentration in polySi and to detect oxygen contamination at the interface. Backscattering spectrometry (Figure 4.18b) provides the depth scale and the composition of the sputter deposited silicide — in this case 1830 Å of $WSi_{2.7}$. In the SIMS data the phosphorus concentration in polySi was calibrated from a standard obtained by implanting phosphorus into a silicon sample.

The ratio of W to Si signals in SIMS data (Figure 4.18a) does not reflect the silicide composition, and there is an order of magnitude increase in the Si signal going from the silicide into the Si. These effects are due to the influence of the matrix on the yield of secondary ions. The peak in the W signal at the silicide/Si interface is due to the enhancement of the W ion yield because of oxygen at the interface (note also the enhancement of the phosphorus yield at the Si/SiO_2 interface). The RBS spectrum (Figure 4.18b) shows that there is no peak in the W distribution at the interface and that the composition of the silicide is $WSI_{2.7}$. The increase in the Si signal around 950 keV is due to the increase in Si concentration going from silicide to Si and the decrease around 800 keV is due to the presence of a 700 Å thick layer of SiO_2 at the interface between the polySi layer and substrate Si. The signal from the 1 atom % of phosphorus in the polySi cannot be detected in the RBS spectrum but is easily detected in the SIMS data. The amount of interface broadening is minimal in the depth profile of the SIMS data and the concentration of light mass elements (oxygen and phosphorus) can easily be detected. The strong influence of the matrix on the ion yield does not allow an accurate determination of the relative concentration of the major constituents (tungsten and

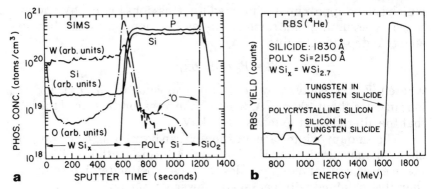

Figure 4.18 (a) Secondary ion mass spectroscopy (SIMS) and (b) Rutherford backscattering spectrometry (RBS) analysis of a tungsten silicide layer sputter deposited on a phosphorus-doped polycrystalline silicon layer on a layer of SiO_2 on silicon. [Data from C. Magee, RCA Research Laboratories, private communication.]

silicon); these quantities are found along with layer thicknesses from the Rutherford backscattering spectrum. By the use of two complementary analytical techniques (RBS and SIMS) a rather accurate picture of the composition of the sample can be obtained.

4.9 Thomas–Fermi Statistical Model of the Atom

In low-velocity collisions the impact parameter is sufficiently large so that the nuclear charge is screened by the electrons. This leads to a modification of the scattering potential from that of the unscreened Coulomb potential, $V = Z_1 Z_2 e^2/r$. The modified potential is found from the Thomas–Fermi description which treats all atoms as identical aside from scaling factors.

The Thomas–Fermi model assumes that the electrons can then be treated by statistical mechanics, Fermi–Dirac statistics, in which they behave as an ideal gas of particles which fills the potential well around the positively charged core. The density of states, $n(E)$, of a free electron gas is obtained by applying periodic boundary conditions and box normalization to a cell of length L to give

$$n(E) = \frac{L^3}{2\pi^2 \hbar^3} (2m)^{3/2} E^{1/2}. \tag{4.36}$$

The energy of the gas increases as the number of electrons increases. For a collection of electrons, the number at a point r, $Z(r)$, is given by

$$Z(r) = \int\limits_{0}^{E_F(r)} n(E)\, dE = \frac{L^3 (2m)^{3/2}}{2\pi^2 \hbar^3} \int\limits_{0}^{E_F(r)} E^{1/2}\, dE = \frac{L^3 (2m)^{3/2}}{3\pi^2 \hbar^3} E_F(r)^{3/2}, \quad (4.37)$$

where $E_F(r)$ is associated with the maximum energy of the ensemble of electrons at r.

The Fermi energy is simply the energy of the highest filled state. In the many-electron atom which we are treating, the total energy E_r of an electron is $E_r = E_K + V(r)$, where E_K is the kinetic energy. For a bound electron $E_r \leq 0$ which requires that for the maximum kinetic energy electron $E_F = -V(r)$. From Eq. (4.37)

$$\rho(r) = \frac{Z(r)}{L^3} = \frac{(2m)^{3/2}}{3\pi^2 \hbar^3} [-V(r)]^{3/2}. \quad (4.38)$$

The self-consistency condition is that the potential due to the electron density in Eq. (4.38), as well as that due to the nuclear charge, properly reproduce the potential energy, $-V(r)$. Consequently, the charge density, $-e\rho$, and the electrostatic potential, $-[V(r)/e]$, must satisfy Poisson's equation

$$-\frac{1}{e} \nabla^2 V = -4\pi(-e\rho),$$

or

$$\nabla^2 V = \frac{1}{r^2} \frac{d}{dr}\left(r^2 \frac{dV}{dr}\right) = 4\pi e^2 \rho = \frac{4e^2 [-2mV(r)]^{3/2}}{3\pi \hbar^3}. \quad (4.39)$$

Equations (4.38) and (4.39) may be solved simultaneously for ρ and V, with the boundary conditions: As $r \to 0$, the leading term in the potential energy must be due to the nucleus, so that $V(r) \to -Ze^2/r$, and as $r \to \infty$, there must be no net charge inside the sphere of radius r, so that V falls off more rapidly than $1/r$, and $rV(r) \to 0$.

Equation (4.39) and the boundary conditions given above are conveniently expressed in a dimensionless form in which Z, E, m, and h appear only in scale factors. We put

$$V(r) = -\frac{Ze^2}{r} \chi, \quad r = ax$$

and

$$a = \frac{1}{2}\left(\frac{3\pi}{4}\right)^{2/3} \frac{\hbar^2}{me^2 Z^{1/3}} = \frac{0.885 a_0}{Z^{1/3}}, \quad (4.40)$$

where $a_0 = \hbar^2/me^2$, the Bohr radius. Equation (4.8) indicates that the scaling parameter to describe the size of an atom is inversely proportional to the cube root of the atomic number. (For electron spectroscopies where transitions are between core levels the 1s radius is approximated by a_0/Z.) With these substitutions, Eq. (4.39) becomes

$$x^{1/2} \frac{d^2\chi}{dx^2} = \chi^{3/2}. \tag{4.41}$$

In this dimensionless Thomas–Fermi (TF) equation, the potential behaves like a simple Coulomb interaction in the extreme case as $r \to 0$. The accurate solution of Equation (4.41) is carried out numerically and there are also analytical approximations represented in series expansions or exponentials. The Molière approximation to the Thomas–Fermi screening function shown in Figure 4.4 is most often used in computer simulations and is given by

$$\chi(x) = 0.35e^{-0.3x} + 0.55e^{-1.2x} + 0.10e^{-6.0x}, \tag{4.42}$$

where $x = r/a$.

Problems

4.1. The maximum value of the nuclear energy loss occurs at reduced energy value of 0.3 for the Thomas–Fermi potential. What energy in keV does $\epsilon' = 0.3$ correspond to for Ar incident on Si, oxygen on Si?

4.2. Assuming nuclear energy loss dominates and the stopping cross section is energy independent, what is the range of 10 keV Ar ions incident on Cu.

4.3. For a screened Coulomb collision with $\chi = a/2r$ use the impulse approximation to show that b, the impact parameter, is proportional to $(a/E\theta)^{1/2}$ and derive $\sigma(\theta)$, the cross section.

4.4. Calculate the ratio of the unscreened to screened nuclear cross section $d\sigma/dT$ for the following cases: 2.0 MeV He^+ on Au, 0.1 MeV He^+ on Au, and 1 KeV Ar on Cu.

4.5. For a scattering potential $V(r) \propto r^{-3}$, what is the energy dependence of the energy loss dE/dx.

4.6. Calculate the sputtering yield for 45 keV Ar ions incident on Si ($U_0 = 4.5$ eV) using a screened potential. Compare your answer with the data given in Figure 4.2.

4.7. If the sputtering yield of species A is twice that of species B in a matrix AB, what is the ratio A to B of the flux of sputtered species at **(a)** the initial time and **(b)** the steady state time and what is the ratio A to B of surface composition at **(c)** the initial time and **(d)** the steady state time?

4.8. Determine the time in seconds required to sputter 500 Å of Si using a $10\mu A/cm^2$ beam of 45 keV ions of **(a)** Ne, **(b)** Kr, and **(c)** Xe. (Use data given by the solid line in Figure 4.2.)

References

1. E. E. Anderson, *Modern Physics and Quantum Mechanics* (W. B. Saunders Co., Philadelphia, 1971).
2. H. H. Andersen and H. L. Bay, "Sputtering Yield Measurements," in *Sputtering by Particle Bombardment I*, R. Behrisch, Ed. (Springer-Verlag, New York, 1981), Chapter 4.
3. R. Behrisch, Ed., *Sputtering by Particle Bombardment I and II* (Springer-Verlag, New York, 1981 and 1983).
4. G. Carter and J. S. Colligon, *Ion Bombardment of Solids* (Elsevier Science Publishing Co., New York, 1968).
5. G. Carter, B. Narvinsek, and J. L. Whitton, Heavy Ion Sputtering Induced Surface Topography Development," in *Sputtering by Particle Bombardment II*, R. Behrisch, Ed. (Springer-Verlag, New York, 1983).
6. W. K. Chu, "Energy Loss of High Velocity Ions in Matter," in *Atomic Physics*, P. Richard, Ed., Methods of Experimental Physics, Vol. 17 (Academic Press, New York, 1980), Chapter 2.
7. Z. L. Liau and J. W. Mayer, "Ion Bombardment Effects on Material Composition," in *Ion Implantation*, J. K. Hirvonen, Ed., Treatise on Materials Science and Technology, Vol. 18, N. Herman, Ed. (Academic Press, New York, 1980), Chapter 2.
8. J. Lindhard, M. Scharff, and H. E. Schiott, Range Concepts and Heavy Ion Ranges (Notes on Atomic Collision, II), *Mat. Fys. Medd. Dan. Vid. Selsk.* **33**(14), (1963).
9. C. Magee and E. M. Botnick, *J. Vac. Sci. Technol.* **19**(47) (1981).
10. J. A. McHugh, "Secondary Ion Mass Spectrometry," in *Methods of Surface Analysis*, A. W. Czanderna, Ed. (Elsevier Science Publishing Co., New York, 1975), Chapter 6.
11. J. M. McCrea, "Mass Spectrometry," in *Characterization of Solid Surfaces*, P. F. Kane and G. B. Larrabee, Eds. (Plenum Press, New York, 1974), Chapter 2.

12. H. Oechsner, Ed., *Thin Film and Depth Profile Analysis* (Springer-Verlag, New York, 1984).

13. P. Sigmund, "Sputtering Processes: Collision Cascades and Spikes," in *Inelastic Ion-Surface Collisions*, N. Tolk et al., Eds. (Academic Press, New York, 1977), pp. 121–152.

14. P. Sigmund, "Sputtering by Ion Bombardment; Theoretical Concepts," in *Sputtering by Particle Bombardment I*, R. Behrisch, Ed. (Springer-Verlag, New York, 1981), Chapter 2.

15. I. M. Torrens, *Interatomic Potentials* (Academic Press, New York, 1972).

16. P. D. Townsend, J. C. Kelley, and N. E. W. Hartley, *Ion Implantation, Sputtering and Their Applications* (Academic Press, New York, 1976).

17. H. W. Werner, "Introduction to Secondary Ion Mass Spectrometry (SIMS)," in *Electron and Ion Spectroscopy of Solids*, L. F. Ermans et al., Eds. (Plenum Press, New York, 1978).

5

CHANNELING

5.1 Introduction

The arrangement of atoms in a solid determines the properties of a material and in single crystals determines the magnitude of incident ion–target atom interactions. The influence of the crystal lattice on the trajectories of ions penetrating into the crystal is known as channeling—a term that visualizes the atomic rows and planes as guides that steer energetic ions along the "channels" between rows and planes. The steering action is effective and can lead to hundred-fold reductions in the yield of backscattered particles. In this chapter we describe channeling of high energy ions in single crystals and show how use of this technique can improve the "depth resolution" of the ion scattering technique and improve its sensitivity to light impurities. The combination of Rutherford backscattering and channeling plays an important role in the materials analysis of a variety of thin film/single crystal problems. Single crystals are not rare in modern day technology as all electronic components are based on single crystal semiconductors.

5.2 Channeling in Single Crystals

Channeling of energetic ions occurs when the beam is carefully aligned with a major symmetry direction of a single crystal. (Lind-

99

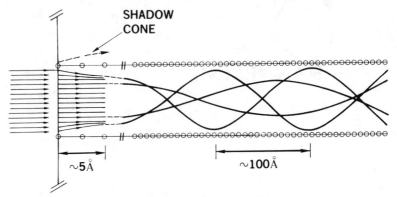

Figure 5.1 Schematic of particle trajectories undergoing scattering at the surface and channeling within the crystal. The depth scale is compressed relative to the width of the channel in order to display the trajectories.

hard, 1965). By a major symmetry direction we mean one of the open directions as viewed down a row of atoms in a single crystal. Figure 5.1 shows a side view of this process in which most of the ion beam is steered (channeled) through the channels formed by the string of atoms. Channeled particles cannot get close enough to the atomic nuclei to undergo large angle Rutherford scattering—hence scattering from the substrate is drastically reduced by a factor of ~ 100. This improves the ion scattering sensitivity to light impurities on the surface. There is always a full interaction with the first monolayers of the solid. This "surface interaction" results in an improved "depth resolution" in these experiments.

The trajectory of a channeled ion is such that the ion makes a glancing angle impact with the axes (axial channeling) or planes (planar channeling) of the crystal and is steered by small angle scattering collisions at distances greater than 0.1 Å from the atomic cores. Since the steering of the channeled particle involves collisions with many atoms, one may consider in a continuum model that the nuclear charge of the atoms in a row (or plane) is uniformly averaged along the row (or plane). The interaction of a channeled particle with an atomic row is described in terms of a single continuum potential $U_a(r)$, where r is the perpendicular distance from the row. $U_a(r)$ is the value of the atomic potential averaged along the atomic row with atomic spacing d. For the axial case,

$$U_a(r) = \frac{1}{d} \int_{-\infty}^{\infty} V\left(\sqrt{z^2 + r^2}\right) dz, \qquad (5.1)$$

where $V(\hat{r})$ is the screened Coulomb potential and $\hat{r}$ is the spherical radial coordinate, $\hat{r}^2 = z^2 + r^2$. Rather than using the Molière potential, we use a more convenient form of the screened Coulomb potential, extensively used in channeling theory because it permits analytical treatments of channeling parameters without significant loss of precision. This "standard potential" is given by

$$V(\hat{r}) = Z_1 Z_2 e^2 \left(\frac{1}{\hat{r}} - \frac{1}{\sqrt{\hat{r}^2 + C^2 a^2}} \right), \qquad (5.2)$$

where C^2 is usually taken equal to 3 and a is the Thomas–Fermi screening distance [Eq. (4.5)]. Then we obtain for the axial continuum potential

$$U_a(r) = \frac{Z_1 Z_2 e^2}{d} \ln \left[\left(\frac{Ca}{r} \right)^2 + 1 \right], \qquad (5.3)$$

where d is the average distance between atoms in the rows. The magnitude of this potential is the order of atomic potentials, that is, 223 eV at $r = 0.1$ Å for He along the $\langle 110 \rangle$ rows of Si ($d = 3.84$ Å).

The continuum potential and conservation of energy allows us to find the critical angle for channeling. The total energy E of a particle inside the crystal is

$$E = \frac{p_{\parallel}^2}{2M} + \frac{p_{\perp}^2}{2M} + U_a(r), \qquad (5.4)$$

where $p_{\parallel}$ and $p_{\perp}$ are the parallel and perpendicular components of the momentum with respect to the string direction (Figure 5.2). Then

$$p_{\parallel} = p \cos \psi, \quad p_{\perp} = p \sin \psi$$

and

$$E = \frac{p^2 \cos^2 \psi}{2M} + \frac{p^2 \sin^2 \psi}{2M} + U_a(r). \qquad (5.5)$$

Channeling angles are small and we use (1) a small angle approximation and (2) equate the last two terms with the transverse energy:

$$E_{\perp} = \frac{p^2 \psi^2}{2M} + U_a(r), \qquad (5.6)$$

i.e., a kinetic energy contribution and a potential energy contribution. The total energy is conserved, and, in this approximation, the transverse energy is conserved. Then the critical angle ψ_c is defined by equating the transverse energy at the turning point $U(r_{min})$ to the transverse energy at the midpoint:

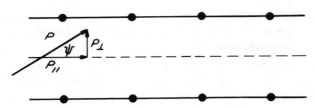

Figure 5.2 Components of the initial momentum vector for a particle incident to an atomic string at an angle ψ.

$$E\psi_c^2 = U(r_{min}) \tag{5.7}$$

or

$$\psi_c = \left(\frac{U(r_{min})}{E}\right)^{1/2}. \tag{5.8}$$

The thermal smearing of the atom positions sets a lower limit to the minimum distance for which a row can provide the necessary correlated sequence of scatterings required for the channeling condition. The most useful first approximation to the critical angle is obtained by substituting $r_{min} = \rho$, in Eqs. (5.3) and (5.8) where ρ^2 is $\frac{2}{3}$ of the mean square thermal vibration amplitude [Eq. (7.5)]:

$$\psi_c(\rho) = \frac{\psi_1}{\sqrt{2}} \left| \ln\left[\left(\frac{Ca}{\rho}\right)^2 + 1\right]\right|^{1/2}, \tag{5.9}$$

where

$$\psi_1 = \left(\frac{2Z_1Z_2e^2}{Ed}\right)^{1/2}. \tag{5.10}$$

(Thermal vibrations and the definition of ρ are discussed in Section 7.3.)

The values of $\psi_c(\rho)$ are within 20% of experimental measurements and follow the measured temperature dependence. For 1.0-MeV He incident on Si⟨110⟩ at room temperature $\psi_c(\rho) = 0.65°$ while the experimentally measured value is 0.55°.

The concept of a distance of closest approach, r_{min}, allows a simple geometric derivation of the fraction of channeled particles for incidence parallel to a crystal axis, $\psi = 0$. In Figure 5.3 we are looking head-on into the crystal. Around each string of atoms is an area πr_{min}^2 in which particles cannot channel; particles incident at $r > r_{min}$ can channel. Then the fraction not channeled is simply $\pi r_{min}^2 / \pi r_0^2$, where r_0 is the radius associated with each string,

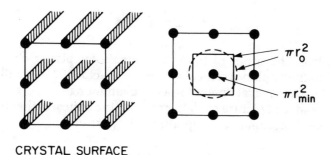

CRYSTAL SURFACE

Figure 5.3 View of an ideal crystal showing continuous rows of atoms and defining the area/row, πr_0^2, and the area associated with the distance of closest approach, πr_{min}^2.

$$\pi r_0^2 = \frac{1}{Nd}; \qquad (5.11)$$

here N is the atomic concentration of atoms and d is the atomic spacing along the string. The ratio $\pi r_{min}^2 / \pi r_0^2$ is usually referred to as the minimum yield, χ_{min} in the literature. In a backscattering experiment, for example, χ_{min} is the yield of close-encounter events as a result of the channeling process. Since $r_{min} \sim 0.1$ Å, the minimum yield is the order of 1% or the fraction of particles channeled is $\sim 99\%$.

The continuum description can be applied to planar channeling as well as axial channeling. For planes, two-dimensional averaging of the atomic potential results in a sheet of charge with the corresponding planar continuum potential $U_p(y)$ being defined:

$$U_p(y) = Nd_p \int V\left(\sqrt{y^2 + r^2}\right) 2\pi r \, dr, \qquad (5.12)$$

where Nd_p is the average number of atoms per unit area in the plane, d_p is the spacing between planes, and y is the distance from the plane. For the standard potential, $U_p(y)$ is given by

$$U_p(y) = 2\pi Z_1 Z_2 e^2 a N d_p \left\{ \left[\left(\frac{y}{a}\right)^2 + C^2 \right]^{1/2} - \frac{y}{a} \right\}. \qquad (5.13)$$

Similar to axial channeling, a critical angle can be defined as

$$\psi_p = \left(\frac{U_p(y_{min})}{E} \right)^{1/2}, \qquad (5.14)$$

where $y_{min} \cong \rho/\sqrt{2}$ is the one-dimensional vibrational amplitude. We define the characteristic angle for planar channeling,

$$\psi_2 = \left(\frac{2\pi Z_1 Z_2 e^2 a N d_p}{E} \right)^{1/2} , \tag{5.15}$$

which is of the order of the critical angle for planar channeling. Experimentally planar critical angles are a factor of 2 – 4 smaller than characteristic critical angles for axial channeling.

A geometric picture of planar channeling indicates that the minimum yield, χ_{min}(planar), is approximately given by

$$\chi_p = \frac{2y_{min}}{d_p} , \tag{5.16}$$

which is substantially larger than the corresponding value for axial channeling. The value of the minimum yield for good planar channeling directions is typically on the order of 10 – 25%.

5.3 Lattice Location of Impurities in Crystals

One application of channeling is the determination of the lattice site (substitutional or nonsubstitutional) of an impurity in a single crystal. For small concentrations of impurities, <1%, the presence of the impurities does not affect the channeling properties of the host lattice. Therefore, the close-encounter probability of a substitutional impurity follows the same angular dependence as that of the host lattice. Impurities that occupy an ensemble of different sites, such as might exist in an impurity cluster, show no angular dependence in the yield curve. The angular yield curves for the substitutional case and for the "random" cluster of nonsubstitutional impurity atoms are shown schematically in Figure 5.4.

Rutherford backscattering is typically used to investigate the lattice site location of impurities that have a larger atomic mass than the host atoms. Scattering kinematics separates the signal of the impurity and host. For lighter mass impurities, the yield from nuclear reactions (Chapter 12) or ion-induced X-ray interactions (Chapter 10) is used to monitor particle-impurity interactions. The angular yield curve, shown in Figure 5.4, is obtained by monitoring the yield of the impurity and the host lattice. The simultaneous accumulation of both angular yield curves provides a sensitive experimental test of the actual substitutionality of the impurity. For substitutional impurities (i.e., those that replace a host atom on its lattice site) the angular scan and minimum yield for scattering from impurities follow that of the host crystal (Figure 5.4). Impurities located at interstitial sites show a

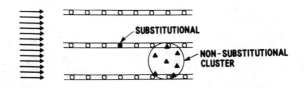

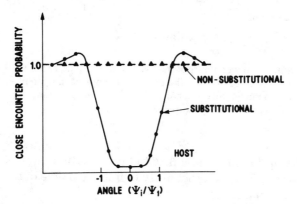

Figure 5.4 Schematic of the close-encounter probability curve expected for substitutional impurities and for a nonsubstitutional cluster.

different angular scan. The interpretation of these angular scans requires a knowledge of the flux distribution of channeled particles.

5.4 Channeling Flux Distributions

A channeled particle is confined within equipotential contours, U_T, such as the continuum contours shown in Figure 5.5. Here $U_T = \Sigma\, U_i$ is the sum of the individual potentials U_i of the nearby rows or planes. Thus a particle with a given transverse energy $E_\perp$ must always lie within a region given by $U_T(r) \leq E_\perp$.

For silicon the 3 eV potential contour closes within the center of the channel (Figure 5.5), and hence particles with a transverse energy less than 3 eV will have their trajectories confined to a particular channel. Channeled particles with higher transverse energies, $E_\perp \gtrsim 10\,\mathrm{eV}$, are not confined within one channel but are guided by the cylindrically symmetric potentials around the axial rows (Figure 5.5).

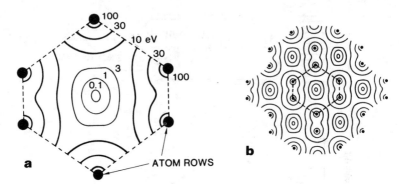

Figure 5.5 (a) Equipotential contours of the axial continuum potential for the case of He in the ⟨110⟩ direction of Si. Note the change in shape of the potential contours corresponding to the geometry of the channel. (b) The potential contours for an array of the channels of the type shown in Figure 5.5a.

Normalizing to unit probability to find a particle somewhere in its allowed area $A(E_\perp)$, the probability to find a particle of transverse energy $E_\perp$ at any point $\mathbf{r}$ is

$$P(E_\perp, \mathbf{r}) = \begin{cases} \dfrac{1}{A(E_\perp)}, & E_\perp \geq U_T(\mathbf{r}), \\ 0, & E_\perp < U_T(\mathbf{r}). \end{cases} \qquad (5.17)$$

The area $A(E_\perp)$ is defined by an equipotential contour such as that shown in Figure 5.5. For example, particles with $E_\perp = 10$ eV (i.e., 1-MeV particles entering the center of channel at an angle $0.18°$) have equal probability of being found at any point within the area defined by the equipotential contour $U_T(\mathbf{r}) = 10$ eV.

In the following, the flux distribution of ions for a channeled beam ($\psi = 0$) is calculated using:

1. Conservation of transverse energy: A particle that enters at r_{in} cannot get closer than r_{in} to the string.

2. For two-dimensional axial channeling a particle has a uniform probability of being found in its "allowed" area; for cylindrical symmetry the allowed area $= \pi r_{in}^2$, where r_{in} is the initial distance from the string at the crystal surface, i.e.,

$$P(r_{in}, r) = \begin{cases} \dfrac{1}{\pi r_0^2 - \pi r_{in}^2}, & r > r_{in}, \\ 0, & r < r_{in}. \end{cases} \qquad (5.18)$$

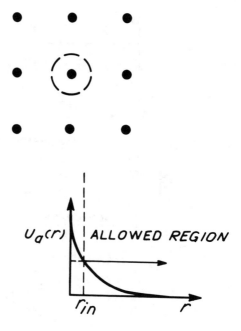

Figure 5.6 Geometry associated with the flux distribution calculation showing the cylindrical geometry and the "allowed area."

3. We assume cylindrical symmetry for simplicity (Figure 5.6).

The flux distribution inside the crystal, $f(r)$, then corresponds to integrating over all initial impact parameters:

$$f(r) = \int_0^{r_0} P(r_{in}, r) 2\pi r_{in} \, dr_{in}$$

$$= \int_r^{r_0} \frac{1}{\pi r_0^2 - \pi r_{in}^2} 2\pi r_{in} \, dr_{in} \qquad (5.19)$$

$$= \ln \frac{r_0^2}{r_0^2 - r^2}.$$

The effect of channeling is to transform a spatially uniform distribution to the peaked distribution shown in Figure 5.7. This flux distribution displays the most prominent feature of channeling, namely, that the flux intensity and hence the close-encounter probability approaches zero near the atomic rows (as $r \to 0$).

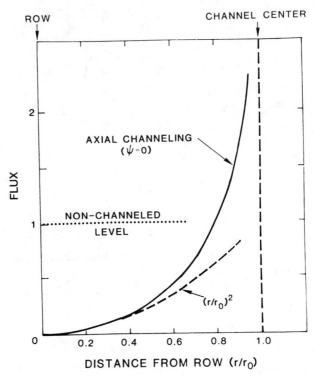

Figure 5.7 Flux distribution of the channeled beam as a function of distance from the row for the case of parallel incidence, $\psi = 0$, and assuming statistical equilibrium and cylindrical symmetry. The small r/r_0 approximation [Eq. (5.20)] is shown for comparison.

Expansion of $f(r)$ for small r, near the atom rows, gives

$$f(r) \approx \frac{r^2}{r_0^2}, \tag{5.20}$$

where the approximation is shown by the dashed line in Figure 5.7. This simple flux distribution is useful for estimating the scattering intensity from substitutional impurities.

Examination of Figure 5.7 indicates another property of the channeling flux distribution: the peaking of the particle density in the center of the channels ($r \simeq r_0$). The intensity is far in excess of unity, the value corresponding to the nonchanneling particle density. Thus the yield from interstitial impurities located near the channels ($r \approx r_0$) will be greater than the nonchanneling value.

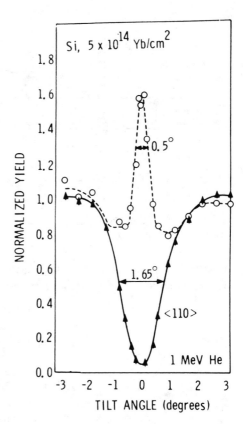

Figure 5.8 Close-encounter probability (normalized yield) as a function of the angle between the beam and ⟨110⟩ symmetry direction for the case of 1.0-MeV He in Si implanted with 5×10^{14} Yb/cm^2 (60 keV, 450°C). The high yield for the Yb (open data points) was the first indication of the flux peaking effect. [From J. U. Andersen et al. (1971).]

The "flux peaking" effect at $r = r_0$ was first shown in lattice location experiments of Yb implanted into Si (Figure 5.8) (Andersen et al., 1971). A substitutional impurity would have an angular scan equivalent to that of the host. Clearly, the Yb is not substitutional but is located at a position where the flux intensity is appreciably greater than that for a nonchanneling direction.

5.5 Surface Interaction via a Two-Atom Model

The two-atom model is a simple and illustrative example of the interaction of the ion beam with the atomic structure at a surface. We calculate the shadow behind a repulsive scattering center, the topmost atom on a surface. Essentially, we calculate the flux distribution $f(r_2)$ at the second atom as a result of scattering interactions with the

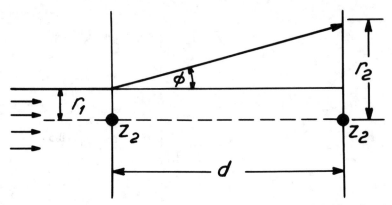

Figure 5.9 Geometry for the calculation of the "shadow cone" and the flux distribution at the second atom.

first atom. The small angle approximation to pure Coulomb scattering is used so that $\phi = Z_1 Z_2 e^2 / E r_1$, as shown in Figure 5.9. Then

$$r_2 = r_1 + \phi d,$$

$$r_2 = r_1 + \frac{Z_1 Z_2 e^2 d}{E r_1}, \tag{5.21}$$

$$r_2 = r_1 + \frac{R_c^2 / 4}{r_1},$$

where the Coulomb shadow cone radius R_c is defined as

$$R_c = 2 \left(\frac{Z_1 Z_2 e^2 d}{E} \right)^{1/2}. \tag{5.22}$$

This quantity is the distance of closest approach to the second atom (Figure 5.10). The flux distribution at the second atom, $f(r_2)$, is given by

$$f(r_2) 2\pi r_2 \, dr_2 = f(r_1) 2\pi r_1 \, dr_1, \tag{5.23}$$

where $f(r_1)$, the incident flux distribution, is uniform and normalized to unity. So that

$$f(r_2) = \frac{r_1}{r_2} \left| \frac{dr_1}{dr_2} \right|. \tag{5.24}$$

As shown in Figure 5.10, r_2 has a minimum, i.e., $dr_2/dr_1 = 0$. It is simplest to evaluate $f(r_2)$ in two steps, for $r_1 < r_{1_c}$ and for $r_1 > r_{1_c}$; $r_2(r_{1_c}) = R_c$. The final result is

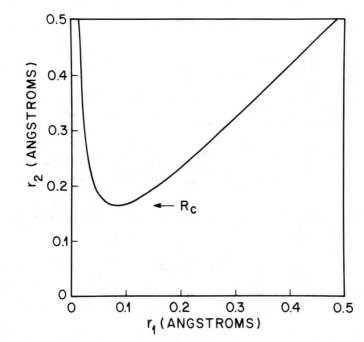

Figure 5.10 The dependence of r_2 on the initial impact parameter r_1 for the case of 1.0 MeV He ions on $W\langle 100\rangle$; r_2 is calculated for $d = 3.16$ Å corresponding to the atom spacing in the $\langle 001\rangle$ direction. The minimum value of r_2 is the Coulomb shadow cone radius R_c.

$$f(r_2) = \begin{cases} 0, & r_2 < R_c \\ \dfrac{1}{2}\left[\dfrac{1}{\sqrt{1 - R_c^2/r_2^2}} + \sqrt{1 - R_c^2/r_2^2}\right], & r_2 > R_c. \end{cases} \quad (5.25)$$

The function $f(r_2)$ is sketched in Figure 5.11.

The flux distribution is so sharp that the curvature occurs within a distance that is small compared to the thermal vibration amplitude of atoms in a crystal. Thus we approximate $f(r_2)$ to a delta function, i.e.,

$$f(r_2) = \begin{cases} 0, & r_2 < R_c, \\ 1 + \dfrac{R_c^2}{2}\dfrac{\delta(r_2 - R_c)}{r_2}, & r_2 \geq R_c. \end{cases} \quad (5.26)$$

The intensity of scattering from the second atom, I_2, which we treat as a measure of the close-encounter probability, is given by the overlap

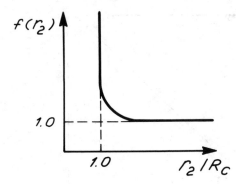

Figure 5.11 Flux distribution $f(r_2)$ as a function of r_2/R_c.

of the flux distribution with the Gaussian position distribution $\tilde{P}(r_2)$ of the second atom.

$$I_2 = \int \tilde{P}(r_2)f(r_2)2\pi r_2 \, dr_2, \qquad (5.27)$$

where

$$\tilde{P}(r) = \frac{1}{\pi\rho^2} e^{-r^2/\rho^2}. \qquad (5.28)$$

This treatment assumes that the close-encounter process has a characteristic interaction distance which is very small compared to thermal vibration amplitudes or R_c; such processes are nuclear reactions, Rutherford backscattering, and inner-shell X-ray excitation. Using (5.26)–(5.28),

$$I_2 = \left[1 + \frac{R_c^2}{\rho^2}\right] e^{-R_c^2/\rho^2} \qquad (5.29)$$

and the total surface peak intensity I is given by the unit contribution for the first atom and I_2 (Figure 5.12). Note that the intensity I_2 is determined by a single parameter ρ/R_c, which defines the ratio of the thermal vibration amplitude to the shadow cone.

$$I = I_2 + 1 \qquad (5.30)$$

For values of $\rho < R_c$, the topmost atom does indeed shadow the underlying atoms from direct, close encounters with the analysis beam. This surface shadowing effect is most vividly revealed in the backscattering spectrum from clean, single crystals. The spectrum is dominated by the surface peak corresponding to interactions with the first few monolayers of the solid; at lower energies there is a continuum of

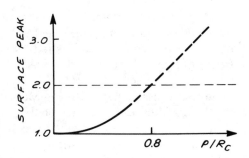

Figure 5.12 Surface peak intensity as a function of ρ/R_c from the "two-atom" calculation.

scattering corresponding to scattering by the relatively few nonchanneled particles. As described previously, the intensity of the surface peak is sensitive to the arrangement of surface atoms.

5.6 The Surface Peak

The application of ion beams to surface structure determination depends upon (1) accurate measurement of a surface peak in monolayers and (2) the ability to predict the surface peak for a given surface structure. For example, the aligned spectra in Figure 5.13 differ in their surface peak intensity for different surface structures: The reconstructed surface shows a higher yield than the ideal surface due to surface displacements.

More sophisticated treatments of the surface peak take into account that more than two atoms may contribute to the surface peak,

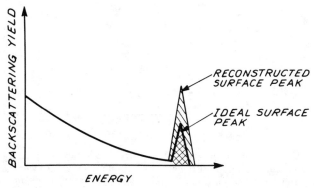

Figure 5.13 Channeling spectrum showing the surface peak for an ideal surface and a "reconstructed" surface.

and the governing potential is not a pure Coulomb description but a screened potential, for example, the Molière potential. The intensity of the surface peak is established by numerical techniques. The results of many calculations demonstrate that the surface peak scaling still holds, the scaling parameter is ρ/R_M, where R_M is the shadow cone radius associated with the Molière potential, roughly $R_M \simeq R_C$.

The values of the surface peak in terms of atoms/row have been calculated for a broad range of cases in which an ideal surface with a "bulklike" structure is assumed. The results are plotted in Figure 5.14, as a "universal curve" for the intensity of the surface peak as a

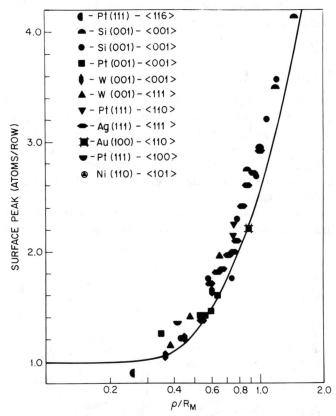

Figure 5.14 Comparison of the "universal" curve with experimental values for a number of different "bulklike" surfaces. The experimental values were determined from backscattering measurements. The notation Pt(111)-⟨116⟩ indicates a Pt crystal with a (111) surface plane; the backscattering measurement is in the ⟨116⟩ axial direction. [From Feldman et al. (1982).]

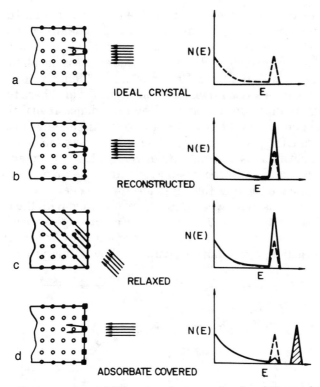

Figure 5.15 Representations of different surfaces on a simple cubic crystal. The back-scattering spectra shown on the right-hand side represent the expected signal from the different structures. The dashed line represents the signal from the "bulklike" crystal.

function of ρ/R_M, where ρ is the two-dimensional vibrational amplitude. Values of R_M may be found in Feldman et al. (1982). Also shown is a number of experimental measurements. The agreement between experiment and theory is extraordinarily good.

Four simple cases of surface structure are shown with their corresponding surface peak spectra in Figure 5.15. The dashed spectra represent the scattering yield from a crystal with an ideal surface for the case where the thermal vibration amplitude ρ is much less than the shadow cone radius R_M. This is the condition that assures that the surface peak intensity corresponds to one atom/row in this ideal case. The crystal with a reconstructed surface where the surface atoms are displaced in the plane of the surface (Figure 5.15b) represents a situation where the second atom is not shadowed. The surface peak intensity in this case is twice that of the ideal crystal. To test for relaxation where the surface atoms are displaced normal to the surface plane (Figure 5.15c), one must use nonnormal incidence, so that the shadow cone established by the surface atoms is not aligned with the atomic rows in the bulk. Here normal incidence would yield a surface peak intensity equivalent to one monolayer. These two measurements, then, at normal and oblique incidence reveal the presence of relaxation. A surface adsorbate atom can shadow the atoms of the substrate if $R_{adsorbate} > \rho_{substrate}$. The atomic mass sensitivity of ion scattering permits discrimination between substrate and adsorbate. In Figure 5.15d the adsorbate is positioned exactly over the surface atoms and hence reduces the substrate surface peak.

The comparison between a spectrum obtained with the incident beam aligned with the ⟨100⟩ axis of W, an aligned spectrum and with the spectrum obtained with the beam oriented away from a crystallographic direction, a random spectrum, is shown in Figure 5.16. The

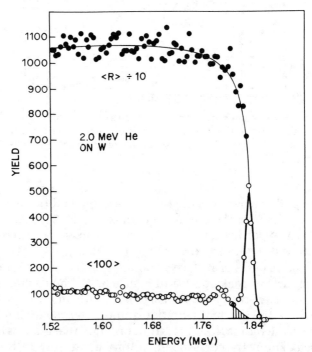

Figure 5.16 Backscattering spectra for 2.0 MeV He incident on a clean W(001) surface for the beam aligned with the ⟨100⟩ axis (O) and for the beam aligned away from any major crystallographic direction (●). Note that the nonchanneling spectrum ⟨R⟩, termed *random incidence*, has been reduced by a factor of 10. [From Feldman et al., 1982.]

surface peak is clearly visible and corresponds to about two atoms per row ($\rho/R_M = 0.65$). Scattering from the bulk of the crystal in this aligned geometry is two orders of magnitude smaller than the scattering yield under random incidence due to the bulk channeling effect. It is this suppression of the scattering from the bulk in the aligned spectrum that permits measurement of the surface peak.

5.7 Substrate Shadowing: Epitaxial Au on Ag(111)

An important application of ion scattering is the study of the initial stages of epitaxy. The ability to monitor epitaxial growth from the very first monolayer is indicated by Figure 5.15d. If the deposited atoms are in perfect registry with the substrate, the shadow cones established by these adsorbed atoms will shield the substrate atoms from the incident beam.

The shielding concept is demonstrated in Figure 5.17 for the deposition of monolayer coverage of Au on Ag(111). The upper portion of the figure shows a cross-sectional view of the (111) surface.

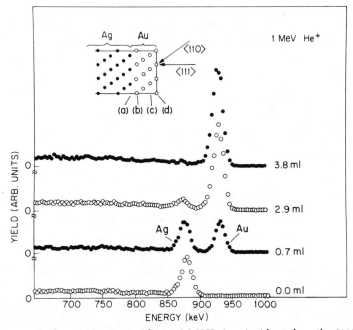

Figure 5.17 Backscattering spectra for 1.0-MeV He ions incident along the ⟨110⟩ axis of (a) a clean Ag(111) surface and Au-covered surfaces (b), (c), and (d). The Ag surface peak decreases because of the Au adsorption. [From Feldman et al., 1982.]

The plane of the figure is the (011) plane, which contains the ⟨111⟩ normal direction and the off-normal ⟨011⟩ direction. In the ⟨111⟩ direction the ion beam sees the first three monolayers of the uncovered Ag surface, whereas in the ⟨011⟩ direction only the first monolayer is visible. Figure 5.17 shows the backscattering spectra for 1.0-MeV ⁴He incident along the ⟨011⟩ direction of a clean Ag surface and a surface covered with approximately one, three, and four monolayers of Au. For the covered surface, the decrease in the Ag surface peak is direct evidence that the Au overlayer is registered with respect to the Ag substrate; that is, the Au is epitaxial on Ag. The determination of registry, monolayer by monolayer, is more sensitive for the ⟨011⟩ direction in which only one monolayer of Au can cover all the available Ag sites.

The decrease of the Ag surface peak as a function of Au coverage is shown in Figure 5.18 for low temperature (140° K) deposition and analysis in the ⟨011⟩ direction. The solid line is the result of computer simulations which assume that Au atoms form uniform coverage monolayer by monolayer. The agreement between the data and calculated curve shows that the Au has good epitaxy and uniform coverage.

The registration of the first monolayer of Au with the Ag substrate can be monitored through the reduction of the Ag surface peak. The registration of subsequent Au layers with the initial Au layer can

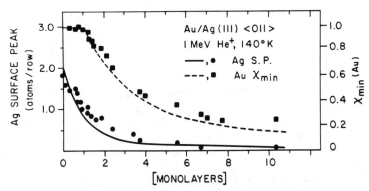

Figure 5.18 Intensity of the Ag surface peak as a function of Au coverage in terms of monolayers on a (111) surface for 1.0-MeV He incident along the ⟨110⟩ direction of a Ag(111) surface at 140° K. Also shown (right-hand scale) is the ratio of the Au signal in the aligned and random direction as a function of Au coverage. The solid line and dashed line are calculated assuming pseudomorphic, monolayer-by-monolayer growth. Bulk vibrations are used and correlations in thermal vibrations are not included. [From R. J. Culbertson, L. C. Feldman, and P. J. Silverman, *Phys. Rev. Lett.*, **47**, 657 (1981).]

be monitored with the Au–Au shadowing effect. The ratio of the Au signal for ⟨011⟩ incidence to random incidence [χ_{min} (Au)] exhibits a break at one monolayer coverage and a pronounced decrease at higher coverage (Figure 5.18). The dashed curve is a computer simulation. The agreement between the calculated curve and the data shows that the Au is indeed epitaxial. The decrease of the Ag surface peak with Au coverage measures the registry of the epitaxial layer to the substrate; the decrease of χ_{min}(Au) measures the quality of the epitaxial film.

5.8 Epitaxial Growth

The Au/Ag case is a good example of the growth of a high quality epitaxial film. This combination of materials satisfies the most im-

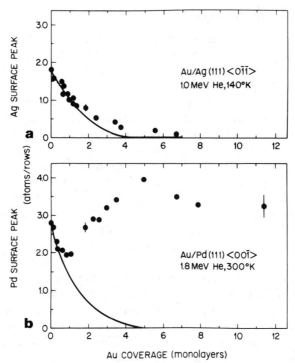

Figure 5.19 Shadowing of the substrate surface peak as a function of Au coverage. (a) Ag(111) along the ⟨110⟩ direction with deposition and analysis at 140° K using 1.0-MeV He$^+$; (b) For Pd(111) along the ⟨100⟩ direction with deposition and analysis at 300 K using 1.8-MeV He$^+$. The full curve is a theoretical line assuming a pseudomorphic, monolayer-by-monolayer growth and includes correlations in thermal vibration. [From Y. Kuk, L. C. Feldman, and P. J. Silverman, *J. Vac. Sci. Technol.* A **1**, 1060 (1983).]

portant criteria for epitaxy, namely, a good lattice constant match; in this case the mismatch is < 0.2% Figure 5.19 compares the reduction of the substrate surface peaks for the case of Au epitaxy on Ag(111) and Pd(111). In this latter case the mismatch is poor, ~4.7%. Notice that in both cases there is an immediate decrease in the surface peak corresponding to pseudomorphic growth. However, in Au/Pd this pseudomorphic growth is disrupted after only two layers. This is consistent with epitaxy theory which considers the strain in a mismatched, epitaxial film and the eventual onset of a misfit dislocations. The calculated thickness corresponding to this onset of dislocations for Au/Pd is ~2 monolayers. In a film with dislocations the overlayer atoms are not registered and hence do not shadow the substrate.

5.9 Thin Film Analysis

An important use of channeling is the suppression of scattering from the single crystal substrate. Scattering from amorphous overlayers such as oxides are not suppressed. The net result is an increased sensitivity to light impurities as well as structural information for the near surface region. Here we discuss an experiment which reveals the stoichiometry and subsurface strain in the SiO_2/Si system.

The sensitivity of these types of experiments has been improved by use of a grazing exit angle geometry as shown by the spectra in Figure 5.20 for a Si crystal with a small overlayer of SiO_2. The placement of the detector has no influence on the close-encounter interactions and channeling of the incident beam, but does influence the relation between detected energy widths and depth intervals. At grazing exit angles (Figure 5.20a) the outgoing path length of the emergent particles can be five times that for the path length near 180° scattering (Figure 5.20b). The stretching of the depth scale spreads the total number of detected scattering events in a given thickness over a greater energy interval and hence decreases the number of counts per energy channel in the bulk crystal. In the sample the surface layer (~13 Å of oxide) is thin compared to the depth resolution, so that the energy width of the signal is determined by the energy resolution of the detector system.

The scattering spectrum from the aligned Si crystal with a thin oxide layer consists of (1) a Si peak with scattering contributions from Si in the oxide, nonregistered Si in the surface region, and the intrinsic surface peak from the underlying single crystal and (2) at

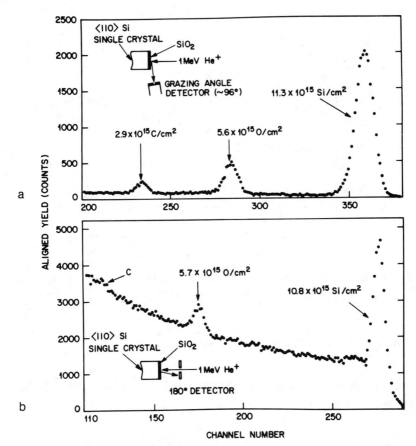

Figure 5.20 Energy spectra taken in channeling configuration from an Si(110) single crystal with a ~ 15 Å of SiO$_2$. Top figure shows a detector placed at grazing exit angle geometry, bottom figure with the detector at ~ 180°.

lower energy, an oxygen peak corresponding to the oxygen coverage. These peak areas can be converted to atoms/cm^2 with approximately 5% accuracy.

The Si versus oxygen intensities are shown in Figure 5.21 for a range of oxides up to ~ 40 Å. Over most of the range, the data are well fitted by a line expected for stoichiometric SiO$_2$ plus an additional offset of 8.6×10^{15} Si/cm^2. Most of this offset is the expected contribution from the intrinsic Si surface peak. The results from Figure 5.21 show that the oxide is primarily stoichiometric SiO$_2$ and the interface is sharp. The data suggest that the interface consists of either two

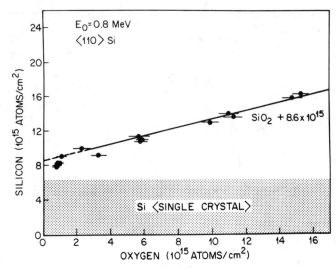

Figure 5.21 Si versus oxygen areal densities for a range of oxides up to ~40 Å. The data points are extracted from spectra of the type shown in Figure 5.20. The shaded portion represents the silicon surface peak intensity expected for clean Si(110) with a "bulk-like" surface structure. [From L. C. Feldman, I. Stensgaard, P. J. Silverman, and T. E. Jackman, in *Proceedings of the International Conference on Physics of SiO₂ and Its Interfaces*, S. T. Pantelides, Ed. (Pergamon Press, Oxford, 1978).]

monolayers of Si not registered to the bulk or a thin region (<5 Å) of nonstoichiometric oxide.

Problems

5.1. Calculate and compare values of the "standard" potential, the Thomas–Fermi potential, and the $1/r^2$ potential at $r/a = 0.01$, 0.1, and 1.0, for He ions incident on Si.

5.2. Cu is an fcc metal with a lattice constant of 3.615 Å. Calculate the axial and planar critical angles and minimum yield for 2 MeV ^{4}He ions incident along 100 axial and planar directions of Cu with a thermal vibration amplitude $\rho = 0.12$ Å.

5.3. Calculate the shadow cone radius R_c and the surface peak in the two atom and "universal" model (assume $R_c = R_M$) for 1.0 MeV ^{4}He ions incident along the 100 axis of Al (lattice constant = 4.05 Å and $\rho = 0.14$ Å). What energy is required for a unity surface peak intensity $(\rho/R_M = 0.4)$?

5.4. Use the r^2/r_0^2 approximation to the flux distribution Eq. (5.20), and calculate the minimum yield of an impurity displaced 0.1, 0.3, and 0.5 Å from the atomic row. Take the vibration amplitude as 0.1 Å and use an r_0 value appropriate to the Si $\langle 110 \rangle$ channeling direction.

5.5. Channeling occurs not only along the strings of atoms but between the sheets of atoms which make up atomic planes. In this case the potential governing the transverse notion can be described by a parabolic potential of the form

$$V(y) = \tfrac{1}{2}ky^2, \qquad 0 \le |y| \le d_p/2$$

where d_p is the planar spacing and y is the distance to the planar wall measured from the midpoint. Using the concept of motion in a harmonic potential derive a formula for the wavelength of the oscillatory motion. Evaluate this wavelength for 1.0 MeV ions channeling along the W(100) planes. The "spring constant" k can be estimated by noting that $V(d_p/2) = U_p(0)$ from Eq. (5.13).

References

1. J. U. Anderson, O. Andreason, J. A. Davies, and E. Uqgerhøj, *Rad. Eff.* **7**, 25 (1971).
2. B. R. Appelton and G. Foti, "Channeling," in *Ion Beam Handbook for Material Analysis*, J. W. Mayer and E. Rimini, Eds. (Academic Press, New York, 1977).
3. W. K. Chu, J. W. Mayer, and M.-A. Nicolet, *Backscattering Spectrometry* (Academic Press, New York, 1978).
4. G. Dearnaley, J. H. Freeman, R. S. Nelson, and J. Stephen, *Ion Implantation* (North-Holland, Amsterdam, 1973).
5. L. C. Feldman, J. W. Mayer, and S. T. Picraux, *Materials Analysis by Ion Channeling* (Academic Press, New York, 1982).
6. D. S. Gemmell, Channeling and Related Effects in the Motion of Charged Particles Through Crystals, *Rev. Mod. Phys.* **46**(1), 129–227 (1974).
7. J. Lindhard, *Mat. Fys. Medd. Dan. Vid. Selsk.* **34**(14), 1 (1965).
8. D. V. Morgan, Ed., *Channeling* (John Wiley and Sons, New York, 1973).
9. R. S. Nelson, *The Observation of Atomic Collisions in Crystalline Solids* (North-Holland, Amsterdam, 1968).
10. P. D. Townsend, J. C. Kelly, and N. E. W. Hartley, *Ion Implantation, Sputtering and Their Applications* (Academic Press, New York, 1976).

6

ELECTRON – ELECTRON INTERACTIONS AND THE DEPTH SENSITIVITY OF ELECTRON SPECTROSCOPIES

6.1 Introduction

Detection of an element from the near surface region of a solid often involves the measurement of an electron energy characteristic of a particular atom. The depth resolution of these techniques is then determined by the thickness of the material that an emitted electron can traverse without undergoing an inelastic event and thus altering the electron energy. We consider these inelastic electron processes in order to obtain a quantitative understanding of the thickness of the analyzed layer. An understanding of these phenomena is particularly useful in the design of surface studies, as an experimenter can often choose the electron energy and thus determine the depth probed.

In this chapter we use the particle scattering concepts developed in Chapters 2 and 3 to derive classical relations for electron – electron collisions. These provide a guide for useful approximations for electron escape depths and impact ionization cross sections.

6.2 Electron Spectroscopies: Energy Analysis

The surface analytical tools, photoelectron and Auger electron spectroscopy, discussed in the following chapters use photons or electrons to excite electrons that escape from the solid with sharply defined energies. Emitted electrons in the 100 eV range have escape depths of

125

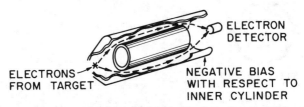

Figure 6.1 Schematic diagram of a cylindrical mirror analyzer used for electron energy detection in various electron spectroscopies.

the order of 10 Å. By choosing the appropriate incident beam parameters and detection systems these electron spectroscopies become extremely surface sensitive. Electron spectroscopies have found, by far, the widest application in surface analysis for a number of reasons. Electrons are easily focused into beams, are efficiently detected and counted, and may be analyzed with respect to angular and energy distribution using electrostatic lenses and deflection systems.

The electron spectroscopies are based on an analysis of the energy distribution of electrons emitted from the surface. The different features of the emission spectra and the requirements imposed by the analytical technique have led to the development of a variety of analyzers for measuring the electron energy distributions (see also Chapter 9). In the cylindrical mirror analyzer (CMA) (Figure 6.1), the emitted electrons are focused electrostatically in such a way that only those electrons with energies within a certain small range pass through the analyzer and arrive at the collector. Focussing is achieved by applying a potential V_a to create a cylindrical electric field between the two coaxial electrodes. The outer cylinder is held at a negative potential with respect to the inner cylinder. Electrons entering the analyzer through the annular entrance are deflected toward the inner cylinder. The analyzer design allows electrons with energy $E = eV_a$ and energy spread ΔE to pass through the exit slit to the collector; the energy resolution $\Delta E/E$ is usually between 0.1 and 1%. The transmission of the analyzer is high since it accepts electrons over a large solid angle. A modulating ac voltage is superimposed upon the cylinder potential to allow single and double differentiation so that the sharp characteristic energy peaks can be pulled out of the background.

6.3 Escape Depth and Detected Volume

For quantitative analysis it is important to determine the escape depth: the distance that electrons of a well-defined energy, E_c, can travel without losing energy (Figure 6.2). The incident radiation,

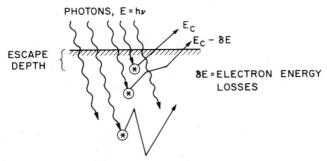

PHOTONS, E = hν

E_c

$E_c - \delta E$

ESCAPE DEPTH

δE = ELECTRON ENERGY LOSSES

Figure 6.2 Schematic of energetic photons incident on a surface and creating characteristic electrons relatively deeply in the solid. Only those electrons created near the surface escape with no loss of energy.

whether photons or electrons, is sufficiently energetic so that it penetrates deeply into the solid, well beyond the escape region for characteristic-energy electrons. The electrons that undergo inelastic collisions and lose energy δE in transport from the point of excitation to the surface leave the solid with a lower energy and contribute to a background signal or tail that can extend several hundred eV below the main signal peak.

In analogy with the experimental methods used to determine the escape depth, we consider the substrate as a source of a flux I_0 of electrons of well-defined energy E_c and deposit a thin film on the substrate. Any inelastic collisions within the thin film will remove electrons from the group of electrons of energy E_c. Consider that the cross section for the inelastic collision is σ and that there are N' scattering centers/cm³ in the deposited film. The number dI of electrons removed from initial group is σI per scattering center and the number of electrons removed per thickness increment dx is

$$-dI = \sigma I N' \, dx,$$

which gives

$$I = I_0 e^{-\sigma N'x}. \tag{6.1}$$

The mean free path λ is related to the cross section by definition as

$$1/\lambda = N'\sigma \tag{6.2}$$

so that Eq. (6.1) can be written as

$$I = I_0 e^{-x/\lambda}. \tag{6.3}$$

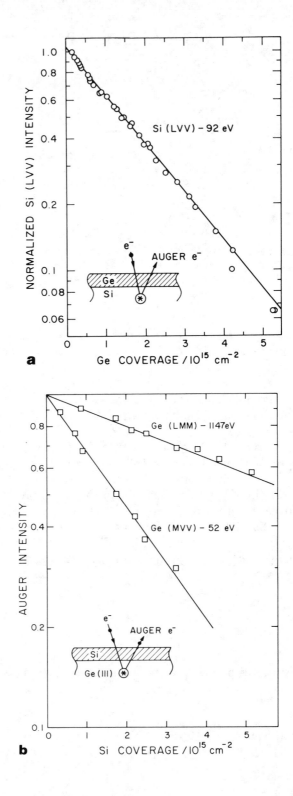

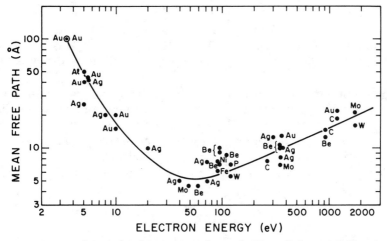

Figure 6.4 Universal curve for electron mean free path. [From G. Somerjai, *Chemistry in Two Dimensions: Surfaces* (Cornell University Press, Ithaca, N.Y., 1981), by permission.]

The number of electrons that can escape from the surface of the deposited absorber film then decreases exponentially with film thickness. In this discussion we treat the mean free path to be synonymous to the escape depth and use the same symbol λ. The yield of electrons from a solid excited uniformly in depth is given by $\int I(x)\,dx = I_0\lambda$, so that a thick substrate appears as a target of thickness λ.

A method used to characterize the attenuation of electrons is to monitor the signal from electrons generated in a substrate as a function of the thickness of a deposited overlayer of different metal. Figure 6.3 shows the relative intensity of Auger electrons (92 eV) from Si as a function of the thickness of an overlayer of Ge. The Ge thickness was determined from Rutherford backscattering analysis (Chapter 3). The data of Figure 6.3 shows that the intensity decreases exponentially with Ge film thickness with a decay length λ equivalent to 2.5×10^{15} Ge atoms/cm^2 (~ 5 Å). The attenuation of electrons depends upon the characteristic energy of the outgoing electrons. Figure 6.3b illustrates the energy dependence of the attenuation for two

Figure 6.3 (a) Attenuation of the Si LVV (92 eV) as a function of Ge coverage. The mean free path corresponds to ~ 5 Å of Ge. In this example the Ge deposition is at 300° K. (b) Attenuation of the Ge LMM (1147 eV) and Ge MVV (52 eV) Auger lines as a function of Si coverage. [From H. J. Gossmann, Ph.D. thesis, State University of New York, Albany, 1984.]

different characteristic energies of electrons emitted from Ge (LMM, 1147 eV; MVV, 52 eV) as a function of Si overlayer thickness. The mean free paths correspond to $9.81 \times 10^{15}/cm^2$ and $2.34 \times 10^{15}/cm^2$, respectively. The terminology for Auger transitions is discussed in Chapter 11.

These values of mean free paths are in agreement with other measurements of electron mean free paths shown in Figure 6.4. The data show that the mean free path is energy dependent with a broad minimum centered around 100 eV. The mean free path is relatively insensitive to the material traversed by the electrons. Such curves of mean free path versus energy have come to be called "universal curves."

6.4 Inelastic Electron-Electron Collisions

The cross section for an inelastic collision can be derived using the impulse approximation (Chapter 3) for scattering in a central force field. For an electron of velocity v the amount of momentum transferred to a target electron is [Eq. (3.2)]

$$\Delta p = \frac{2e^2}{bv},$$ (6.4)

where b is the impact parameter. Here we have taken the small angle scattering results from Chapter 3 with $Z_1 = Z_2 = 1$ and $M_1 = M_2 = m$ (Figure 6.5). Let T denote the energy transferred by an electron of energy $E = \frac{1}{2}mv^2$, then

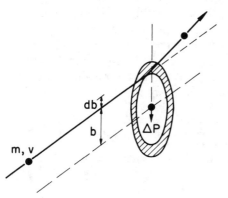

Figure 6.5 Schematic of an electron of momentum mv interacting with a free electron at an impact parameter b.

$$T = \frac{(\Delta p)^2}{2m} = \frac{e^4}{Eb^2}. \tag{6.5}$$

The differential cross section $d\sigma(T)$ for an energy transfer between T and $T + dT$ is

$$d\sigma(T) = -2\pi b \, db. \tag{6.6}$$

From Eq. (6.5), $2b \, db = -(e^4/ET^2) \, dT$, so that

$$d\sigma(T) = \frac{\pi e^4}{E} \cdot \frac{dT}{T^2}. \tag{6.7}$$

The cross section for an electron to transfer an energy between T_{min} and T_{max} is

$$\sigma_e = \int_{T_{min}}^{T_{max}} d\sigma(T), \tag{6.8}$$

$$\sigma_e = \pi \frac{e^4}{E} \left(\frac{1}{T_{min}} - \frac{1}{T_{max}} \right). \tag{6.9}$$

For energetic electrons with energy E of several hundred eV or greater, the maximum energy transfer $(T_{max} = E$ for $M_1 = M_2)$ is very much greater than T_{min}; therefore,

$$\sigma_e \cong \frac{\pi e^4}{E} \frac{1}{T_{min}} = \frac{6.5 \times 10^{-14}}{E T_{min}} \text{ cm}^2 \tag{6.10}$$

with E and T_{min} in eV and using the values $e^2 = 14.4$ eV Å.

6.5 Electron Impact Ionization Cross Section

The value of the cross section can be estimated from Eq. (6.10) with $T_{min} = E_B$,

$$\sigma_e = \frac{\pi e^4}{E E_B} = \frac{\pi e^4}{U E_B^2}, \tag{6.11}$$

where $U = E/E_B$ and E_B is the binding energy of an orbital electron. For incident energies less than E_B, $U < 1$, the cross section must be equal to zero. The actual shape of the cross section as a function of U is shown in Figure 6.6. The value of the cross section has a maximum near reduced energy values of $U \cong 3$ to 4. For $E_B = 100$ eV and $U = 4$, the value of the cross section is 1.6×10^{-18} cm^2. The value is in rea-

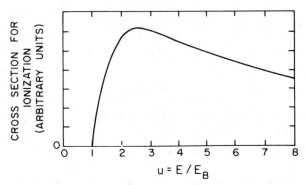

Figure 6.6 Ionization cross section versus reduced energy U for interactions within a solid. [From Chang, in Kane and Larrabee, Eds., *Characterization of Solid Surfaces* (Plenum Press, New York, 1974).]

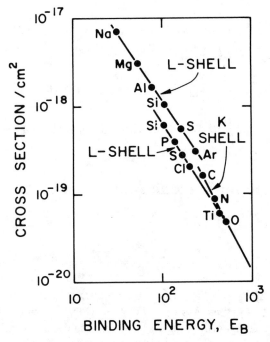

Figure 6.7 The maximum electron impact ionization cross section per electron (near $U = 4$) versus binding energy for various electron shells. [From Kirschner, in Ibach, 1977.]

sonable agreement with measured values of the maximum electron impact ionization cross section (measured near $U = 4$) that are shown in Figure 6.7.

6.6 Plasmons

In solids, the collective excitation of the conduction electron gas leads to discrete peaks in the energy loss of electrons. The plasmon is a quantum of a plasma oscillation and has an energy $\hbar\omega_p$ of about 15 eV.

From a classical viewpoint, the plasma frequency is determined by oscillations of the valence electrons in a metal with respect to the positively charged cores (Figure 6.8). Consider the fluctuation δr in radial distance r from a positive core of a free electron gas containing a concentration n of electrons. If the gas expands from its equilibrium radius δr, the number of electrons in the shell, $\delta n = 4\pi nr^2\delta r$, establish an electric field $\mathscr{E}$

$$\mathscr{E} = \frac{e}{r^2} \cdot \delta n = 4\pi ne\delta r. \tag{6.12}$$

The retarding force F created by the expansion is

$$F = -e\mathscr{E} = -4\pi e^2\delta rn. \tag{6.13}$$

The solution for the frequency of a harmonic oscillator with a force given in Eq. (6.13) is

$$\omega_p = \left(\frac{4\pi e^2 n}{m}\right)^{1/2}, \tag{6.14}$$

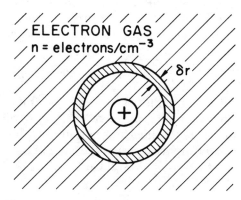

Figure 6.8 Electron gas with $\frac{4}{3}\pi r^3 n$ electrons around a positive core and undergoing a radical contraction δr.

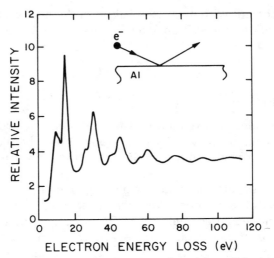

Figure 6.9 Energy loss spectrum for electrons reflected from Al for incident primary electron energy of 2 keV. The loss peaks are made up of a combination of surface and bulk plasmon losses.

where m is the mass of the electron. For metals, the value of $n \cong 10^{23}/cm^2$ gives an oscillator frequency $\omega_p = 1.8 \times 10^{16}$ rad/s and an energy $\hbar\omega_p = 12$ eV. The plasma frequency may be thought of as a "natural" frequency of the electron – ion system which is excited by an incoming charged particle.

The measured value of the plasmon energy for Mg is 10.6 eV and for Al is 15.3 eV. Figure 6.9 shows energy loss spectra for electrons reflected from a film of Al. The loss peaks are made up of combinations of the bulk plasmon $\hbar\omega_p = 15.3$ and the surface plasmon at an energy of 10.3 eV. The surface plasmon frequency $\omega_p(s)$ has the following relationship to the bulk plasmon:

$$\omega_p(s) = \frac{1}{\sqrt{2}}\,\omega_p; \tag{6.15}$$

this equation is found to hold for many metals and semiconductors. The calculated plasmon energies of Si and Ge are 16.0 eV based on four valence electrons per atom with the entire valence electron sea oscillating with respect to the ion cores. The measured values for Si are 16.4 – 16.9 eV and Ge are 16.0 – 16.4 eV.

6.7 The Electron Mean Free Path

The mean free path can be estimated using the general formulation for electron energy loss in a solid of n electrons/volume [see Eq. (3.10)]:

$$-\frac{dE}{dx} = \frac{4\pi e^4 n}{mv^2} \ln B, \tag{6.16}$$

where B represents a ratio of particle energy to excitation energy. The excitation of plasma oscillations, plasmons, in distant collisions is the dominant mode of energy loss of electrons in solid. The electron energy loss for 80 keV electrons traversing a thin NiAl foil is shown in Figure 6.10. The dominant feature in the figure is the energy distribution of electrons which have lost 17.8 eV, the bulk plasmon energy,

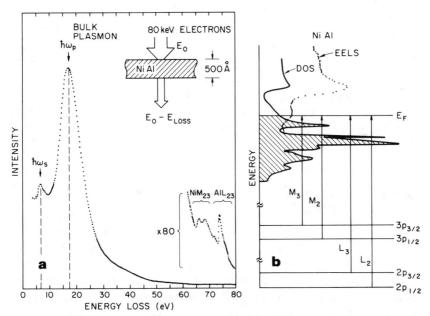

Figure 6.10 (a) Electron energy loss spectrum for 80 keV electrons transmitted through a 500 Å NiAl film. The dominant peak in the loss spectrum is the single plasmon loss at $\hbar\omega_p \sim 17.8$. The surface plasmon ($\hbar\omega_s = 7.1$ eV) and core excitation (Ni M_{23} and Al L_{23}) are indicated. (b) The calculated density of states (DOS) for NiAl (solid line) and Al L_{23} core excitation (data points) from EELS spectra similar to that shown in (a). Core level threshold transitions are shown by arrows. [From Grunes, Barbour, Hung, Mayer and Ritsko, *J. Appl. Phys.* **56**, 168 (1984).]

in transmission through the film. This suggests that the energy losses occur in quantum jumps of $\hbar\omega_p$, and we set

$$B = \frac{2mv^2}{\hbar\omega_p} \tag{6.17}$$

and from $\omega_p = (4\pi e^2 n/m)^{1/2}$ [Eq. (6.14)] write the energy loss in terms of ω_p,

$$-\frac{dE}{dx} = \frac{\omega_p^2 e^2}{v^2} \ln\frac{2mv^2}{\hbar\omega_p}. \tag{6.18}$$

A similar formulation is presented by Ibach (1977).

If we treat the plasmon as the major source of energy loss in determining the mean free path λ for electrons, we can write λ as

$$\frac{1}{\lambda} = \left(-\frac{dE}{dx}\right) \cdot \frac{1}{\hbar\omega_p}, \tag{6.19}$$

which gives

$$\frac{1}{\lambda} = \frac{\omega_p e^2}{\hbar v^2} \ln\frac{2mv^2}{\hbar\omega_p}. \tag{6.20}$$

For example, we calculate the value of λ to be 9.2 Å for 350 eV electrons in Al ($\hbar\omega_p = 15$ eV, $v^2 = 2E/m = 1.23 \times 10^{18}$ cm^2/s^2 and $\hbar = 6.6 \times 10^{-16}$ eV s). This value is in reasonable agreement with the data presented in Figure 6.4.

6.8 Influence of Thin Film Morphology on Electron Attenuation

One of the many prominent uses of electron spectroscopies is the characterization of different modes of film growth. Questions of the growth mode are particularly important in the creation of uniform layered films in which one desires extreme uniformity in composition for films thinner than 100 Å. Common types of growth include:

1. *Layer-by-layer*, in which the deposited film completes one monolayer of coverage, then the second, etc. This mode is commonly referred to as *Frank–van der Merwe growth*.

2. *Layer plus islanding*, in which the first layer completely covers the surface of the substrate and subsequent layers form islands of deposited material. This mode is referred to as *Stranski–Krastanov growth*.

3. *Complete islanding*, in which the material immediately forms islands on the surface. This mode is commonly referred to as *Volmer–Weber growth.*

4. *Statistical deposition*, in which the growth corresponds to the random occupancy of surface sites according to Poisson statistics.

6.8.1 Layer-by-Layer Growth

In the following discussions we continually make use of the exponential probability of inelastic scattering for a characteristic electron emerging from a solid. As a first example, consider the attenuation of electrons from a substrate as a result of an overlayer which deposits in a layer by layer mode. Although the derivation here is for attenuation of substrate electrons, very similar considerations can be used to derive formulae for the *increase* of the electron yield from the overlayer.

A typical experiment involves a measurement of the intensity I_s of a characteristic Auger electron from the substrate as a function of overlayer thickness t. (Naturally, the overlayer is a different material.) The mean free path of the substrate Auger electron in the material of the overlayer is λ. Then for coverages up to one monolayer the Auger intensity can be written as

$$I_s/I_{s_0} = (1 - x) + xe^{-l/\lambda}, \quad 0 \le x \le 1, \tag{6.21}$$

where x is the fraction of the surface covered by the overlayer, I_{s_0} is the intensity from the clean surface, and l is the thickness of a monolayer. In this regime, up to one monolayer, the dependence of the intensity on coverage is linear in x.

A similar formula can be written for the region from 1 to 2 monolayers as

$$I_s/I_{s_0} = (1 - x)e^{-l/\lambda} + xe^{-2l/\lambda}, \quad 0 \le x \le 1, \tag{6.22}$$

where x is now the fraction of the surface covered with two layers and $(1 - x)$ is the fraction covered with one layer. Again in the region of coverage of one to two layers, the attenuation is linear in x.

In general, a similar formula can be written for the transition from the n to the $(n + 1)$ layer as

$$I_s/I_{s_0} = (1 - x)e^{-nl/\lambda} + xe^{-(n+1)l/\lambda}, \quad 0 \le x \le 1, \quad n = 0, 1, \cdots, \tag{6.23}$$

where $(1 - x)$ is the fraction of the surface covered with n layers and x is the fraction covered with $(n + 1)$ layers. The characteristic shape of this curve is a series of straight lines (on a linear plot) with breaks at

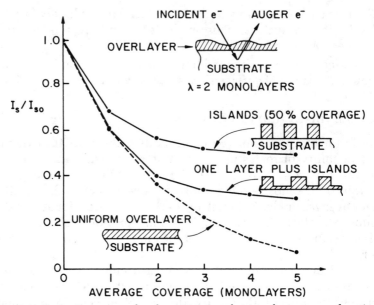

Figure 6.11 Extinction curves for characteristic substrate electrons as a function of average coverage of an overlayer. In these examples the mean free path, λ, is taken as two monolayers, close to the minimum mean free path achievable. In the case of the uniform overlayer the growth is assumed to be layer-by-layer and the extinction curve is a series of straight lines with the envelope of points at integral coverage corresponding to an exponential decay. The other curves correspond to a "single layer plus islanding" growth mode and a case of pure islanding. For the cases involving islanding it is assumed that 50% of the surface is covered.

coverages corresponding to an integral number of monolayers. The envelope of points corresponding to integral coverages describes an exponential decay of the form $e^{-nt/\lambda}$ (Figure 6.11).

6.8.2 Single Layer plus Islanding

A second type of growth mode corresponds to the deposition of a single uniform layer which is then followed by islanding. From the discussion in the previous section we can write the attenuation up to one monolayer coverage as

$$I_s/I_{s_0} = (1 - x) + xe^{-1/\lambda}, \quad 0 \le x \le 1, \tag{6.24}$$

where x represents the fraction of the covered surface. In the second stage of growth the deposited material forms islands of unspecified dimensions and we cannot write a simple analytical formula as the

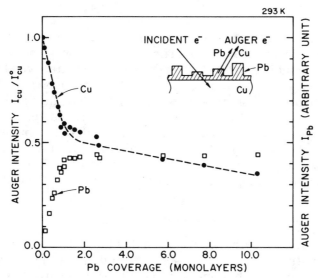

Figure 6.12 Auger yield curve for the deposition of Pb on Cu(100) at room temperature. The solid circles represent the attenuation of the Cu substrate signal and the open squares the growth of the Pb signal. This type of growth corresponds to the "one layer plus islanding" mode. In this example the (average) Pb coverage is measured by RBS. [From R. J. Culbertson, unpublished data.]

actual attenuation will depend on the fraction of covered surface. As an example, however, we take as a very simple case that the islands cover 50% of the surface. Then for the second monolayer's worth of material the coverage formula can be written as

$$I_s/I_{s_0} = (1 - x)e^{-1/\lambda} + xe^{-3l/\lambda}, \quad 0 \leq x \leq 0.5, \qquad (6.25)$$

where x is the fraction of the surface covered with "2 monolayer high" islands. A similar formula can be written for any coverage regime but there will always be a term of magnitude $0.5e^{-1/\lambda}$ in the formula, plus additional positive contributions. The net result is that the Auger yield from the substrate is always finite and does not approach the zero yield associated with layer by layer growth. The decay curve for this type of growth is shown schematically in Figure 6.11. An actual case is shown in Figure 6.12 for the growth of Pb on Cu(100). Note the almost constant level of the Cu Auger yield beyond 1 monolayer of coverage. Such curves can be fitted with simple models of islands as described above. The Auger signal from the Pb overlayer increases with coverage but saturates again consistent with islanding.

6.8.3 Islanding

As described in the preceding section, the substrate yield curve as a function of coverage does not show a simple exponential decay in cases of islanding. In this type of growth the substrate yield remains high, since some fraction of the substrate is not covered by the overlayer. Figure 6.11 shows the expected curve for the simple case of islands covering 50% of the surface. Obviously the yield can never have values below 0.5 in this scheme. The differences in the extinction curves (Figure 6.11) reveal the different modes of growth. The interpretation of these curves requires an accurate measurement of the absolute coverage *and* knowledge of the mean free path.

6.8.4 Distribution of Deposited Atoms

By a statistical distribution we mean that the deposited atoms simply reside on the surface in a Poisson distribution — as if they impinged randomly and simply stuck at the landing site. For an average coverage θ, the probability of finding a structure k atom layers high is

$$P_k = \frac{\theta^k e^{-\theta}}{k!} \qquad (6.26)$$

so that

$$P_0 = e^{-\theta},$$

the usual coverage formula for the fraction of the surface not covered by the adsorbate.

Then the Auger intensity factor is given for layers of thickness l by

$$I_s/I_{s_0} = \sum_k P_k e^{-kl/\lambda} \qquad (6.27)$$

or

$$I_s/I_{s_0} = e^{-\theta} \sum_k \frac{\theta^k}{k!} e^{-kl/\lambda}. \qquad (6.28)$$

This expression can be rewritten as

$$I_s/I_{s_0} = e^{-\theta} \sum_{k=0} \frac{(\theta e^{-l/\lambda})^k}{k!} \qquad (6.29)$$

or

$$I_s/I_{s_0} = e^{-\theta(1-e^{-l/\lambda})}. \qquad (6.30)$$

Note that this attenuation curve is a pure exponential for all coverages! The effective decay length is now $(1 - e^{-l/\lambda})^{-1}$ rather than λ.

6.9 Range of Electrons in Solids

In materials analysis energetic electrons are used to generate inner-shell vacancies which decay by Auger emission or X-ray emission. In measurements of the emitted characteristic X-rays (the electron microprobe), the concern is with the depth over which the X-rays are generated. The situation is more complex for incident electrons with energies between 1 and 50 keV than for heavy ions whose path is relatively straight over most of the range. For electrons significant deviations from the incident direction occur due to elastic scattering. Monte Carlo calculations of the paths of electrons in Fe ($E_0 = 20$ keV) are shown schematically in Figure 6.13. The elastic scattering is composed of both large-scale scattering events and multiple small-angle

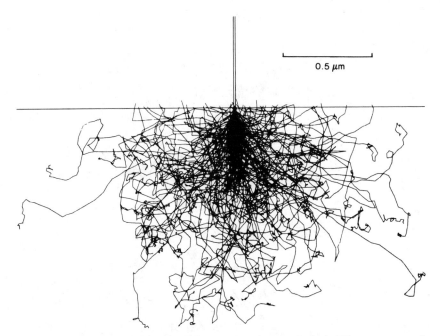

0.5 μm

Figure 6.13 Monte Carlo electron trajectory simulation of a 20 keV beam at normal incidence on Fe. The density of trajectories gives a visual impression of the interaction volume. [From Goldstein et al. (1981).]

scattering events which can also lead to a large change in the direction of the electrons.

The electron range R is defined as the total distance that an electron travels in the sample along a trajectory and can be written as

$$R = \int_{E_0}^{0} \frac{dE}{dE/dx},$$ (6.31)

where the energy loss expression has been discussed previously. The energy loss formula, dE/dx, is of the form

$$\frac{dE}{dx} \propto \frac{NZ_2}{E} \ln \frac{E}{I}$$ (6.32)

or

$$\frac{dE}{dx} \propto \frac{\rho}{E} \ln \frac{E}{I},$$ (6.33)

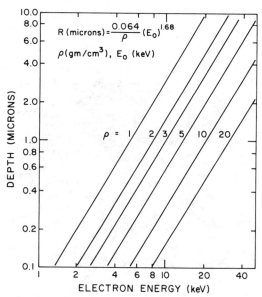

Figure 6.14 The electron range R (μm) versus incident electron energy for different density, ρ, materials. The lines are calculated from Eq. (6.24) with $K = 0.064$ and $\gamma = 1.68$.

where N is the atomic density so that NZ_2 is proportional to ρ and I is the average ionization energy, $I \sim 10Z_2$ (eV). From experimental results, the dependence of the range on incident energy has the form

$$R = \frac{K}{\rho} E_0^\gamma, \tag{6.34}$$

where ρ is the density (g/cm^3), K is a material independent constant, and γ varies from 1.2 to 1.7. It is convenient to use the mass range ρR since to a first approximation ρR is independent of target material. The electron range R as a function of energy is given in Figure 6.14 for $K = 0.064$ and $\gamma = 1.68$.

The mass range ρR_x for characteristic X-ray production is smaller than the range of electrons since characteristic X-rays can only be produced at energies above the critical excitation energy, or binding energy E_B, for a given element. The mass range for characteristic X-ray production is given by

$$\rho R_x = K(E_0^\gamma - E_B^\gamma). \tag{6.35}$$

As discussed by Goldstein et al. (1981), the mass range equation for X-ray production can be expressed as

$$\rho R_x = 0.064(E_0^{1.68} - E_B^{1.68}) \tag{6.36}$$

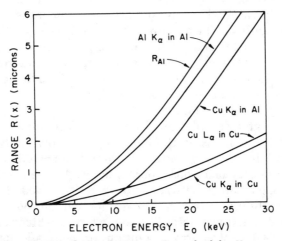

Figure 6.15 The effective electron ranges in Cu and Al for X-ray production. The ranges denote the end point of penetration of electrons for generation of characteristic K_α and L_α X-rays. [From Goldstein et al., 1981.]

where E_0 and E_B are in keV, ρ is in g/cm^3, and R_x is in microns (μm). Figure 6.15 shows the electron range R in Al and R_x for the Al K$_\alpha$ and Cu K$_\alpha$ lines generated in Cu-doped Al and the Cu K$_\alpha$ and Cu L$_\alpha$ lines in pure Cu. The ranges for X-ray production depend of course on the density of the matrix (Al = 2.7 g/cm^3, Cu = 8.9 g/cm^3) and on the value of E_B (Cu K$_\alpha$, E_B = 8.98 keV; Cu L$_\alpha$, E_B = 0.93 keV).

6.10 Electron Energy Loss Spectroscopy (EELS)

The characteristic energy losses of electron beams penetrating through a film or reflected from a surface can give important information on the nature of the solid and the relevant binding energies. Electron energy loss spectroscopy (EELS) is carried out from ≲ 1 eV to ~ 100 keV. The choice is based on a variety of experimental considerations and the energy range of interest. The low energy regime is used primarily in surface studies where the investigation centers on the energies of vibrational states associated with absorbed molecules. The energy loss spectrum contains discrete peaks corresponding to the vibrational states of absorbed molecules.

At higher energies, as shown in Section 6.7, the dominant peak corresponds to a plasmon loss or losses. A detailed examination of the energy loss spectrum would also show discrete edges in the spectra corresponding to excitation and ionization of atomic core levels. These features represent a means of element identification, particularly useful in cases where the spatial resolution of the electron microscope is required. The features tend to be broad since the incident electron can transfer a continuum of energies to the bound electron. For example, a core electron may be excited to the unoccupied states within the solid (excitation) or actually ejected from the solid (ionization). The cross section tends to strongly favor small energy transfers, thus making excitation dominant. An examination of the loss features under high resolution can then yield information on the unoccupied density of states. In the following we show EELS spectra of Ni$_x$Si$_y$ films using the inelastic spectrum from an ~ 100 keV electron beam. High energy electrons, approximately 100 keV, are used because the long distance between collisions, about 500–1000 Å, permits examination of self-supporting films that can be mounted on conventional electron microscopy sample grids. Electrostatic analyzers of 0.1–0.5 eV energy resolution are generally used so that changes in the density of states can be monitored.

   The energy loss of 80 keV electrons transmitted through 500 Å films of crystalline NiAl was given earlier in Figure 6.10. The dominant feature in the spectrum is the large bulk plasmon peak, labeled $\hbar\omega_p$, that is centered at 17.8 eV. This resonance involves all the valence electrons and is displaced to higher energy loss in NiAl than the bulk plasmon found at 15.0 eV in Al. The sensitivity of the bulk plasmon to the composition of the sample is in Figure 6.16 which gives spectra for samples of Si, $NiSi_2$ (the most Si rich nickel silicide), Ni_2Si, and Ni. The bulk plasmon peaks become broader and are centered at increasingly higher energies with increasing Ni concentration. Silicon, $NiSi_2$, and Ni_2Si have bulk plasmon energies, $\hbar\omega_p$, of 16.7, 17.2, and 21.8 eV, respectively. The spectra in Figure 6.16 were scaled such that the plasmon peak heights are equivalent, even though the absolute intensity of the Si plasmon peak is much greater than the corresponding peaks in the Ni spectrum.

   The bulk plasmon may also be used to estimate the relative amount of plural scattering. Scattering of the incident electron beam by two sequential bulk plasmon events produces a peak in the loss function at twice the bulk plasmon energy, as can be seen in the peak in the Si spectrum at 33.4 eV in Figure 6.16. From the figure we can

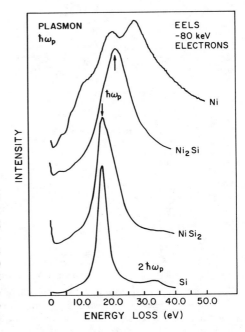

Figure 6.16 Electron energy loss spectra for 80 keV electrons incident on thin (about 400 Å) self-supporting films of Si, $NiSi_2$, Ni_2Si and Ni films. The positions of the bulk plasmon peaks, $\hbar\omega_p$, are shown for the two nickel silicides. [From Barbour et al., in *Thin Films and Interfaces II*, Baglin, Campbell and Chu, Eds., North-Holland, Amsterdam, 1984).]

see that the ratio of the double to single plasmon intensity is quite small, which indicates that the sample thickness is less than the mean free path for excitation of bulk plasmons.

In the EELS spectrum of NiAl (Figure 6.10), weak but sharp peaks appear at high energy loss values of ~70 eV. These correspond to excitation of individual deeply bound core electrons to the unfilled conduction band states. The Al L_{23} transition of about 75 eV in the spectrum of Figure 6.10 corresponds to excitations of the Al 2p core

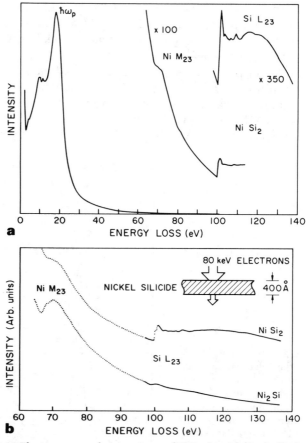

Figure 6.17 (a) Electron energy loss spectrum of $NiSi_2$ showing the bulk plasmon peak and the characteristic Ni M_{23} and Si L_{23} core level excitations. (b) A comparison of EELS spectra in the energy region of the characteristic Ni and Si core level excitations. [From Barbour et al., *Ultramicroscopy* **14**, 79 (1984).]

electrons to unfilled states above the Fermi level. In this figure the measured Al L_{23} data points are aligned with calculated values (solid line) of the density of states (DOS). The experimental points mirror the DOS shape indicating that EELS measurements can be used to determine the density of states above the conduction band.

The energy loss spectra for nickel silicides also display features due to core excitations. Figure 6.17a is an electron energy loss spectrum of a 400 Å thick $NiSi_2$ self-supporting film taken over the energy region from 0 to 138 V. The largest peak is the bulk plasmon ($\hbar\omega_p$) with the Ni M_{23} and Si L_{23} core excitation peaks at higher energies, magnified by 100 and 350 times, respectively. No multiple scattering events are discernible in the EELS spectrum (no higher order plasmon loss peaks are present), indicating that the background on the low energy loss side of the Ni before the Ni M edge is primarily due to the tail of the plasmon peak. The heights of the steps at the Ni M and Si L edges can be used to determine the composition.

At energies below the bulk plasmon peak, energy loss values 0–15 eV in Figure 6.17a, there are peaks in the spectrum which correspond to interband transitions. The interband transitions involve a convolution of the valence and conduction band densities of states and hence are more difficult to interpret than core level spectra where the initial states are sharp.

The differences in the spectrum step heights at the Ni and Si edges for different Ni to Si concentration ratios can be seen in Figure 6.17b which gives spectra taken from samples of Ni_2Si and $NiSi_2$. The yield Y_A for detection of an incident electron which loses energy E_A when passing through a material of thickness t containing a concentration N_A of atoms is given by

$$Y_A = QN_A t\sigma_A \eta \Omega, \tag{6.37}$$

where Q is the integrated incident electron current density, $N_A t$ is the number of A atoms contributing to the inelastic scattering events, σ_A is the cross-section for excitation of an electron in a given core energy level of atom A, η is a collection efficiency, and Ω is the detector collection angle. Equation (6.37) assumes the collected electrons experience only single inelastic scattering events. Provided that the collection efficiency in scattering from A atoms is equal to that in scattering from B atoms, then the atomic ratio of A atoms to B atoms is

$$(N_A/N_B) = (Y_A/Y_B)(\sigma_B/\sigma_A), \tag{6.38}$$

where Y_A and Y_B can be experimentally measured as the areas above background in an energy window above the edge. Therefore, the accuracy of the atomic ratio is sensitive to the cross section calculations and the accuracy in determining the relative Ni M to Si L areas by fitting the background after the edges.

Electron energy loss spectroscopy is not the most straightforward or sensitive method of detecting average composition or trace impurities. Its major advantage is analysis of small areas (<1000 Å) for detecting microprecipitates and composition variations.

6.11 Bremsstrahlung

The energy loss of electrons passing through matter contains an additional significant component corresponding to the radiation loss. Classical physics tells us that an accelerated charge emits radiation. This acceleration is caused, for example, by the deflection of an electron in the field of an atom as the charged particle penetrates matter. Since the acceleration is essentially the ratio of the electrostatic force to the mass, this radiation component is significantly more important for electrons than heavy projectiles. As we will show, this bremsstrahlung (German for "braking radiation") yields a continuum of photons up to the energy of the incident electron. It is of interest to the materials analyst since bremsstrahlung may provide a significant background in analysis techniques that use incident electrons and detect characteristic photons. It is most obvious in electron-μ-probe analysis as a background underlying the characteristic X-ray spectrum. ·

The elastic scattering cross section of a charged particle, Z_1, by a nucleus of charge, Z_2, is given by Eq. (2.17):

$$\frac{d\sigma}{d\Omega} = \left(\frac{Z_1 Z_2 e^2}{4E}\right)^2 \frac{1}{\sin^4 \theta/2}, \qquad (6.39)$$

where θ is the scattering angle. It is convenient to express this cross section in terms of the associated momentum transfer Δp (Chapter 2).

$$\Delta p = 2p \sin \theta/2 \qquad (6.40)$$

and

$$d\Omega = 2\pi \sin \theta \, d\theta = \frac{2\pi \Delta p \, d\Delta p}{p^2}$$

Then the cross section for a momentum transfer Δp is given by

$$\frac{d\sigma}{d\Delta p} = 8\pi \left(\frac{Z_1 Z_2 e^2}{v}\right)^2 \frac{1}{(\Delta p)^3}. \qquad (6.41)$$

Classical electromagnetic theory shows that the total energy radiated per unit frequency interval per collision is (Jackson, 1975):

$$\frac{dI}{d\omega} = \frac{2}{3\pi} \frac{(Z_1 e)^2}{m^2 c^3} \Delta p^2, \tag{6.42}$$

where m is the mass of the deflected particle. This formula is derived for the nonrelativistic case and in the limit of low ω; both conditions are of interest here. Then the differential radiation cross section is

$$\frac{d^2\chi}{d\omega \, d\Delta p} = \frac{dI}{d\omega} \cdot \frac{d\sigma}{d\,\Delta p},$$

that is, the probability of emitting a photon of energy $\hbar\omega$ associated with a momentum transfer Δp times the probability of a momentum transfer, Δp. Explicitly

$$\frac{d^2\chi}{d\omega \, d\Delta p} = \frac{16}{3} \frac{Z_2^2 e^2 (Z_1^2 e^2)^2}{m^2 v^2 c^3} \frac{1}{\Delta p}. \tag{6.43}$$

We integrate over all possible energy transfers Δp_{min} to Δp_{max} to find the frequency spectrum,

$$\frac{d\chi}{d\omega} = \frac{16}{3} \frac{Z_2^2 e^2 (Z_1^2 e^2)^2}{m^2 v^2 c^3} \ln \frac{\Delta p_{max}}{\Delta p_{min}}.$$

To determine the ratio, $\Delta p_{max}/\Delta p_{min}$, we consider in detail the kinematics of the process. Energy conservation and momentum conservation can be written as:

$$E = E' + \hbar\omega$$

and

$$(\Delta p)^2 = (\mathbf{p} - \mathbf{p}' - \mathbf{k})^2 \simeq (\mathbf{p} - \mathbf{p}')^2, \tag{6.44}$$

where E, p and E', p' refer to the energy and momentum of the particle before and after the collision and $\hbar\mathbf{k}$ is the momentum of the emitted Bremsstrahlung photon. Here we have neglected the momentum associated with the photon. Then

$$\frac{\Delta p_{max}}{\Delta p_{min}} = \frac{p + p'}{p - p'} = \frac{(\sqrt{E} + \sqrt{E - \hbar\omega})^2}{\hbar\omega}$$

so that

$$\frac{d\chi}{d\omega} = \frac{16}{3} \frac{Z_2^2 e^2 (Z_1^2 e^2)^2}{m^2 v^2 c^3} \ln\left(\frac{2 - \hbar\omega/E + 2\sqrt{1 - \hbar\omega/E}}{\hbar\omega/E}\right). \tag{6.45}$$

This is a function which falls off as $(\hbar\omega/E)$ for small $\hbar\omega/E$ and flattens out with a sharp cut-off at $E = \hbar\omega$ (Figure 6.18). This form of the Bremsstrahlung spectrum was first derived by Bethe and Heitler in 1934.

The radiation cross section, $d\chi/d\omega$ is proportional to $Z_2^2 Z_1^4 / m^2$ showing that the emission is most important for light particles (electrons) in materials of high atomic number.

The total energy lost by radiation for a particle traversing a material of N nuclei per volume is

$$\frac{dE_{\text{rad}}}{dx} = N \int_0^{\omega_{\text{max}}} \frac{d\chi(\omega)}{d\omega}\, d\omega$$

Letting $x = \hbar\omega/E$ and noting that

$$\left(\frac{1 + \sqrt{1-x}}{\sqrt{x}}\right)^2 = \frac{2 - x + 2\sqrt{1-x}}{x}$$

we have

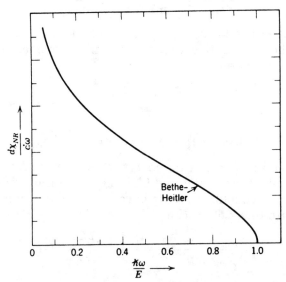

Figure 6.18 Radiation cross section (energy · area/unit frequency) for nonrelativistic Coulomb collisions as a function of $\hbar\omega/E$, Eq. (6.45).

$$\frac{dE_{rad}}{dx} = \frac{16}{3} \frac{NZ_2^2 e^2 (Z_1^2 e^2)^2}{c^3 \hbar} \int_0^1 \ln\left(\frac{1 + \sqrt{1-x}}{\sqrt{x}}\right) dx \qquad (6.46)$$

The dimensionless integral has the value of unity. The radiative loss is essentially energy independent.

The ratio of radiative loss (for electrons) to nonradiative loss, dE_{rad}/dE_{nr}, is approximately given by

$$\frac{dE_{rad}}{dE_{nr}} \simeq \frac{4}{3\pi} \frac{Z_2}{137} \left(\frac{v}{c}\right)^2 \qquad (6.47)$$

and is small for $v < c$.

For electron probe analysis we are concerned that the Bremsstrahlung photon may obscure the characteristic energy of interest. As shown by Evans, the ratio of the ionization cross section to the radiative cross section is roughly $Z_2(v/c)^2/137$, or ~ 0.01 for $Z_2/137 = \frac{1}{4}$ and 100 keV electrons. This is what sets the limit for detecting approximately 1% of an impurity in a solid matrix.

Problems

6.1. Draw the curve of Auger yield versus overlayer thickness for the signal from the overlayer atoms in the case of
(a) Uniform growth in a layer by layer fashion and
(b) Island growth with 50% coverage.
Assume that the mean free path λ is equivalent to two monolayers.

6.2. Typically electron impact ionization cross sections have a maximum at $E/E_B = 3$. Calculate the value of E at σ_{max} for the K shell and L shell of Si for incident electrons. As discussed in the text, ionization cross sections depend primarily on the velocity of the incident charged particles; calculate the energies of incident protons where the protons have the same velocities as that of the electrons in the examples for the ionization of K and L shell electrons.

6.3. You deposit an Al film on a copper substrate, and use 3 keV electrons to excite Cu L_α X-rays for the determination of Cu penetration into Al. What thickness of the deposited Al film is required so that the electrons do not excite Cu L_α X-rays in the Cu substrate. Use values of $K = 0.064$ and $\gamma = 1.68$, and Eq. (6.34).

6.4. You are carrying out a transmission electron energy loss experiment with 100 keV electrons incident on an Al film where the bulk plasmon energy is 15.3 eV. Estimate the value of λ for 100 keV electrons. Compare the values of the range of 100 keV electrons using Eq. (6.34) with that of Eq. (6.31) where you assume that dE/dx is a constant evaluated at 50 keV.

6.5. The Si LVV Auger electron is emitted with 92 eV. Assuming that the only mechanism for energy loss is ionization of Si atoms (Appendix 4), calculate the mean free path for these electrons:
(a) With the relations given in Section 6.7 and
(b) With $\lambda = 1/N\sigma$, where σ is the cross section given in equation 6.11.

6.6. In a vacuum system, the flux of gas atoms impinging on a surface is $Nv/4$, where N is the density of atoms per cm^3 and v is the thermal velocity of the gas atoms. Assuming every gas atom sticks, calculate the value of N such that an absorbed layer of oxygen atoms is thinner than two monolayers after one hour. Express N in Torr noting that one atmosphere (760 torr) is approximately 2×10^{19} atoms/cm^3. It is these basic requirements that determine the need for good vacuum in thin film analysis.

6.7. Consider a semiconductor structure consisting of 25 Å of Si, one monolayer of Ge, 25 Å of Si, and one monolayer of Ge on a thick Si substrate. Imagine an Auger/sputter profiling analysis of this structure in which one detects the Ge MVV Auger line ($\lambda = 10$ Å in Si). Assume that the sputtering process removes one monolayer of material at a time with no interface broadening or other mixing effects. Write an equation for the yield of Ge as a function of material removed and plot the Ge profile expected in this kind of analysis. (Ignore the effect of the single monolayers of Ge in the extinction path.)

References

1. C. R. Brundle and A. D. Baker, Eds., *Electron Spectroscopy: Theory, Techniques and Applications* (Academic Press, New York, 1981).
2. M. Cardona and L. Ley, Eds., *Photoemission in Solids, I, II*, Topics in Applied Physics, Vols. 26 and 27 (Springer-Verlag, New York, 1978 and 1979).
3. T. A. Carlson, *Photoelectron and Auger Spectroscopy* (Plenum Press, New York, 1975).
4. A. W. Czanderna, Ed., *Methods of Surface Analysis* (Elsevier Science Publishing Co., New York, 1975).

5. G. Ertl and J. Kuppers, *Low Energy Electrons and Surface Chemistry* (Verlag Chemie International, Weinheim, 1974).

6. R. D. Evans, *The Atomic Nucleus* (McGraw-Hill, New York, 1955).

7. J. I. Goldstein, D. E. Newbury, P. Echlin, D. C. Joy, C. Fiori, and E. Lifshin, *Scanning Electron Microscopy and X-ray Microanalysis* (Plenum Press, New York, 1981).

8. H. Ibach, Ed., *Electron Spectroscopy for Surface Analysis*, Topics in Current Physics, Vol. 4 (Springer-Verlag, New York, 1977).

9. J. D. Jackson, *Classical Electrodynamics*, 2nd Edition (John Wiley and Sons, New York, 1975).

7

SURFACE STRUCTURE

7.1 Introduction

Throughout this book we concentrate on elemental analysis in the near surface region of a solid. The techniques yield elemental concentration as a function of depth from which compound phases can be inferred. Most materials tend to form distinct crystalline phases — i.e., ordered arrangements of atoms — which can be identified through diffraction techniques. A particular lattice type and lattice constant will produce a distinct diffraction pattern which identifies that compound. Thus diffraction probes are an additional important component in determining composition at a surface. Furthermore, such probes provide information on surface crystal structure — an important component in characterizing and explaining the properties of solids.

X-ray diffraction is the classic and well-developed technique for determining crystalline order in three-dimensional solids. Here three-dimensional implies a bulk measurement determined essentially by the X-ray absorption length — typically 10 to 100 μm in most experiments. For near surface analysis we require a geometry or interaction which establishes the diffraction condition in or near the surface. Surface sensitivity usually comes about through the use of grazing angle geometries and/or electron diffraction probes which are strongly interacting and reveal structure for depths less than ~1 μm.

155

Diffraction techniques represent a large and distinct field of materials analysis. Major review articles and textbooks on this subject are available. Our goal here is to provide some insight into the mechanisms of these techniques and to illustrate their use in surface analysis. The reader is referred to other sources for detailed accounts.

7.2 Diffraction Parameters

The crystal structure of materials is usually determined by diffraction techniques in which the wavelength of the incident radiation is of the order of atomic spacing in crystal lattices. Hence diffraction analyses with neutrons of thermal energies and photons of kiloelectron volt energies probe the crystallinity of solids. For surface crystallography electrons of ~100 eV energy are used. In all diffraction analyses the ordered array of atoms in the crystal lattice acts coherently as a diffraction grating to establish interference minima and maxima.

The existence of diffraction processes is essential to the understanding of the wavelike nature of atomic particles. The early experiments of Davisson and Germer demonstrated diffraction of electrons; confirmation of the wavelike nature of neutrons and other particles soon followed. These experiments provided justification for the use of atomic wave equations—the Schrödinger equation—and gave direct confirmation of the de Broglie relation which relates the wavelength λ of a particle to its momentum p:

$$\lambda = h/p = h/\sqrt{2mE}. \tag{7.1}$$

If we consider atomic spacings of about 1 Å, the associated particle energy is 150 eV for electrons and 0.08 eV for neutrons, that is, thermal neutrons.

Different depth regimes may be sampled depending on the diffracting species. Diffraction is the coherent addition of radiation of the *same* wavelength; therefore, absorption or inelastic scattering—whichever is greater—determines the depth regime. In X-ray diffraction one uses X-rays whose energies correspond to wavelengths comparable to the lattice spacing, i.e., spacings about 1 Å correspond to 12.4 keV X-rays. For X-rays, absorption is determined by the photoelectric absorption (Chapter 8), and relatively thick or bulk samples can be analyzed. The dominant X-ray interaction at ~10 keV is the photoelectric effect, i.e., absorption. Inelastic effects, i.e., Compton scattering, are not significant since the photoelectron absorption is

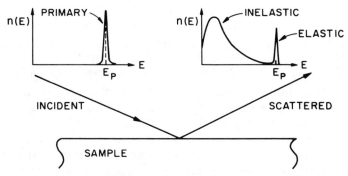

Figure 7.1 Schematic of the incident and scattered electron energy distribution of ~1 keV electrons scattered from a single crystal surface.

much greater than that for Compton scattering. X-ray diffraction extends over depths the order of $\simeq 10 \ \mu$m. The absorption of electrons is given essentially by the escape depth (Chapter 6) and low energy electron diffraction (LEED) is thus used to analyze surface structure. As a pure surface technique, LEED is not perfect since electrons can penetrate a few monolayers.

The energy distribution of backscattered electrons is shown in Figure 7.1. Most of the distribution is associated with inelastic events (Auger electrons, secondary electrons, and plasmon excitations). These inelastic events, primarily the plasmon loss, determine the escape length curve shown in the previous chapter. The sharp peak of elastically scattered electrons is used for diffraction analysis. Experimentally this peak is isolated by energy discrimination against all particles which have undergone inelastic scattering and are not in the elastic group.

It is important to distinguish the diffraction effects associated with 100 eV electrons from the orientation effects (channeling) associated with MeV ions incident on single crystals. For channeling, the wavelength of the incident energetic particle is much less than the lattice spacing. However, the ordered array of atoms in a crystal can also influence the interactions between a million electron volt ions and the atoms in a crystal. In this case, a classical approximation is used, and a wavelike or diffraction picture is not necessary. The parameters of the individual collisions are determined by classical mechanics under the influence of an appropriate screened Coulomb potential. In channeling the collective nature of the interaction comes about because of the correlation of scattering events from the ordered atoms. Simply, as the energetic ion approaches the atomic row at a glancing angle, the small deflection due to the scattering from one atom determines the impact parameter and hence scattering angle of the collision with the second atom; and this in turn determines the impact with the third atom and so forth. The net effect is a steering of the trajectories of the incident ions. The steered or channeled ions are confined to the spaces or channels between the atomic rows and do not make close impact collisions with the lattice atoms. This channeling condition exists when the beam is almost parallel (within about 0.1°) to a crystal axis and leads to a near hundredfold decrease in the number of close encounter collisions such as Rutherford backscattering, nuclear reactions, or inner-shell X-ray production. It is this reduction in yield which is so useful in solid state applications (as discussed in the Chapter 5).

7.3 Thermal Vibrations and the Debye – Waller Factor

In the analysis of atomic positions we must consider the thermal vibrations of atoms about their equilibrium positions which is a disruption of the "perfect" lattice. A measure of the displacement of atoms due to thermal motion is the mean square thermal vibration amplitude u^2. In the harmonic approximation the distribution of amplitudes, $P(u)$, is a Gaussian

$$P(u) = \frac{1}{(2\pi \langle u^2 \rangle)^{3/2}} e^{-u^2/2\langle u^2 \rangle}, \tag{7.2}$$

where $\langle u^2 \rangle$ is the mean value of the distribution. In many cases we are interested in the one-dimensional component $\langle u_x^2 \rangle$ such that

$$P(u_x) = \frac{e^{-u_x^2/2\langle u_x^2 \rangle}}{(2\pi \langle u_x^2 \rangle)^{1/2}}. \tag{7.3}$$

For cubic systems $\langle u_x^2 \rangle = \langle u_y^2 \rangle = \langle u_z^2 \rangle = \frac{1}{3} \langle u^2 \rangle$. In Chapter 5, we defined ρ^2 as the two-dimensional mean square thermal amplitude, i.e., $\rho^2 = \langle u_x^2 \rangle + \langle u_y^2 \rangle$.

In the Debye approximation we define a Debye temperature θ_D, by

$$\hbar \omega_D = k\theta_D \tag{7.4}$$

where ω_D is the Debye cutoff frequency, the maximum frequency allowed in the solid. Then

$$\langle u^2 \rangle = \frac{3\hbar^2 T}{Mk\theta_D^2} \left[\Phi\left(\frac{\theta_D}{T}\right) + \frac{\theta_D}{4T} \right], \tag{7.5}$$

where

$$\Phi(x) = \frac{1}{x} \int_0^x \frac{y \, dy}{e^y - 1}$$

is the Debye function and M is the mass of atoms in the monoatomic solid.

The Debye function is shown graphically in Figure 7.2. Numerically

$$\langle u^2 \rangle = 439.2 \left[\left(\frac{\Phi(x)}{x} + \frac{1}{4} \right) \Big/ M\theta_D \right] (\text{Å})^2 \qquad (7.6)$$

where $x = \theta_D/T$.

The dependence of $\langle u^2 \rangle$ on temperature is shown schematically in Figure 7.3. At high temperature, $T \gg \theta_D$, $\langle u^2 \rangle$ is proportional to the absolute temperature; at low temperature $\langle u^2 \rangle$ approaches a constant and finite value corresponding to the zero point motion of the solid. The Debye temperatures of a variety of materials is given in Table 7.1.

Thermal vibrations are responsible for the temperature dependence of diffracted X-ray intensity as shown in Figure 7.4. The intensity of an X-ray spot decreases with increasing temperature as

$$I = I_0 e^{-2W} \qquad (7.7)$$

where W is the Debye–Waller factor. In X-ray diffraction $2W = \frac{1}{3}\langle u^2 \rangle (\Delta k)^2$, where Δk is the momentum transfer in the X-ray scatter-

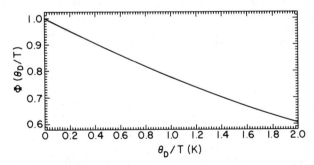

Figure 7.2 The Debye function $\Phi(\theta_D/T)$ used in evaluating the mean square vibration amplitude

$$\Phi(x) = \frac{1}{x} \int_0^x \frac{y \, dy}{e^y - 1}.$$

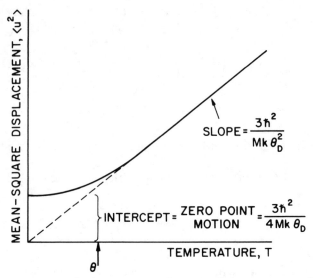

Figure 7.3 Mean square displacement as a function of temperature in the Debye approximation.

Table 7.1 Crystal Parameters at Room Temperature[a]

Z_2	M_2	Name	Structure	θ_D (K)	$\langle u_x^2 \rangle^{1/2}(\text{Å})^b$	Lattice constant (Å)
6	12.01	C	fcc (diam)	2000	0.04	3.567
13	26.98	Al	fcc	390	0.105	4.050
14	28.09	Si	fcc (diam)	543	0.075	5.431
23	50.94	V	bcc	360	0.082	3.024
24	52.00	Cr	bcc	485	0.061	2.884
26	55.85	Fe	bcc	420	0.068	2.867
28	58.71	Ni	fcc	425	0.065	3.524
29	63.54	Cu	fcc	315	0.084	3.615
32	72.59	Ge	fcc (diam)	290	0.085	5.657
41	92.91	Nb	bcc	275	0.079	3.300
42	95.94	Mo	bcc	380	0.057	3.147
45	102.91	Rh	fcc	340	0.061	3.803
46	106.4	Pd	fcc	275	0.074	3.890
47	107.87	Ag	fcc	215	0.093	4.086
73	180.95	Ta	bcc	245	0.064	3.306
74	183.85	W	bcc	310	0.050	3.165
78	195.09	Pt	fcc	225	0.066	3.923
79	196.97	Au	fcc	170	0.087	4.078
82	207.19	Pb	fcc	88	0.164	4.951

[a] From D. S. Gemmel, *Rev. Mod. Phys.* **46**, 129 (1974).
[b] Calculated for $T = 293\,°K$.

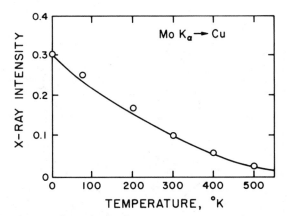

Figure 7.4 Temperature dependence of the integrated X-ray intensity of Mo K_α X-rays reflected from the (800) planes of Cu. The solid line is the calculated curve from the Debye–Waller factor. [From P. A. Flinn et al., *Phys. Rev.* **123**, 809 (1961).]

ing, i.e., $\Delta k = (4\pi/\lambda) \sin \theta$. Here λ is the wavelength of the incident radiation and 2θ is the scattering angle.

7.4 Low Energy Electron Diffraction (LEED)

An electron with a wavelength λ impinges perpendicular to a periodic row of atoms with interatomic spacing a (Figure 7.5). When the electron is scattered, wavelets from one atom will interfere with those from adjacent atoms. Where interference occurs constructively, new wave fronts will be generated. The condition for constructive interference is that the wavelets add rather than cancel. Therefore, they must be in a phase and there must be an integral number of wavelengths along this direction between wave fronts from the different atoms. This condition for constructive interference is

$$n\lambda = a \sin \theta, \qquad (7.8)$$

where $n\lambda$ is the integral number of wavelengths and $a \sin \theta$ is the projection of the interatomic spacing along the new direction of travel and, consequently, the distance between wavelets for adjacent atoms. If this condition for constructive interference is met for wavelets generated from adjacent atoms, it will be met for wavelets generated from all of the atoms in this row owing to the periodic nature of the problem. Depending upon a and λ, there may be several angles θ for which constructive interference can occur. As this row of atoms

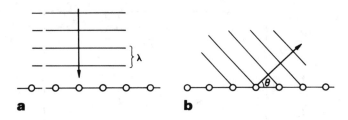

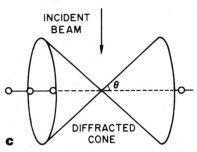

Figure 7.5 Diffraction of a particle from a row of scattering centers: (a) incident plane wave; (b) scattered wave front; (c) cone of constructive interference.

has one-dimensional symmetry, there will be cones around the axis of the row where constructive interference can occur and where we will have a finite probability of finding the electron.

A two-dimensional periodic arrangement with primitive interatomic spacing a and b will generate two sets of diffraction conditions which must be met simultaneously, namely,

$$n_a \lambda_a = a \sin \theta_a \qquad (7.9)$$

and

$$n_b \lambda_b = b \sin \theta_b. \qquad (7.10)$$

The new set of cones are also the only allowed areas of constructive interference. As both conditions must be met simultaneously, the only regions where we may find the electron are in the intersections of these cones. As the intersection of two cones with common origin and nonparallel axes is a set of lines, we see that when an electron is diffracted from a periodic two-dimensional array of atoms, it can only be scattered along a set of lines or rods radiating away from the surface. If we set a detection device across this set of rods, we will perceive them as points or spots. In many LEED experiments (Figure 7.6) these diffraction rods are intercepted by a fluorescent screen, and the resulting diffraction spots are observed visually. These spots

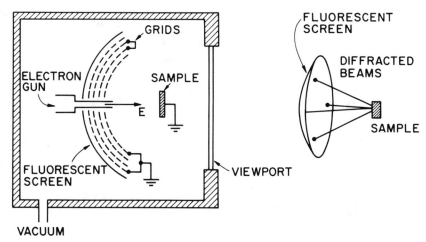

Figure 7.6 Schematic of a vacuum chamber with a low energy electron diffraction apparatus.

may be conveniently indexed as (n_a, n_b) where n_a and n_b are the integral number of wavelengths, in the a and b directions, respectively, that are associated with the rod in question.

Low energy electron diffraction can be practiced at different levels of sophistication. In the simplest scheme the LEED pattern is viewed on a fluorescent screen. The apparatus, shown schematically in Figure 7.6, contains a set of retarding grids to reflect the inelastic electrons. The elastic group is of sufficiently high energy to overcome the retardation; after passage through the grids the elastic electrons are further accelerated so as to fluoresce a phosphor screen. In this mode the LEED pattern is a relatively quick and simple method of determining crystalline order at the very surface of a single crystal. These types of experiments must be carried out under carefully controlled ultrahigh vacuum conditions since even a monolayer of surface contamination can seriously affect the quality of the pattern. More sophisticated uses of electron diffraction measure the electron intensity in a reflected spot. Subsequent analysis can then reveal a more detailed description of the surface structure.

It is important to recognize that a LEED pattern reveals the periodicity of atoms of the surface and the overall symmetry of the surface — but not the detailed atomic position. We can illustrate this with a very simple example. Consider the geometry of Figure 7.5 and the signal corresponding to zeroth order diffraction ($\theta = 0°$) and first order diffraction $n = 1$. The angular difference between the spots is given by

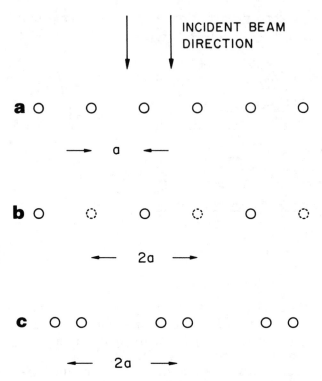

Figure 7.7 Strings of atoms with periodicity (a) a, (b) 2a, and (c) 2a. Examples (b) and (c) give rise to the same LEED pattern of ½ order spots but have different atom configurations.

$$\Delta\theta = \sin^{-1}(\lambda/a).$$

If $\lambda/a = \frac{1}{3}$, then $\Delta\theta = 19.5°$. Assume now that $a \rightarrow 2a$, i.e., we halve the atom density along the row. Then $\Delta\theta = 9.59°$ and $\Delta\theta$ (n = 0,2) is 19.5°. The result is an extra set of diffraction spots in between the original set. Any structure which results in the doubling of the periodicity will give rise to "½ order spots." Examples of such structures are shown in Figure 7.7 and include (a) the original structure of period a, (b) a structure of period 2a resulting from a vacancy in every other site, and (c) a structure of period 2a resulting from a pairing of adjacent atoms. Structures (b) and (c) consist of atoms at very different positions but result in the same LEED patterns.

Although the assignment of atom positions on the basis of LEED patterns is not unique, it is possible to predict the symmetry of a LEED pattern from the real space configuration of atoms. Figure 7.8 gives

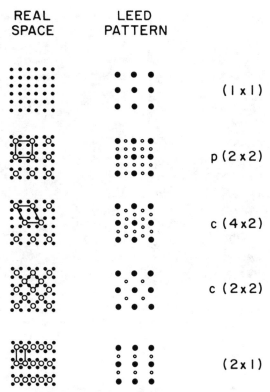

Figure 7.8 Overlayers on the (100) surface of a cubic crystal and the associated LEED patterns in reciprocal space. The notation for the various patterns is indicated on the right. [From Estrup and McRae, 1971.]

examples of overlayers on the (100) surface of a cubic crystal. The letter p in Figure 7.8 indicates the unit cell is primitive and the LEED pattern for $p(2 \times 2)$ has extra, half-order spots. The letter c in Figure 7.8 indicates that the unit cell has an additional scatterer in the center which gives rise to $\frac{1}{2}$, $\frac{1}{2}$ spots in the diffraction pattern.

In general, changes in the periodicity of the surface will result in changes in the diffraction pattern that are easily observable and interpretable in terms of the new two-dimensional symmetry. Such changes are often observed, for example, when gases are adsorbed on crystal surfaces. They frequently arrange themselves in an orderly fashion with periodicities that are integral multiples of the substrate spacings. A standard notation for such structures is $M(hkl)$-$(n \times m)$-C, where M is the chemical symbol for the element whose surface is being studied, (hkl) indicates the particular crystalline plane that the

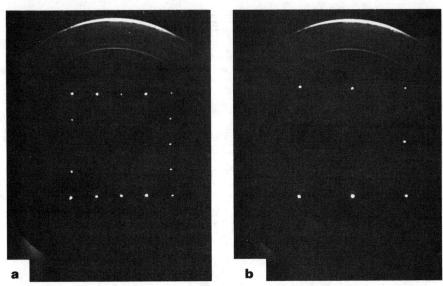

Figure 7.9 LEED patterns from Si(100) for (a) the clean surface and (b) the hydrogen covered surface. In the first case the pattern consists of two (2 × 1) rows rotated 90° with respect to one another. The $\frac{1}{2}$ order spots arise from the double periodicity in this surface. With hydrogen coverage, the periodicity of the substrate is (1 × 1), and the $\frac{1}{2}$ order spots are no longer present. (Note that in this figure the sample obscures some of the LEED spots.)

surface is parallel to, and (n × m) shows that the new surface structure has a periodicity which is n times the original surface periodicity in the a direction and m times that in the b direction. The final symbol C is the chemical symbol for the adsorbed gas or other contaminant on the surface. Frequently, a letter is inserted before the (n × m) term; for example, p means that the new unit cell is primitive, and c means that it is centered. The p is frequently deleted for brevity. If the new unit cell is rotated relative to that of the substrate, an angle is included to show the relative orientation. Examples of such nomenclature are Ni(111)(2 × 2)-0, Pt(100)-C(2 × 2)-C_2H_4, W(110)-C(9 × 5)-CO, and Si(111)-(7 × 7). When the contaminant is unknown, or it is believed that the surface has reconstructed itself into a new periodicity in the absence of any other elements, then the final symbol is deleted.

Diffraction from a clean Si(100) surface gives a pattern which contains the "normal" spots, as well as $\frac{1}{2}$ order spots due to a surface reconstruction (Figure 7.9a). With hydrogen adsorption the surface periodicity is removed and the pattern corresponds to that of the bulklike structure (Figure 7.9b). The high surface sensitivity of LEED arises from the large scattering cross sections of atoms for low-energy electrons. Medium energy electron diffraction (MEED) and reflection high energy electron diffraction (RHEED) extend the energy region up to ~50 keV, thus making these techniques more useful for thin films. These techniques are often carried out in a grazing angle geometry which puts stronger requirements on the surface flatness of the sample.

7.5 Glancing Angle X-ray Diffraction

The analysis of the elemental composition of thin films is often aided through the use of diffraction techniques which can uniquely identify crystalline phases near a solid surface. X-ray diffraction is most suited to films thicker than a few hundred angstroms. To limit the penetration of the beam and enhance the diffraction pattern of the film with respect to the substrate, glancing angle geometries are used. Individual crystalline phases are identified by their characteristic diffraction patterns. Such X-ray patterns also reveal information on the orientation and size distribution of the crystallites.

X-ray techniques provide superior angular resolution and more accurate structural data than available from electron diffraction. On the other hand, diffracted intensities can be much smaller for X-ray than for electrons necessitating long exposures. A major advantage of electron diffraction is in conjunction with electron microscopes which permit diffraction analysis within a small region of the sample.

The basis of X-ray diffraction is the Bragg equation which describes the condition for constructive interference for X-rays scattering from atomic planes of a crystal. The condition for constructive interference is

$$2d \sin \theta = n\lambda,$$

where λ is the wavelength of the incident radiation.

The Bragg law requires that θ and λ be matched for diffraction. The condition may be satisfied by varying λ, or varying the orientation of a single crystal. In thin films the distribution of crystallite orientations is nearly continuous. Diffraction occurs from crystallites which happen to be oriented at the angle to satisfy the Bragg condition.

In evaluation of thin-film (200–1000 Å thick) samples, generally a Seemann–Bohlin X-ray diffractometer or Read camera are used.

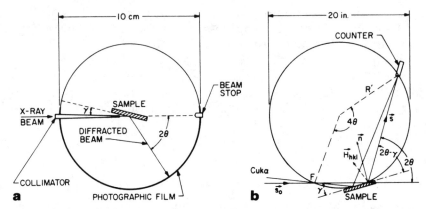

Figure 7.10 (a) Schematic diagram of the geometry of the Read camera; (b) Schematic diagram of the geometry of the Seemann–Bohlin configuration.

Both configurations are basically glancing angle X-ray diffraction setups with a fixed angle of incidence. By use of glancing angles of incidence, a relatively large volume is examined even in thin film samples. For example, an incident angle of 6.4° will increase the length of the X-ray beam path in a thin specimen to about nine times the film thickness.

The experimental geometries of the two configurations are shown in Figure 7.10. In both, X-ray diffraction patterns are obtained from a monochromatic X-ray beam (Cu K_α, for example) impinging at about 6–14° to the sample surface. The beam is focused in the Seemann–Bohlin configuration so that the foci of the incident and diffracted beams lie on the circumference of the diffraction circle. The detector is moved along the diffraction circle, which is 20 in. in diameter, in steps of 0.15° (4θ) with a counting time of 30 s at each step. In the Read camera the incident beam is collimated through two pinholes, and the patterns at all angles are recorded simultaneously on film placed on a 5 cm radius from the sample center.

In order to illustrate the application of X-ray diffraction, we use an example of silicide formation[1] in which a deposited layer of Ni is partially reacted with a silicon substrate to form an Ni_2Si layer between the unreacted Ni and the silicon. The Rutherford backscattering spectrum in Figure 7.11 shows that there is intermixing of Ni and Si (shaded area) that forms a reacted layer with a Ni to Si ratio of 2 to 1. This suggests that Ni_2Si is formed, but conclusive identification of the

---

[1] S. S. Lau, W. K. Chu, J. W. Mayer, and K. N. Tu, *Thin Solid Films* **23**, 205 (1974).

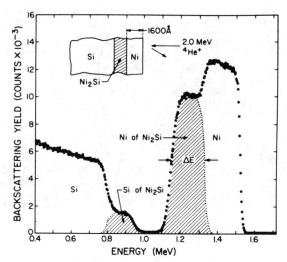

Figure 7.11 Backscattering spectrum at 2 MeV of a sample formed by depositing 2600 Å of Ni on a Si substrate and annealing at 250 °C to form a layer (shaded region) of Ni_2Si that is about 1600 Å thick.

phase requires diffraction techniques because the backscattering spectra provide only atomic ratios of composition.

Phase identification of Ni_2Si was established by X-ray diffraction using a Read camera and a Seemann–Bohlin diffractometer. In the Read camera diffraction pattern shown in Figure 7.12, the spot pattern arises from the single-crystal Si substrate and the ring pattern from the polycrystalline film. The components of the film are identified by measuring the angular positions of the rings and then comparing these positions with data in American Society for Testing and Materials (ASTM) powder pattern compilations. In Figure 7.12, the presence of Ni and Ni_2Si is detected. The same result was obtained by the Seemann–Bohlin technique, as shown in the diffraction pattern of Figure 7.13. In this figure diffraction intensities are plotted as a function of 4θ. This diffraction pattern is similar in concept to a scan along the equator plane AB in Figure 7.12. Reflections from the single crystal Si substrate (the spot pattern in Figure 7.12) are not seen. This is due to the diffraction geometry in the Seemann-Bohlin diffractometer which is insensitive to Bragg reflections of a single crystal since the incident angle is fixed and monochromatic Cu K_α radiation is used. Consequently, two extremes of film crystallinity are difficult to analyze — single crystal epitaxial layers and amorphous layers where there are broad halos rather than sharp Bragg reflections.

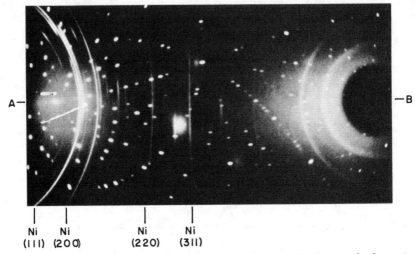

Ni	Ni	Ni	Ni
(111)	(200)	(220)	(311)

Figure 7.12 X-ray diffraction pattern from the Read camera for the sample shown in Figure 7.11. The spot patterns are from the Si substrate and the indexed lines are from Ni. The other lines are from Ni_2Si. [From S. S. Lau, W. K. Chu, J. W. Mayer, and K. N. Tu, *Thin Solid Films* **23**, 205 (1974).]

In the Seemann–Bohlin geometry, where the diffracted intensity is plotted versus 4θ (Figure 7.13), the vector H_{hkl} in Figure 7.10 represents the reciprocal lattice vector of the (hkl) plane which fulfills the Bragg diffraction condition. For fixed angle of incidence γ, the angle $\theta - \gamma$ between H_{hkl} and the surface normal n increases with the Bragg angle. The direction of H_{hkl} with respect to n changes with each (hkl) reflection so that each intensity peak in the reflection spectra measures the lattice parameter in a different direction.

Glancing-angle X-ray diffraction techniques can be used to determine the thickness[2] of a growing intermetallic compound or silicide as well as to provide identification of the compound phase. As an example, consider the growth of polycrystalline Pd_2Si from a deposited layer of palladium on silicon. Figure 7.14 shows the Seeman–Bohlin X-ray diffraction from the as-deposited state of 1000 Å Pd on Si and that of the final state after all the Pd has been consumed in the formation of polycrystalline Pd_2Si. In the as-deposited state, the spectrum in Figure 7.14a is that of a pure Pd film (fcc, $a = 3.8808$ Å) with no reflection from the Si single crystal substrate. The reflections cor-

---

[2] K. N. Tu and B. S. Berry, *J. Appl. Phys.* **43**, 3283 (1972).

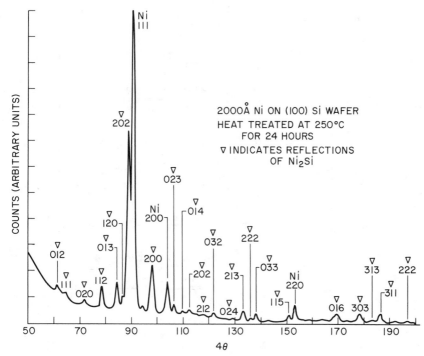

Figure 7.13 X-ray diffraction lines from the Seemann–Bohlin diffractometer from a sample similar to that shown in Figure 7.11. [From K. N. Tu, unpublished.]

respond to Pd lattice spacings of 2.24 Å (111) to 1.12 Å (222). The spectrum from the totally reacted polycrystalline Pd_2Si layer in Figure 7.14b shows the peaks indexed according to the hkl reflections for a hexagonal structure with $a = 6.493$ Å and $c = 3.427$ Å. The lattice spacings range from 3.246 Å (110) to 1.058 Å (203).

The integrated intensity of a given reflection from the Pd_2Si is a measure of the total volume of the silicide phase under irradiation by the X-ray beam. For laterally uniform layers, a condition that is generally met in silicide formation, the total integrated intensity I_{int} is linearly proportional to the thickness of the film since X-ray absorption corrections can be small in thin films. The measured intensity is also determined by the structure and geometrical factors which should be unchanged for a given reflection as long as the degree of crystallinity of the growing silicide layer does not change. Rutherford backscattering can be used to give a calibration point of thickness. Figure 7.15a shows RBS spectra for scattering from a 1000 Å Pd layer deposited on Si and annealed at 200°C. The step in the Pd signal

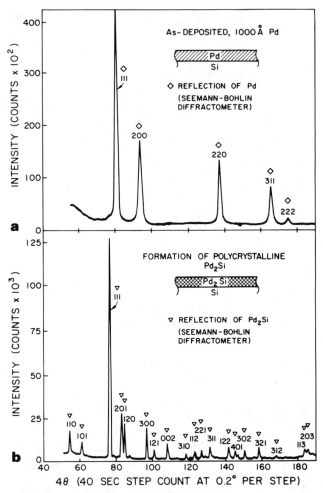

Figure 7.14 Glancing incidence Seemann–Bohlin diffraction spectra of (a) a 1000 Å Pd film deposited on Si and (b) the film after total reaction to form polycrystalline Pd_2Si. [From K. N. Tu, *J. Appl. Phys.* **53**, 428 (1982).]

provides a direct measure of the silicide thickness assuming a bulk density of Pd_2Si. The X-ray measurements were taken on a diffractometer to obtain the integrated intensity from the 00.2 reflection from the Pd_2Si layer. The comparison between the two measurements made on the same set of samples is shown in Figure 7.15c where the X-ray integrated intensity is plotted against silicide thickness determined from RBS analysis. There is a linear relationship with the extrapolated line passing through the origin. The diffraction tech-

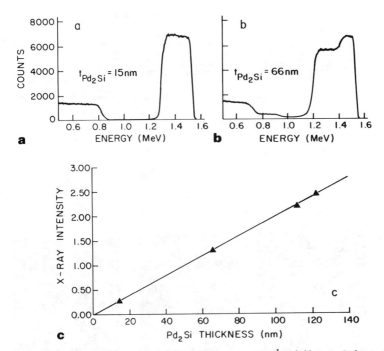

Figure 7.15 Analysis of the growth of Pd_2Si from a 1000 Å Pd film on Si heated at 200°C: (a) and (b) Rutherford backscattering spectra (1.8 MeV ^{4}He) for samples of two thicknesses and (c) integrated X-ray intensity from the Pd_2Si 00.2 reflection versus Pd_2Si thickness determined from backscattering spectra. [From B. Coulman, H. Chen, and L. E. Rehn, *J. Appl. Phys.* **57**, 643 (1985).]

nique can be used to measure film growth as long as variations in film texture or absorption effects do not play a significant role.

7.6 Transmission Electron Microscopy

Electron diffraction is another technique used to identify the structure of solids. In the simplest view, electron diffraction from a crystalline lattice can be described as a kinematic scattering process that meets the wave reinforcement and interference conditions given in the Bragg equation. In many applications the samples are thinned to thicknesses of a few thousand angströms by chemical etching or ion-milling. The patterns are formed by diffraction of an electron beam (typically 50–200 keV) transmitted through the thin sample. As in the case of X-ray diffraction, the electron diffraction patterns illustrated in Figure 7.16 are spot patterns from single crystal films, ring

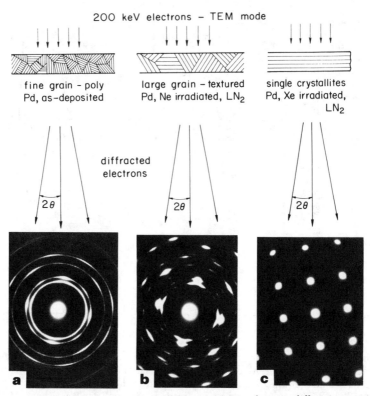

Figure 7.16 Schematic representation of transmission electron diffraction patterns from (a) fine-grain, randomly oriented polycrystalline films, (b) textured polycrystalline, and (c) single crystal films. Ion irradiation causes a change in the structure of the film. [From M. Nastasi, unpublished.]

patterns from fine-grain, randomly oriented crystallites, and super-imposed ring and spot-patterns from larger-grain polycrystalline films containing some orientation (textured films). In the example shown in Figure 7.16, ion irradiation of a thin Pd film causes a change in the structure from fine grain to large single crystallites.

For cubic crystals with planar Miller indices $\{hkl\}$ and lattice parameter a, the interplanar spacing d is given by

$$d_{hkl} = \frac{a}{\sqrt{h^2 + k^2 + l^2}}. \tag{7.11}$$

If the specimen to photographic recording plane distance is L, then the radius R of the diffraction rings is related to the interplanar spacing by

$$\lambda L = R d_{hkl}, \tag{7.12}$$

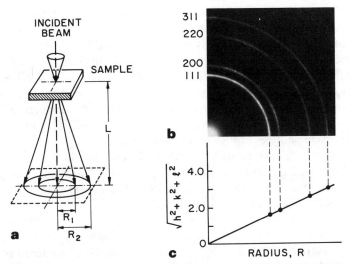

Figure 7.17 Determination of camera constant λL: (a) schematic of experimental geometry; (b) a quadrant of the electron diffraction pattern from a 500 Å thick Ni film with the prominent rings indexed; (c) ring radii versus $(h^2 + k^2 + l^2)^{1/2}$.

where λ is the wavelength of the electrons and a small angle approximation is used. Figure 7.17 illustrates a method of determining the camera constant λL from the diffraction pattern of a polycrystalline film of Ni [lattice parameter $a = 3.524$ Å for face-centered cubic (fcc) Ni]. To prepare the sample, a layer of Ni, about 500 Å thick, was deposited on a NaCl substrate, the NaCl substrate was dissolved in water, the Ni film was placed on a fine-mesh grid TEM sample holder 3 mm in diameter, and the holder was then placed in the electron microscope. As shown in Figure 7.17a, the sample was irradiated with 200 keV electrons whose transmission diffraction pattern was a successive array of reflection circles of radius R on the photographic plane. A quadrant of these circles is shown in Figure 7.17b along with the Miller hkl indices for the different planes. The radii, R, are plotted versus $(h^2 + k^2 + l^2)^{1/2}$ in Figure 7.17c and the slope, $a/\lambda L$, determines the camera constant. For a given specimen, the interplanar spacing d_{hkl} can be found from the measured values of the radii and the camera constant. The structure can be identified by use of ASTM index files or by use of computer programs as described by Murr (1982).

We illustrate the application of transmission electron microscopy with an example[3] of ion-irradiation-induced phase transforma-

---

[3] M. Nastasi, et al., *J. Appl. Phys.* **57**, 1050 (1985).

tion of Ni_2Al_3 to NiAl. The structures of the NiAl and Ni_2Al_3 phases are closely related: NiAl has a CsCl structure with Ni atoms occupying the body centers, Ni_2Al_3 is made up of pseudo-NiAl cubes, every other one of which has a vacant Ni body center site. The similarity between these two structures (and the ease with which Ni_2Al_3 can transform to NiAl) is evident upon comparison of their diffraction patterns (Figure 7.18). Peaks *a* to *d* are found in both patterns with the additional peaks *e* and *f* appearing only in the Ni_2Al_3 pattern. The observed *d* spacings are listed in Table 7.2.

If Ni and Al atoms randomly occupied the bcc lattice, the rings *b* and *d* would be found, characteristic of a bcc structure. The placement of Al atoms at cell corners and Ni atoms at body centers in the NiAl phase results in the formation of the CsCl structure. The additional rings *a* and *c* are superlattice reflections due to the ordering of Ni and Al from bcc to CsCl. The Ni_2Al_3 compound can be viewed as a NiAl structure with periodic Ni sites. The addition of vacant Ni sites to the NiAl structure on a random basis would not lead to additional

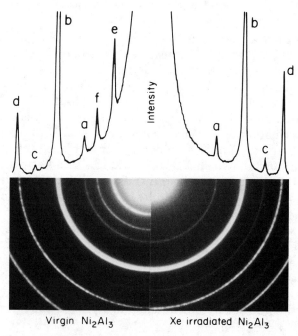

Virgin Ni_2Al_3 Xe irradiated Ni_2Al_3

Figure 7.18 Electron diffraction pattern and densitometer trace of Ni_2Al_3 before and after Xe ion irradiation. Peaks a to d correspond to fundamental CsCl reflections (NiAl) and peaks e and f arise from vacancy ordering in Ni_2Al_3. The ASTM data of d spacings observed in the diffraction from both NiAl and Ni_2Al_3 is listed in Table 7.2. [From Nastasi et al., *J. Appl. Phys.* **57**, 1050 (1985).]

Table 7.2 Observed d Spacings for NiAl and Ni_2Al_3

Disordered NiAl[a] (bcc)	NiAl[b]	Ni_2Al_3[b]
		4.90 (e)
		3.52 (f)
	2.87 (a)	2.863 (a)
2.02	2.02 (b)	2.028 (b)
	1.655 (c)	1.652 (c)
1.43	1.434 (d)	1.428 (d)

[a] Calculated using the NiAl lattice parameter of 2.88 Å for $h + k + l$ = even.
[b] ASTM data.

reflections in the diffraction pattern but would affect the relative intensities. However, the periodic ordering of vacancies does contribute the extra superlattice reflections (e and f) found in the Ni_2Al_3 diffraction pattern.

On an experimental basis, thin films of Ni_2Al_3, ~350 Å thick, were produced by thermally reacting samples containing alternating layers of Ni and Al. Samples with the Ni_2Al_3 composition were placed on electron microscope grids, and diffraction patterns were obtained. The samples were then irradiated with Xe or Ne ions whose ranges were greater than the sample thickness. The diffraction patterns showed that the structure had transformed from Ni_2Al_3 to NiAl with a dose of 2×10^{15}/cm² Xe ions. As in the example shown in Figure 7.16, the data in Figure 7.18 illustrates changes in structure that occur without changes in composition.

Transmission electron microscopy as an analytical technique for structural determination differs from the other structural techniques (channeling, LEED, glancing angle X-ray diffraction) in that sample preparation is generally destructive during the thinning process. Consequently, TEM analysis is usually viewed as the last step in the history of a sample rather than as one part of a succession of measurements. In this regard, one can take the sample preparation a step beyond the conventional thinning technique and prepare cross sectional TEM samples (Marcus and Sheng, 1983). In this case, one cuts the sample in mm thick slices and mounts them sideways in a holder for polishing to about 50 μm and then ion mills to thicknesses of about 500–1000 Å as shown schematically in Figure 7.19. Sample preparation involves care as different materials erode in the ion miller at different rates and the final sample is fragile. However, the furnished cross section allows direct examination of the thin film reaction zone edge-on rather than a top view through the sample.

$\Longrightarrow$

Si

ION MILL
OR ETCH

b CROSS-SECTION TEM SAMPLE

1) Si
SLICE SECTIONS

2) Si
$\approx$ 1 mm

3) Si
EPOXY
POLISH TO 50 μm

4) Si
ION MILL
$\approx$ 500 Å
EPOXY

Figure 7.19 Sample preparation for transmission electron microscopy (TEM): (a) Preparation of standard planar samples involves polishing and a final etch or ion mill to thin the sample; (b) samples for cross section TEM involve (1) and (2) slicing sections about 1 mm thick, (3) using epoxy to hold two sections together and then polishing to about 50 μm thickness, and then (4) ion milling to produce the final sample thickness of about 500 Å.

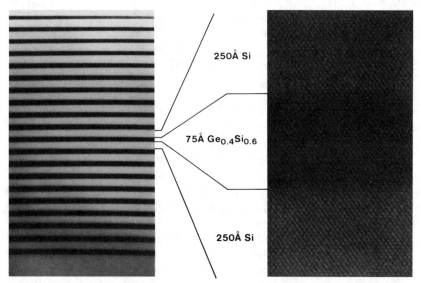

250Å Si

75Å Ge$_{0.4}$Si$_{0.6}$

250Å Si

Figure 7.20 A transmission electron cross section micrograph of a Ge$_{0.4}$Si$_{0.6}$/Si multilayer structure. The left side indicates a "low resolution" view showing the planarity of the multilayer structure. The right side is a "high resolution" view showing the high degree of epitaxy in the structure. [From J. C. Bean, *J. of Cryst. Growth* **70**, 444 (1984).]

Cross-sectional TEM samples show the planarity of the interface and the structure of the various reacted layers. Figure 7.20 shows a cross section TEM micrograph of a $Ge_{0.4}Si_{0.6}$/Si superlattice. There are 20 layers of 75 Å of the alloy interspersed with 250 Å of Si. The micrograph shows the planarity of the interface and the uniformity of the layers. The right hand view of Figure 7.20 shows a high resolution imaging of the lattice planes. The continuation of the lattice planes across the interface show that a high degree of epitaxy has been achieved. The multilayer structure is only apparent when using these cross sectional thinning techniques.

Problems

7.1. Al is a fcc cubic metal with a density of 2.7 g/cm³. Calculate the lattice parameter a, the number of atoms/cm² on a {100} planar surface, and the interlattice spacing between {111} planes.

7.2. The Debye temperature θ_D for Al is 390°K. Calculate the one-dimensional thermal vibration amplitude $\langle u_x^2 \rangle^{1/2}$, and compare the value of $\langle u_x^2 \rangle^{1/2}$ with the value of the lattice parameter a.

7.3. Sketch the real space and LEED pattern for a (3 × 1) overlayer on a 100 surface of a cubic crystal.

7.4. Ni is an fcc crystal with a density of 8.91 g/cm³. In the glancing-angle Read camera configuration of Figure 7.10 with Cu K_α radiation incident on a Ni thin film, what are the Bragg angles for the Ni (220) and (311) planes and at what distance along the film (Figure 7.12) would these patterns appear.

7.5. For TEM measurements with 120 kV electrons and camera length L, calculate the camera constant λL and the diffraction pattern radii for the Ni (220) and (311) planes.

7.6. Al has a Debye temperature of 390°K. Calculate $\langle u^2 \rangle$ at 293°K and compare your answer with the value given in Table 7.1. The melting temperature of Al is 933°K. Calculate $\langle u^2 \rangle$ at 933°K and compare with the nearest neighbor distance in the fcc crystal. What is the ratio of X-ray intensities at these two temperatures I_{933}/I_{293} for incident radiation at a wavelength of 1.5 Å and a scattering angle of 30°.

7.7. Consider radiation with momentum k incident on two atoms a distance a apart. The diffracted beam can be represented by $\psi = e^{ikr} + e^{ik(r+a)}$. If the intensity of the diffracted beam is given by $|\psi^*\psi|^2$ where ψ^* is the complex conjugate of ψ, show that the intensity has a form 2 sin(ka). In real lattices such scattering conditions give rise to the structure factor which defines the intensity of the diffracted beam from different crystal planes.

7.8. You are given a GeSi/Si superlattice as shown in Figure 7.20, but with 100 Å layers of Si between 10 Å layers of GeSi. In order to observe diffraction effects from the 100 Å spacing, you use X-rays and grazing incidence (10°). At approximately 10° exit angle, what wavelength radiation is required to see a peak in the diffraction intensity? With that wavelength, what accuracy is required in the measurement of the exit angle in order to evaluate the Si layer thickness within ± one atomic layer (5.4 Å)?

References

1. G. Ertl and J. Kuppers, *Low Energy Electrons and Surface Chemistry* (Verlag Chemie International, Weinheim, 1974).
2. P. J. Estrup, "LEED Studies of Surface Layers," in *Characterization of Metal and Polymer Surfaces* (Academic Press, New York, 1977).
3. P. J. Estrup and E. G. McRae, Surface Studies by Electron Diffraction, *Surface Sci.* **25**, 1 (1971).
4. R. D. Heidenreich, *Fundamentals of Transmission Electron Microscopy* (Wiley-Interscience Publishers, New York, 1964).
5. M. Henzler, "Electron Diffraction and Surface Defect Structure," in *Electron Spectroscopy for Surface Analysis*, H. Ibach, Ed., (Springer-Verlag, New York, 1977).
6. J. J. Hren, J. I. Goldstein, and D. C. Joy, Eds., *Introduction to Analytical Electron Microscopy* (Plenum Press, New York, 1974).
7. R. B. Marcus and T. T. Sheng, *Transmission Electron Microscopy of Silicon VLSI Circuits and Structures* (Wiley-Interscience Publishers, New York, 1983).
8. L. E. Murr, *Electron and Ion Microscopy and Microanalysis* (Marcel Dekker, New York, 1982).
9. G. Pinsker, *Electron Diffraction* (Butterworth Scientific, London, 1953).
10. G. A. Somorjai, *Chemistry in Two Dimensions: Surfaces* (Cornell University Press, Ithaca, NY, 1981).
11. M. A. van Hove and S. Y. Tong, *Surface Crystallography by LEED* (Springer-Verlag, New York, 1979).
12. D. B. Williams, *Practical Electron Microscopy in Materials Science* (Verlag Chemie International, Weinheim, 1984).

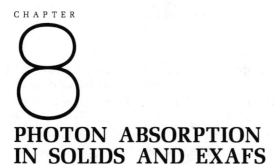

CHAPTER

8

PHOTON ABSORPTION IN SOLIDS AND EXAFS

8.1 Introduction

In surface and thin film analysis the most important photon absorption process is the photoelectric effect. In this process an incident photon of energy $\hbar\omega$ transfers *all* of its energy to a bound electron in an atom. The energy E_e of the outgoing electron is

$$E_e = \hbar\omega - E_B,$$

where E_B is the binding energy of the electron in the atom. The binding energies of electrons are well known and distinct for each element so that the measurement of E_B for the atomic constituents in a solid represents a technique for materials analysis. In this chapter we consider the binding energy of electrons in atoms and the cross section for photon absorption by a bound electron.

The photoelectric effect was instrumental in the early development of the quantum theory of matter. Today, it has been developed to a high degree of sophistication to yield not only an elemental analysis of materials but also a detailed description of the energies and momenta of electrons in solids.

The proper treatment of the photoelectric process requires a knowledge of the wave functions of the electrons within an atom. These wave functions result from solving the basic equation of quantum mechanics, the Schrödinger equation, which describes the prop-

181

erties of a quantum system through a wave equation. In this chapter we review the Schrödinger equation and its solutions in order to arrive at a quantitative estimate of the cross section for the photo effect.

8.2 The Schrödinger Equation

The wave particle duality of matter is expressed mathematically by the Schrödinger equation

$$\left[-\frac{\hbar^2}{2m}\nabla^2 + V(\mathbf{r})\right]\psi(\mathbf{r},t) = i\hbar\,\frac{\partial\psi(\mathbf{r},t)}{\partial t}, \tag{8.1}$$

where $\psi(\mathbf{r},t)$ is the "wave function" which describes the motion of the particle under the influence of the potential $V(\mathbf{r})$. As in classical mechanics, a physics problem is "solved" when, under the influence of a given potential, the coordinates of the particle are expressed as a function of time. In quantum mechanics a problem is solved when ψ is known as a function of $\mathbf{r}$ and t. There are very few interesting physical problems that can be solved exactly. Almost all processes which are necessary for materials analysis are approximate solutions to the Schrödinger equation using "perturbation theory."

In Cartesian coordinates the Schrödinger equation is written explicitly as

$$-\frac{\hbar^2}{2m}\left(\frac{\partial^2}{\partial x^2} + \frac{\partial^2}{\partial y^2} + \frac{\partial^2}{\partial z^2}\right)\psi + V(x,y,z)\psi = i\hbar\,\frac{\partial\psi}{\partial t}, \tag{8.2}$$

where ψ may be a function of x,y,z, and t.

For many examples it is sufficient to consider the one-dimensional Schrödinger equation

$$-\frac{\hbar^2}{2m}\frac{\partial^2\psi(x,t)}{\partial x^2} + V(x)\psi(x,t) = i\hbar\,\frac{\partial\psi(x,t)}{\partial t}. \tag{8.3}$$

The solution $\psi(x,t)$ can be written as the product of two functions:

$$\psi(x,t) = u(x)T(t). \tag{8.4}$$

This can be substituted into Eq. (8.3) to yield

$$\frac{1}{u}\left(-\frac{\hbar^2}{2m}\frac{\partial^2 u}{\partial x^2} + V(x)u\right) = \frac{i\hbar}{f}\frac{dT}{dt}. \tag{8.5}$$

Using the mathematical device of separation of variables, we note that the left-hand side depends only on x and the right-hand side only on t; therefore, both sides are proportional to a separation constant E. Then

$$T(t) = Ce^{-iEt/\hbar}, \tag{8.6}$$

where C is an arbitrary constant and the equation for u(x) is

$$\left[-\frac{\hbar^2}{2m} \frac{\partial^2}{\partial x^2} + V(x) \right] u(x) = Eu(x). \tag{8.7}$$

The total solution is then

$$\psi(x,t) = Au(x)e^{-iEt/\hbar}, \tag{8.8}$$

where A is a normalization constant. Equation (8.7) is known as the time-independent Schrödinger equation; the separation is valid provided that the potential V is not a function of time.

The philosophy underlying the Schrödinger equation is discussed in other books. Note however, that $-(\hbar^2/2m) \cdot (\partial^2/\partial x^2)$ is associated with the kinetic energy, V with the potential energy, and Eq. (8.7) is often written in short-hand as

$$H\psi = E\psi, \tag{8.9}$$

where H $(=$ kinetic energy $+$ potential energy) is associated with the classical Hamiltonian. The meaning of $\psi(\mathbf{r},t)$ is given through $|\psi(\mathbf{r},t)|^2$, the probability of finding the particle at position $\mathbf{r}$ at time t. Equation (8.7) is an example of an "eigenvalue equation;" u is said to be the eigenfunctions of the operator H $[=-(\hbar^2/2m) \nabla^2 + V]$ and E the eigenvalue. E is the energy of the system. The solution of Eq. (8.7) for many real potentials involves only discrete values of E, thus verifying the quantization assumptions in the Bohr theory given in Chapter 1.

In central force problems the potential $V(x,y,z) = V(r)$ and we convert the Schrödinger equation to spherical coordinates (r,θ,ϕ):

$$-\frac{\hbar^2}{2m} \frac{1}{r^2} \frac{\partial}{\partial r} \left(r^2 \frac{\partial u}{\partial r} \right) - \frac{\hbar^2}{2mr^2} \left[\frac{1}{\sin \theta} \frac{\partial}{\partial \theta} \left(\sin \theta \frac{\partial u}{\partial \theta} \right) \right.$$
$$\left. + \frac{1}{\sin^2 \theta} \frac{\partial^2 u}{\partial \phi^2} \right] + Vu = Eu. \tag{8.10}$$

Separation of variables is again possible,

$$u(r,\theta,\phi) = R(r)f(\theta)g(\phi), \tag{8.11}$$

and the radial part becomes

$$-\frac{\hbar^2}{2m}\frac{1}{r^2}\frac{d}{dr}\left(r^2\frac{dR}{dr}\right)+\left[\frac{l(l+1)\hbar^2}{2mr^2}+V(r)\right]R=ER, \qquad (8.12)$$

where l is the "orbital" quantum number.

8.3 Wave Functions

There are two kinds of wave functions that we will need to illustrate the main points of this chapter: (1) a free particle wave function $(V = 0)$ and (2) hydrogenic wave functions $(V = Z_1Z_2e^2/r)$.

8.3.1 Plane Waves $(V = 0)$

Many analysis problems require an incident beam or outgoing radiation. The incident particle, of energy E directed along the x direction say, is not under the influence of a potential. Then the solution to the appropriate Schrödinger equation

$$-\frac{\hbar^2}{2m}\frac{d^2u}{dx^2}=Eu$$

is

$$u(x)=Ae^{ikx}, \qquad (8.13)$$

where

$$\frac{\hbar^2k^2}{2m}=E. \qquad (8.14)$$

This describes a particle moving in the positive x direction with momentum $|\mathbf{p}|=\hbar k$. The total wave function $\psi(x,t)$ is given by

$$\psi(x,t)=Ae^{i(kx-\omega t)}, \qquad (8.15)$$

where $E = \hbar\omega$. In both Eqs. (8.13) and (8.15) A is an arbitrary constant to be determined by the beam parameters.

8.3.2 Hydrogenic Wave Function $[V(r) = Z\,e^2/r]$

The solution to the Schrödinger equation in radial coordinates may be written as

$$u(r,\theta,\phi)=R(r)Y(\theta,\phi), \qquad (8.16)$$

where $Y(\theta,\phi)$ is known as a spherical harmonic. The solution to the angular part of the differential equation results in quantum numbers,

Table 8.1 Hydrogenlike Wave Functions $u_{nlm}(r,\theta,\phi)$ $(\rho = r/a$ and $a = a_0/Z)$

$n = 1, l = 0, m = 0$	$u_{100} = \left[\dfrac{1}{\pi a^3}\right]^{1/2} e^{-\rho}$
$n = 2, l = 0, m = 0$	$u_{200} = \dfrac{1}{8}\left[\dfrac{2}{\pi a^3}\right]^{1/2} (2 - \rho)e^{-\rho/2}$
$n = 2, l = 1, m = 0$	$u_{210} = \dfrac{1}{8}\left[\dfrac{2}{\pi a^3}\right]^{1/2} \rho e^{-\rho/2} \cos\theta$
$n = 2, l = 1, m = \pm 1$	$u_{211} = \dfrac{1}{8}\left[\dfrac{1}{\pi a^3}\right]^{1/2} \rho e^{-\rho/2} \sin\theta e^{i\phi}$
	$u_{21-1} = \dfrac{1}{8}\left[\dfrac{1}{\pi a^3}\right]^{1/2} \rho e^{-\rho/2} \sin\theta e^{-i\phi}$
$n = 3, l = 0, m = 0$	$u_{300} = \dfrac{1}{243}\left[\dfrac{2}{\pi a^3}\right]^{1/2} (27 - 18\rho + 2\rho^2)e^{-\rho/3}$
$n = 3, l = 1, m = 0$	$u_{310} = \dfrac{1}{81}\left[\dfrac{2}{\pi a^3}\right]^{1/2}\rho(6 - \rho)e^{-\rho/3} \cos\theta$
$n = 3, l = 1, m = \pm 1$	$u_{311} = \dfrac{1}{81}\left[\dfrac{1}{\pi a^3}\right]^{1/2} \rho(6 - \rho)e^{-\rho/3} \sin\theta e^{i\phi}$
	$u_{31-1} = \dfrac{1}{81}\left[\dfrac{1}{\pi a^3}\right]^{1/2} \rho(6 - \rho)e^{-\rho/3} \sin\theta e^{-i\phi}$
$n = 3, l = 2, m = 0$	$u_{320} = \dfrac{1}{486}\left[\dfrac{6}{\pi a^3}\right]^{1/2} \rho^2 e^{-\rho/3}(3\cos^2\theta - 1)$
$n = 3, l = 2, m = \pm 1$	$u_{321} = \dfrac{1}{81}\left[\dfrac{1}{\pi a^3}\right]^{1/2} \rho^2 e^{-\rho/3} \sin\theta \cos\theta e^{i\phi}$
	$u_{32-1} = \dfrac{1}{81}\left[\dfrac{1}{\pi a^3}\right]^{1/2} \rho^2 e^{-\rho/3} \sin\theta \cos\theta e^{-\phi}$
$n = 3, l = 2, m = \pm 2$	$u_{322} = \dfrac{1}{162}\left[\dfrac{1}{\pi a^3}\right]^{1/2} \rho^2 e^{-\rho/3} \sin^2\theta e^{i2\phi}$
	$u_{32-2} = \dfrac{1}{162}\left[\dfrac{1}{\pi a^3}\right]^{1/2} \rho^2 e^{-\rho/3} \sin^2\theta e^{-i2\phi}$

i.e., $Y(\theta,\phi)$ is not always a solution; it only satisfies physical conditions for certain integer values of the parameters l and m.

To see how this arises, consider the ϕ dependence of the Schrödinger equation:

$$\frac{d^2g}{d\phi^2} = -m^2g(\phi), \tag{8.17}$$

where m is a constant that arises in the separation of variables. Then

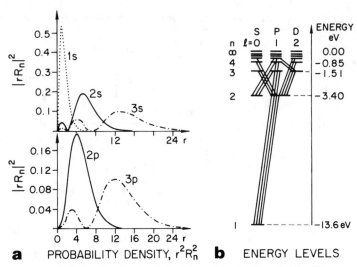

a   PROBABILITY DENSITY, $r^2R_n^2$ **b**   ENERGY LEVELS

Figure 8.1 (a) Probability density for the hydrogenic wave function for different values of the quantum numbers n and l. (b) Energy levels in the hydrogen atom; the vertical lines represent transitions that satisfy the selection rule $\Delta l = \pm 1$.

$$g(\phi) = A e^{im\phi}, \tag{8.18}$$

but ϕ is the azimuthal angle and the function must have the same value at $\phi = 0$ or $\phi = 2\pi$. This is true if

$$m = 0 \quad \text{or} \quad \text{an integer.} \tag{8.19}$$

The separation of the three variables leads to three quantum numbers n, l, and m and the solutions to the hydrogen atom problem are

$$u_{nlm}(r,\theta,\phi) = R_{nl}(r)f_{lm}(\theta)e^{im\phi}. \tag{8.20}$$

R_{nl} and f_{lm} are functions that can be written explicitly for a particular value of n, l, and m. The first few wave functions are

$$u_{100} = \frac{1}{\sqrt{\pi a^3}} e^{-\rho},$$

$$u_{200} = \frac{1}{\sqrt{4\pi(2a)^3}} (2 - \rho)e^{-\rho}, \tag{8.21}$$

where $\rho = Zr/a_0$ and $a = a_0/Z$. Table 8.1 lists the lowest energy wave functions for the hydrogen atom, and Figure 8.1 gives the radial distribution function in terms of $r^2R_n^2$. This quantity gives the probability density of finding a particle within a shell of radius r. Figure 8.1 also shows the energy levels for the hydrogen atom; the same values are found in the Bohr treatment.

The above wave functions are properly normalized; i.e., the probability of finding a particle somewhere in space is unity. That is,

$$\int u^*u \, dr = 1, \tag{8.22}$$

where u^* is the complex conjugate of u.

8.4 Quantum Numbers, Electron Configuration, and Notation

Quantum mechanics requires the assignment of four quantum numbers to specify the state of the electron in an atom:

Principal quantum number $\quad n = 1, 2, \cdots .$
Orbital quantum number $\quad l = 0, 1, 2, \ldots , n - 1.$
Magnetic quantum number $\quad m = 0, \pm 1, \pm 2, \ldots , \pm l.$
Spin quantum number $\quad m_s = +\frac{1}{2}, -\frac{1}{2}.$

The internal angular momentum of the electron, the spin, leads to the fourth quantum number, m_s, which has values $\pm\frac{1}{2}$. The existence of spin is responsible for the fine structure seen in high resolution measurements of spectral lines. From the Pauli exclusion principle, only one electron can have a given set of quantum numbers; that is, no two electrons in an atom can have the same set of values of the quantum numbers n, l, m, and m_s. Another set of quantum numbers can be assigned when the intersection between the orbital and spin is taken into account (i.e., the spin orbit interaction).

An electron has both orbital angular momentum (the quantum number l) and spin angular momentum s. The resultant spin plus orbital angular momentum, $j = l + s$, has the value

$$A_j = \sqrt{j(j + 1)} \, \hbar \quad \text{with } j = |(l \pm \tfrac{1}{2})|,$$

and the projection of the component around the polar axis has a quantized value of

$$(A_j)_z = m_j\hbar,$$

where m_j takes on integrally spaced values of $j, j - 1, \ldots , -j$. Thus for $j = \frac{3}{2}, m_j = \frac{3}{2}, \frac{1}{2}, -\frac{1}{2}, -\frac{3}{2}$. In spectroscopic notation, the total angular momentum quantum number of an atomic state is written as a subscript so that a state with principal quantum number 2, $l = 1$, and $j = \frac{1}{2}$ is denoted as $2p_{1/2}$.

When spin-orbit splitting is considered, an appropriate set of quantum numbers is:

Principal quantum number $\quad\quad\quad n = 1, 2, 3, \cdots$.

Orbital quantum number $\quad\quad\quad\quad l = 0, 1, 2, \ldots, n - 1$.

Angular momentum quantum
number $\quad\quad\quad\quad\quad\quad\quad\quad\quad j = |l \pm \frac{1}{2}|$.

Z-component of j quantum number $\quad m_j = j, j - 1, \ldots, -j$.

Here the quantum numbers j and m_j are always half-integral.

The periodic table can be built by assigning electrons to quantum states characterized by the four quantum numbers n, l, m, and m_s with no two electrons in any one atom having the same four quantum numbers. The Z electrons in the atom occupy the lowest energy states with the energy primarily determined by the principal quantum number n and to a lesser degree by the orbital quantum number l with no appreciable difference due to different values of the spin

Table 8.2 Atomic versus Ionic Energy Levels

		Atomic levels		Singly ionized atom
n	l	Electron shell	Number of electrons	X-ray symbol
1	0	1s	2	K
2	0	2s	2	L_1
	1	2p	6	L_2
				L_3
3	0	3s	2	M_1
	1	3p	6	M_2
				M_3
	2	3d	10	M_4
				M_5
4	0	4s	2	N_1
	1	4p	6	N_2
				N_3
	2	4d	10	N_4
				N_5
	3	4f	14	N_6
				N_7
5	0	5s	2	O_1

substates m_s (spin-orbit splitting ignored). The filling of electron levels within an atom is often designated in terms of the principal quantum numbers, n, and the historical notation for angular momentum where $l = 0$ corresponds to the letter s (for sharp), $l = 1$ to p (for principal), $l = 2$ to d (for diffuse), and $l = 3$ to f (fundamental). Thus He has an electronic configuration given by $1s^2$ (2 electrons in the $n = 1$, $l = 0$ shell) and neon has a configuration $1s^2$, $2s^2$, $2p^6$ (2 electrons in the $n = 1$, $l = 0$ shell; 2 electrons in the $n = 2$, $l = 0$ subshell and 6 electrons in the $n = 2$, $l = 1$ subshell). Table 8.2 lists the atomic levels, the electron shells, and X-ray symbols. When spin-orbit interactions are included we find that the p, d, $\cdots$ shells can be further split (in energy) to give configuration of the sort 1s, 2s, $2p_{1/2}$, $2p_{3/2}$ where the subscript denotes the angular momentum $l \pm \frac{1}{2}$ which is a result of summing the orbital angular momentum and spin angular momentum. The $2p_{3/2} - 2p_{1/2}$ splitting is approximately 1.5 eV for chlorine and is easily resolved in a standard XPS spectrometer. The splitting increases with increasing Z. Appendix 4 gives electron configurations of atoms along with ionization potentials (the energy to remove one electron from neutral atom) and Appendix 5 lists binding energies E_B.

8.5 Transition Probability

Analysis techniques involve one or more atomic transitions. Auger analysis involves the creation of a core hole, one atomic transition, and a subsequent Auger decay, the second atomic transition. Similarly, X-ray fluorescence (X-ray in, characteristic X-ray out) and the electron-microprobe (electron in, X-ray out) are examples of two transition processes. X-ray photoelectric spectroscopy is an example of a process that involves one atomic transition, the creation of an inner electron hole and an energetic photoelectron.

The most useful formula to calculate the probability of a transition comes from time-dependent perturbation theory. In first order perturbation theory the Hamiltonian acting on the system is written as

$$H = H_0 + H', \tag{8.23}$$

H_0 is a Hamiltonian for which the Schrödinger equation can be solved, and H' contains an additional potential, i.e., an applied electric field.

There is a set of eigenfunctions which are the solutions to H_0 such that $H_0 u_n = E_n u_n$. As discussed in Section 8.10, the transition probability per unit time for a transition to the state k from the initial state m is given by

$$W = \frac{2\pi}{\hbar} \rho(E) |\langle \psi_k | H' | \psi_m \rangle|^2. \qquad (8.24)$$

In this equation $\rho(E)$ is the density of final states per unit energy,

$$\langle \psi_k | H' | \psi_m \rangle = \int \psi_k^* H' \psi_m \, d\tau = |H'_{km}|, \qquad (8.25)$$

where ψ^* is the complex conjugate of ψ, $d\tau$ is a three-dimensional volume element (i.e., $r^2 \, dr \sin \theta d\theta d\phi$), and the wave function ψ_m is

$$\psi_m = e^{it\omega_m} u_m. \qquad (8.26)$$

Equation (8.24) is famous in quantum mechanics and is known as Fermi's Golden Rule. Its great advantage is that one need not know the wave functions for the true potential, $H_0 + H'$, rather only the solutions to H_0 are required. Note that W has dimensions of $(\text{time})^{-1}$.

$$W = \frac{1}{\text{energy} \cdot \text{time}} \cdot \frac{1}{\text{energy}} \cdot (\text{energy})^2 = \frac{1}{\text{time}}.$$

The derivation of Eq. (8.24) is given in Section 8.10.

8.6 Photoelectric Effect—Square Well Approximation

The goal in this chapter is to make a quantitative calculation of the cross section for the photoeffect: i.e., an electron with binding energy E_B is irradiated with light of energy $\hbar\omega$ and is ejected with energy $\hbar\omega - E_B$. In this section we calculate the relatively simple problem of the transition probability and then the cross section for an electron bound in a one-dimensional square well. In the next section we consider the more realistic case of the photoeffect in a hydrogenic atom.

A flux of incident photons is represented by an electromagnetic field and the perturbation is

$$H'(x,t) = H'(x)e^{i\omega t} = e\mathscr{E}xe^{-i\omega t}, \qquad (8.27)$$

where the electric field $\mathscr{E}$ which acts on the particle is uniform in space (wavelength greater than atomic dimensions) but harmonic in time. This perturbation represents the potential energy of the electron in the field of a photon of frequency ω.

The transition rate between the initial and final state is given in Eq. (8.24) as

$$W = \frac{2\pi}{\hbar} \rho(E)|H'_{fi}|^2,$$

where the initial state is the wave function ψ_i of a particle bound in the well and the final state is an outgoing one-dimensional plane wave, e^{-ikx}.

To calculate the density of final states $\rho(E)$, we consider the system to have dimension of length L and require periodic boundary conditions so that $\psi(x_0) = \psi(x_0 + L)$. The normalized states are

$$\psi_E = \frac{1}{\sqrt{L}} e^{ikx}, \qquad (8.28)$$

where $kL = (2mE)^{1/2}L/\hbar = 2\pi N$ (i.e., periodic boundary conditions).

The density of states is the number of states with energy between E and $E + \Delta E$. Then

$$\rho(E)\Delta E = \Delta N = \frac{L}{2\pi}\Delta k,$$

$$\rho(E) = \frac{L}{2\pi} \cdot \frac{\Delta k}{\Delta E}. \qquad (8.29)$$

For a free particle, $E = \hbar^2 k^2/2m$ and $\Delta E = (\hbar^2 k/m)\cdot \Delta k$ so that

$$\rho(E) = \frac{L}{2\pi\hbar}\left(\frac{2m}{E}\right)^{1/2}, \qquad (8.30)$$

where we have included a factor of 2 for positive and negative values of N. The matrix element in the transition probability is

$$H'_{fi} = \int \frac{1}{\sqrt{L}} e^{ikx}e\mathscr{E}x\,\psi_i(x)\,dx, \qquad (8.31)$$

where $\hbar k$ is the momentum of the final state.

In order to calculate the matrix element for a specific example, we consider the electron to be weakly bound in a narrow potential well. For the one-dimensional well, the wave function u_0 outside the well must satisfy

$$\frac{-\hbar^2}{2m}\frac{d^2u_0}{dx^2} = -E_Bu_0,$$

which leads to exponentially decaying solutions of the form

$$u_0 = Ce^{\pm k_i x},$$

where k_i is the momentum associated with the bound state, $k_i = (2mE_B)^{1/2}/\hbar$ and $C = \sqrt{k_i}$. For ease of calculation we extrapolate the exterior wave function to the origin, as indicated in Figure 8.2. Now, the normalized initial wave function ψ_i is

$$\psi_i(x) \simeq [\sqrt{k_i}]\, e^{-k_i|x|}. \qquad (8.32)$$

With this simplification we have

$$\langle \psi_f|x|\psi_i \rangle = \frac{1}{\sqrt{L}} \int_{-\infty}^{\infty} e^{ikx} x\, \psi_i(x)\, dx$$

$$= \frac{1}{\sqrt{L}} k_i^{1/2} \left[\frac{4ikk_i}{(k^2 + k_i^2)^2} \right]. \qquad (8.33)$$

The transition rate is then

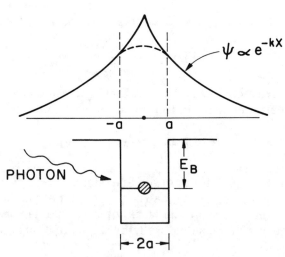

Figure 8.2 Schematic of the photo effect from an electron bound in a square well. The full curve shows the approximate wave function used in the calculation while the dashed line shows the true wave function in the region $x = \pm a$.

$$W = \frac{2\pi}{\hbar} \rho(E) |H'_{fi}|^2$$

$$= \frac{4e^2 \mathscr{E}^2 \hbar}{m} \left(\frac{E_B}{E}\right)^{1/2} \frac{E E_B}{(E + E_B)^4},$$

where $\mathscr{E}$ is the external electric field. We consider the case that $\hbar\omega \gg E_B$, so that $E \cong \hbar\omega$. Then

$$W = \frac{4e^2 \mathscr{E}^2 \hbar}{m} \cdot \frac{E_B^{3/2}}{E^{7/2}}. \tag{8.34}$$

For analytical purposes we are interested in the probability of the event per incident photon, i.e., the cross section for the process. This quantity σ is related to W by determining the flux F of photons in the oscillating electric field.

From classical electromagnetic theory we have that the power density in the field is $c\mathscr{E}^2/2$. The power density is the energy/area/s. Thus the flux F, the number of photons/area/s, is given by $c\mathscr{E}^2/2\hbar\omega$. Dimensionally the cross section is

$$\sigma = \frac{W}{F} = \frac{\text{Transition probability/time}}{\text{No. of photons/area/time}}$$

or

$$\sigma = \frac{8e^2 \hbar}{mc} \frac{E_B^{3/2}}{E^{5/2}}. \tag{8.35}$$

The cross section decreases with increasing photon energy ($E = \hbar\omega$) as $E^{5/2}$.

8.7 Photoelectric Transition Probability for a Hydrogenic Atom

In this section we describe a calculation of the cross section for the photoelectric effect using hydrogenic wave functions in three-dimensional space. The formula for the transition probability is given by Eq. (8.24). The relevant wave functions for the initial and final state are given by

$$\psi_i = \frac{1}{\sqrt{\pi a^3}} e^{-\rho}$$

and

$$\psi_f = \frac{1}{\sqrt{V}} e^{i\mathbf{k}\cdot\mathbf{r}}, \tag{8.36}$$

where the initial state ψ_i describes a ground state hydrogenic wave function in an atom of atomic number Z and the final state is the usual outgoing plane wave of final energy $E_f = \hbar^2 k^2/2m$ normalized to a volume V. The binding energy of the electron, E_B, is expressed as $Z^2 e^2/2a_0$ [see Eq. (1.17)], and in this calculation we assume the energy of the incoming photon, $\hbar\omega \gg E_B$. Here, the three-dimensional density of states, $\rho(E) = (V/2\pi^2)(2m/\hbar^2)^{3/2}E^{1/2}$, is used.

The transition probability can be calculated explicitly if the perturbation potential used is

$$H' = -e\mathscr{E}ze^{-i\omega t}.$$

In that case the final result for the photoeffect cross section σ_{ph} yields

$$\sigma_{\text{ph}} = \frac{288\pi}{3} \frac{e^2\hbar}{mc} \frac{E_B^{5/2}}{E^{7/2}},$$

a result similar to the square-well one-dimensional calculation carried out explicitly in the previous section.

Using a more sophisticated description of the perturbation potential, but precisely the same wave functions and assumption of $\hbar\omega \gg E_B$, Schiff (1968) shows that

$$\sigma_{\text{ph}} = \frac{128\pi}{3} \frac{e^2\hbar}{mc} \frac{E_B^{5/2}}{E^{7/2}}. \tag{8.37}$$

The value of $e^2\hbar/mc = 5.56 \times 10^{-2}$ eV Å^2, so that, for convenience, we write

$$\sigma_{\text{ph}} = \frac{7.45 \text{ Å}^2}{\hbar\omega} \times \left(\frac{E_B}{\hbar\omega}\right)^{5/2} \tag{8.37'}$$

by setting the incoming photon energy $\hbar\omega$ in eV equal to the energy E of the outgoing electron since $E_B \ll \hbar\omega$ (i.e., $E = \hbar\omega - E_B \approx \hbar\omega$).

As an example, the photoelectric cross section for Fe K_α radiation ($\hbar\omega = 6.4 \times 10^3$ eV, Appendix 6A) incident on K-shell electrons in Al ($E_B^K = 1.56 \times 10^3$ eV, Appendix 5) has a value

$$\sigma_{\text{ph}} = \frac{7.45}{6.4 \times 10^3} \times \left(\frac{1.56}{6.4}\right)^{5/2} \text{Å}^2 = 3.4 \times 10^{-21} \text{ cm}^2.$$

For the total absorption all electrons in all shells must be considered.

The cross section for impact ionization by electrons is given in Chapter 6 [Eq. (6.11)] for $E > E_B$ as $\sigma_e = \pi(e^2)^2/E_B E$, where E is the energy of the incident electron. For the same conditions, $E = 6.4 \times 10^3$ eV and $E_B = 1.56 \times 10^3$ eV, the electron impact ionization cross section σ_e has a value

$$\sigma_e = \frac{\pi \times (14.4 \text{ eV Å})^2}{6.4 \times 1.56 \times 10^6} = 6.5 \times 10^{-21} \text{ cm}^2,$$

which is a factor of 2 greater than that for the photoelectron cross section.

The electron impact cross section depends inversely as the energy of the incident particle while the photo-effect cross section is a strong function, $\sigma \propto (\hbar \omega)^{-7/2}$, of the incident photon energy for cases where $\hbar \omega \gg E_B$. Thus, in most cases, the values of σ_e are significantly greater than that of the photoelectric cross section. The primary advantage of using electrons as a method of creating inner shell vacancies is not the increase in the cross section but that an electron beam can be obtained with orders of magnitude greater intensity than is possible with an X-ray source in a laboratory system. An electron beam can also be focused and scanned for the analysis of submicron regions.

8.8 X-ray Absorption

In the previous section, we have been concerned with photoelectric absorption. This is but one of three processes that lead to attenuation of a beam of high energy photons penetrating a solid: photoelectron production, Compton scattering, and pair production. In the Compton effect, X-rays are scattered by the electrons of an absorbing material. The radiation consists of two components, one at the original wavelength λ and one at a longer wavelength (lower energy). The problem is generally treated as an elastic collision between a photon with momentum $p = h/\lambda$ and a stationary electron with rest energy mc^2. After scattering at an angle θ, the photon wavelength is shifted to larger values by an amount $\Delta \lambda = (h/mc)(1 - \cos \theta)$, where $h/mc = 0.0243$ Å is known as the Compton wavelength of the electron.

If the photon energy is greater than $2mc^2 = 1.02$ MeV, the photon can annihilate with the creation of an electron–positron pair. This process is called pair production. Each of the three processes — photoelectric, Compton scattering, and pair production — tend to dominate in a given region of photon energies as shown in Figure 8.3. For X-ray and low energy gamma rays photoelectric absorption makes the dominant contribution to the attenuation of the photons

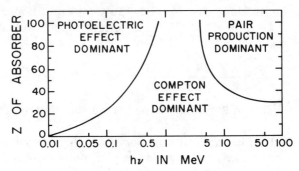

Figure 8.3 The relative importance of the three major types of photon interactions. The lines show the values of Z and $\hbar\omega$ for which the neighboring effects are equal.

penetrating the material. It is this energy regime which is of primary concern for *atomic* processes in materials analysis.

The intensity I of X-rays transmitted through a thin foil of material for an incident intensity I_0 follows an exponential attenuation relation

$$I = I_0 e^{-\mu x} = I_0 \exp\left(-\mu/\rho\right) \rho x, \qquad (8.38)$$

where ρ is the density of the solid (g/cm^3), μ is the linear attenuation coefficient, and μ/ρ is the mass attenuation coefficient given in cm^2/g. Figure 8.4 shows the mass absorption coefficient in Ni as a function of X-ray wavelength. The strong energy dependence of the absorption coefficient follows from the energy dependence of the photoelectric cross section. At the K absorption edge, photons eject electrons from the K shell. At wavelengths longer than the K edge,

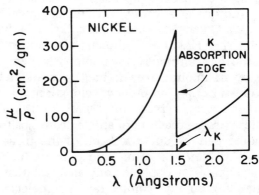

Figure 8.4 The mass absorption coefficient μ/ρ (cm^2/g) of Ni versus λ.

absorption is dominated by the photoelectric process in the L shells; at shorter wavelengths where $\hbar\omega \gtrsim E_B(K)$, photoelectric absorption in the K shell dominates.

Both X-ray photoelectron spectroscopy (discussed in Chapter 9) and X-ray absorption depend on the photoelectric effect. The experimental arrangements are shown in the upper portion of Figure 8.5 (XPS on the left side and X-ray absorption on the right side). In XPS, a bound electron such as the K-shell electron shown in Figure 8.5 is promoted to a free state outside the sample. The kinetic energy of the photoelectron is well defined and sharp photopeaks appear in the photoelectron spectrum. In X-ray absorption spectra, an edge occurs when a bound electron is promoted to the first unoccupied level allowed according to the selection rules. With metallic samples this unoccupied level is at or just above the Fermi level. In X-ray absorption, the absorption is measured as a function of X-ray energy whereas in XPS one irradiates with constant energy photons and measures the kinetic energy of the electrons.

The mass absorption coefficient μ/ρ for electrons in a given shell or subshell can be calculated from the photoelectric cross section σ:

$$\frac{\mu}{\rho} = \frac{\sigma \ (cm^2/electron) \times N \ (atoms/cm^3) \cdot n_s \ (electrons/shell)}{\rho \ (g/cm^3)} \tag{8.39}$$

where ρ is the density, N the atomic concentration, and n_s the number of electrons in a shell. For example, for Mo K_α radiation ($\lambda = 0.711$ Å and $\hbar\omega = 17.44$ keV) incident on Ni, which has a K shell binding energy of 8.33 keV, the value of the photoelectric cross section per K electron is

$$\sigma_{ph} = \frac{7.45 \times 10^{-16} \ cm^2}{17.44 \times 10^3} \left(\frac{8.33}{17.44}\right)^{5/2} = 6.7 \times 10^{-21} \ cm^2.$$

The atomic density of Ni is 9.14×10^{22} atom/cm³ and the density is 8.91 g/cm³. The mass absorption coefficient μ/ρ for K shell absorption ($n_s = 2$ for the K shell) is

$$\frac{\mu}{\rho} = \frac{6.7 \times 10^{-21} \times 9.14 \times 10^{22} \cdot 2}{8.91} = 138 \ cm^2/g.$$

In this calculation the contribution of the L shell electrons was neglected. For photon energies greater than the K shell binding energy the photoelectric cross section for the L shell is at least an order of magnitude smaller than that of the K shell; of course, this is the major factor in the sharp increasing absorption when one crosses the

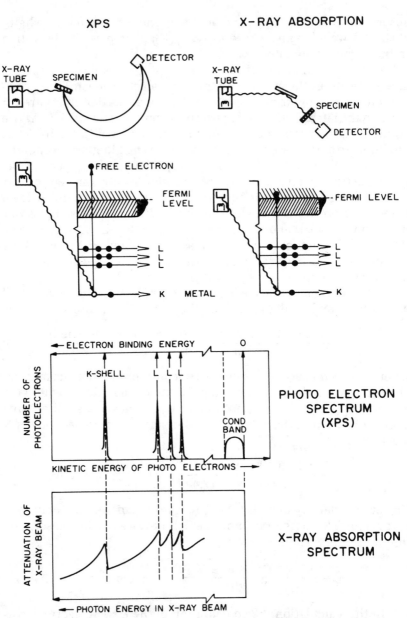

Figure 8.5 Comparison of X-ray absorption and X-ray photoelectron spectroscopy. [From Siegbahn et al., in *ESCA* (Almqvist and Wiksells, Uppsala, Sweden, 1967).]

K absorption edge. In the present case of Mo K_α radiation on Ni, if we assume an average binding energy of 0.9 keV for the L_1, L_2, and L_3 shells, the photoelectric cross section per electron is a factor of 3.8×10^{-3} smaller than that of the K shell electrons due to the $(E_B/\hbar\omega)^{5/2}$ term.

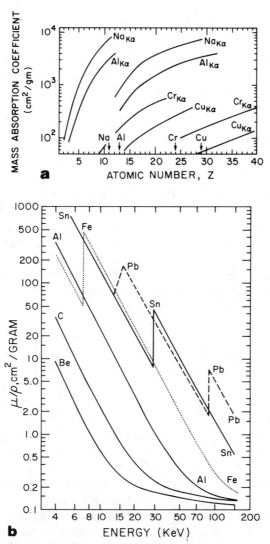

Figure 8.6 Mass absorption coefficient (a) for elements from $Z = 2–40$ for K_α radiation from a variety of sources and (b) as a function of energy for different absorbers.

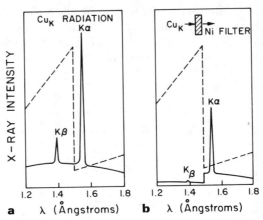

Figure 8.7 Comparison of the Cu radiation before and after passage through an Ni filter. The dashed line represents the mass absorption coefficient of nickel. [Adapted from Cullity, 1978.]

The calculated value, 138 cm²/g, is greater than the measured value of 47.24 (Appendix 6A). The major difficulty in the mass absorption calculation above was that the energy E of Mo K_α radiation is only twice that of the K shell binding energy E_B and the derivation of Eq. (8.37) was based on $\hbar\omega \gg E_B$. For Cu K_α radiation with $E = 8.04$ keV, the photon energy is about 10 times that of the L shell binding energy and the calculated photoelectric cross section ($\sigma = 3.1 \times 10^{-21}$ cm²) for the L shell gives a value of $\mu/\rho = 32$ cm²/g, a value close to the tabulated value of 48.8 cm²/g.

Measured values of the mass absorption coefficient for different radiation is tabulated in Appendix 6B and is displayed in Figure 8.6a for $Z = 2-40$. For a given element the absorption coefficient can vary over 2 orders of magnitude depending on the wavelength of the incident radiation. The strong photon energy dependence ($\hbar\omega^{-7/2}$) of the absorption coefficient is illustrated in Figure 8.6b.

The tenfold change in the absorption coefficient on either side of the K edge as shown in Figure 8.4 represents a major change in transmitted intensity for thin foils because of the exponential nature of the transmission factor I/I_0. If the transmission factor of a particular sheet is 0.1 for a wavelength just longer than λ_K, then for a wavelength just shorter the transmission is reduced by a factor of about $\exp(-10)$. This effect has been used to design filters for X-ray diffraction experiments which require nearly monochromatic radiation. As shown in Figure 8.7a, the characteristic radiation from the K shell contains a strong K_α line and a weaker K_β line (the K_β/K_α emission ratio is discussed in Section 10.10). The K_β line intensity relative to that of the K_α line can be decreased by passing the beam through a filter made of material whose absorption edge lies between the K_α and K_β wavelength of the target material. For metals with Z near 50, the filter will have an atomic number one less than that of the target. As shown in Figure 8.7, a Ni filter has a strong effect on the ratio of the Cu K_α and K_β lines where μ/ρ has a value of 48 for Cu K_α and 282 for K_β radiation.

8.9 Extended X-ray Absorption Fine Structure (EXAFS)

In the previous sections, the emphasis was upon the photoelectric cross section and absorption edges without consideration of the fine structure that is found at energies above the absorption edges. Figure 8.8 is a schematic representation of an X-ray absorption, of μx versus the energy of the incident radiation plotted over an energy region extending about 1 keV above the K absorption edge. In this energy region there are oscillations in absorption. The term extended X-ray absorption fine structure (EXAFS) refers to these oscillations, which

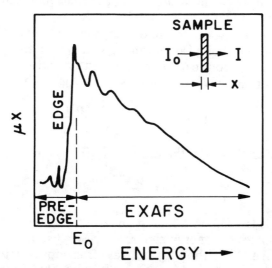

Figure 8.8 Schematic of the transmission experiment and the resulting X-ray absorption μx versus E for an atom in a solid.

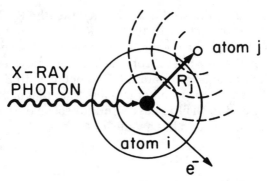

Figure 8.9 Schematic of the EXAFS process illustrating an emitted e^- scattering from a nearby atom at a distance R_j.

may have a magnitude of about 10%, of the absorption coefficient in the energy region above the edge. The oscillations arise from interference effects due to the scattering of the outgoing electron with nearby atoms. From analysis of the absorption spectrum for a given atom one can assess the types and numbers of atoms surrounding the absorber. EXAFS is primarily sensitive to short range order in that it probes out to about 6 Å in the immediate environment around each absorbing species. Synchrotron radiation is used in EXAFS measurements because it provides an intense beam of variable energy, monoenergetic photons.

For an incoming photon of energy $\hbar\omega$, a photoelectron can be removed from a K shell of atom i and have a kinetic energy $\hbar\omega - E_B^K$. The outgoing electrons can be represented as a spherical wave (Figure 8.9) which has a wave number $k = 2\pi/\lambda$ given by

$$k = \frac{p}{\hbar} = \frac{\sqrt{2m(\hbar\omega - E_B^K)}}{\hbar} \tag{8.40}$$

and a wave function ψ of the form

$$\psi = \frac{\psi_0 e^{i\mathbf{k}\cdot\mathbf{r}}}{r}. \tag{8.41}$$

Note that this is a different final state term than used in the calculation of σ_{ph} since low energy electrons are well represented by this spherical wave. When the outgoing wave from atom i arrives at atom j a distance R_j away, it can be scattered through $180°$ so that its wave function is

$$\psi_j = \frac{\psi_0 f e^{i\mathbf{k}\cdot\mathbf{R}_j+\phi_a}}{R_j},$$

where f is an atomic scattering factor and ϕ_a is a phase shift. When the scattered wave arrives back at atom i, it has a wave function ψ_{ij}

$$\psi_{ij} = \psi_0 \frac{f e^{i\mathbf{k}\cdot\mathbf{R}_j+\phi_a}}{R_j} \cdot \frac{e^{i\mathbf{k}\cdot\mathbf{R}_j+\phi_i}}{R_j}.$$

That is the outgoing photoelectron wave from atom i is backscattered with amplitude f from the neighboring atom, thereby producing an incoming electron wave. It is the interference between the outgoing and incoming waves that gives rise to the sinusoidal variation in the absorption coefficient.

The net amplitude of the wave at atom will be $\psi_0 + \psi_{ij}$

$$\psi_0 + \psi_{ij} = \psi_0 \left(1 + \frac{f}{R_j^2} e^{i2\mathbf{k}\cdot\mathbf{R}_j+\phi_j+\phi_a} \right),$$

and the intensity $I = \psi\psi^*$ will have the form

$$I = \psi_0\psi_0^* \left(1 + \frac{2f \sin(2\mathbf{k}\cdot\mathbf{R}_j + \phi_{ij})}{R_j^2} + \text{higher terms} \right),$$

where ϕ_{ij} represents the phase shifts. There are additional terms to account for the fact that the atoms have thermal vibrations and that the electrons which suffer inelastic losses in their path between atoms will not have the proper wave vector to contribute to the interference process. This latter factor is usually accounted for by use of an exponential damping term, $\exp(-2R_j/\lambda)$, where λ is the electron mean free path. The damping term is responsible for the short-range order description while the sinusoidal oscillation is a function of the interatomic distances $(2kR_j)$ and the phase shift (ϕ_{ij}). The important part of this equation is the term proportional to $\sin(2kR_j + \phi)$. By measuring $I(k)$ and taking a Fourier transform of the data with respect to k, one can extract R_j. The data and the transform are illustrated in Figure 8.10.

The ability of EXAFS to determine the local structure around a specific atom has been used in the study of catalysts, multicomponent alloys, disordered and amorphous solids, and dilute impurities and atoms on a surface. In surface EXAFS (SEXAFS), the technique has been used to determine the location and bond length of absorbed atoms on clean single crystal surfaces. EXAFS is an important tool in structural studies; the requirement for strong radiation sources leads

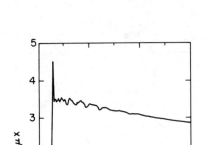
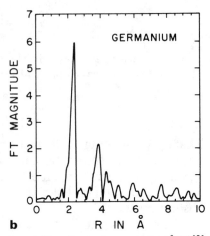

Figure 8.10 (a) X-ray absorption spectrum of crystalline Ge at a temperature of 100°K. The sharp rise near 11 keV is the K edge and the modulation in μx above the edge is the EXAFS. (b) Fourier transform of (a) showing the nearest neighbor and second nearest neighbor distances. [From Lee et al., 1981.]

to these types of experiments being carried out at synchrotron radiation facilities.

8.10 Time–Dependent Perturbation Theory

8.10.1 Fermi's Golden Rule

In this section we give a brief treatment of perturbation theory which leads to the basic formula for the transition probability of a quantum system. It is the formula which is the starting point for many of the derivations of cross sections given in this book.

Consider a system with a Hamiltonian H given by

$$H = H_0 + H', \tag{8.42}$$

where H_0 is a time-independent operator with an eigenvalue ψ_0. H_0 could be, for example, the Hamiltonian which describes a hydrogenic atom while H' may be a time–dependent perturbation, i.e., an oscillating electric field. The wave function ψ_0 satisfies the Schrödinger equation such that

$$i\hbar \frac{\partial \psi_0}{\partial t} = H_0 \psi_0. \tag{8.43}$$

Since H_0 is time independent, we can write

$$\psi_0 = u(x,y,z)e^{-iE_0 t/\hbar} \tag{8.44}$$

or

$$\psi_0 = \sum_n a_n^0 u_n^0 e^{-iE_n^0 t/\hbar}, \tag{8.45}$$

where

$$H_0 u_n^0 = E_n^0 u_n^0 \tag{8.46}$$

with the u_n^0 being an orthonormal set of eigen vectors and the constants a_n^0 independent of time.

For the perturbed Hamiltonian we write

$$H\psi = i\hbar \frac{\partial \psi}{\partial t} = (H_0 + H')\psi \tag{8.47}$$

and

$$\psi = \sum_n a_n(t)u_n^0 e^{-iE_n^0 t/\hbar}, \tag{8.48}$$

where the coefficients a_n are now a function of time.

Substitution of (8.48) into (8.47), multiplying by the complex conjugate u_s^{0*} and using the orthonormality relation yields

$$\frac{da_s}{dt} = \dot{a}_s = -\frac{i}{\hbar} \sum_n a_n(t)H'_{sn} e^{i(E_s^0 - E_n^0)t/\hbar}, \tag{8.49}$$

where

$$H'_{sn} = \int u_s^{0*}H'u_n^0 \, d\tau,$$

the integral being over all space. We approximate the solution by noting that if the perturbation is small, the time variation of $a_n(t)$ is slow; then $a_n(t) \simeq a_n(0)$ and

$$a_s(t) - a_s(0) \cong -\frac{i}{\hbar} \sum_n a_n(0) \int_0^t H'_{sn}(t)e^{i\omega_{sn}t} \, dt, \tag{8.50}$$

where $\hbar\omega_{sn} = E_s^0 - E_n^0$.

A special case is if the system is in state n at $t = 0$; then $a_n(0) = 1$ and all other a's are zero. Equation (8.50) then gives

$$a_s(t) = \frac{i}{\hbar} \int_0^t H'_{sn}(t) e^{i\omega_{sn}t} \, dt \tag{8.51}$$

for $s \neq n$.

The perturbation $H'(t)$ can induce transitions from the state n to any state s, and the probability of finding the system in the state s at time t is $|a_s(t)|^2$.

If H' is independent of time, then

$$a_s(t) = - H'_{sn} \frac{(e^{i\omega_{sn}t} - 1)}{\hbar \omega_{sn}}, \tag{8.52}$$

and

$$|a_s(t)|^2 = 4 \frac{H'^2_{sn} \sin^2(\omega_{sn}t/2)}{\hbar^2 \omega_{sn}^2} \tag{8.53}$$

and is valid if $|a_s(t)| < 1$.

In many applications the result of a perturbation is a particle in the continuum, i.e., a free particle. Then an explicit final state is not appropriate but the density of final states is relevant. We call $\rho(E)$ the density of final states (number of energy levels/per unit energy interval) and assume that H'_{sn} is the same for all final states.

The transition probability $P(t)$ is then given by

$$P(t) = \sum_s |a_s(t)|^2 = 4|H'_{sn}|^2 \sum_s \frac{\sin^2(\omega_{sn}t/2)}{\hbar^2 \omega_{sn}^2}. \tag{8.54}$$

For a continuum we transform the sum to an integral and note that the number of states in energy interval dE_s is $\rho(E_s) \, dE_s$; then

$$P(t) = 4|H'_{sn}|^2 \int_{-\infty}^{\infty} \rho(E_s) \frac{\sin^2(\omega_{sn}t/2)}{\hbar^2 \omega_{sn}^2} \, d\hbar \omega_{sn}. \tag{8.55}$$

The major contribution of the integral comes from $\omega_{sn} = 0$ (it is a bit like a delta function) and noting that $\int (\sin^2 \alpha x/x^2) \, dx = \pi \alpha$, we have

$$P(t) = 4|H'_{sn}|^2 \rho(E_n) \frac{\pi t}{2\hbar} \tag{8.56}$$

so that the rate of the transition, W, is given by

$$W = \frac{2\pi}{\hbar} \rho(E_n)|H'_{sn}|^2. \tag{8.57}$$

This is a famous and important formula denoted the "Golden Rule" of Fermi.

8.10.2 Transition Probability in an Oscillating Electric Field

Another important application of perturbation theory is when the perturbing field has a time dependence of the form $e^{i\omega t}$. Consider for example an oscillating electric field directed along the x axis such that

$$\mathscr{E} = \mathscr{E} \cos \omega t = \frac{\mathscr{E}_0}{2} (e^{i\omega t} + e^{-i\omega t}) \tag{8.58}$$

and

$$H'(t) = \frac{e}{2} \mathscr{E}_0 x (e^{i\omega t} + e^{-i\omega t}). \tag{8.59}$$

Inserting $H'(t)$ into (8.51) gives

$$a_s(t) = \frac{i}{\hbar} e \frac{\mathscr{E}_0 x_{sn}}{2} \int_0^t (e^{i(\omega_{sn}+\omega)t} + e^{i(\omega_{sn}-\omega)t}) \, dt. \tag{8.60}$$

If $\omega_{sn} \neq \omega$, the integral averages to zero. If $\omega_{sn} \simeq \omega$, we have

$$|a_s(t)|^2 = \frac{e^2 \mathscr{E}_0^2}{\hbar^2} |x_{sn}|^2 \frac{\sin^2[(t/2)(\omega - \omega_{sn})]}{(\omega - \omega_{sn})^2}, \tag{8.61}$$

where $|x_{sn}| = \int u_s^* x u_n \, d\tau$. We recognize that $\mathscr{E}_0^2/2$ is the energy density in the field which can be represented by $\rho(\omega)$ to obtain a transition probability

$$T_{ns} = |a_s(t)|^2 = 2e^2 |x_{sn}|^2 \int_0^\infty \frac{\sin^2[(t/2)(\omega - \omega_{sn})]}{\hbar^2(\omega - \omega_{sn})^2} \rho(\omega) \, d\omega. \tag{8.62}$$

Assume that the distribution $\rho(\omega)$ varies much more slowly than the sharply peaked function by which it is multiplied. Then $\rho(\omega)$ is nearly constant over the small range of values of ω for which the integrand is nonzero, so that we can replace $\rho(\omega)$ by its value at $\omega = \omega_{sn}$, and remove it from the integral, with no loss of accuracy. With the further substitution $z = \frac{1}{2}(\omega - \omega_{sn})t'$, the expression becomes

$$|a_{sn}(t')|^2 = \frac{e^2\rho(\omega_{sn})|x_{sn}|^2 t'}{\hbar^2} \int\limits_{-\infty}^{\infty} \frac{\sin^2 z}{z^2}\, dz$$

or, since

$$\int\limits_{-\infty}^{\infty} \frac{\sin^2 z}{z^2}\, dz = \pi,$$

the transition probability is

$$T_{sn} = |a_{sn}(t')|^2 = \frac{\pi e^2 \rho(\omega_{sn})|x_{sn}|^2 t'}{\hbar^2}, \tag{8.63}$$

where $x_{sn} = \int u_s^* x u_n\, d\tau$. Equation (8.63) holds for radiation which is polarized along the x axis. In the general case, when radiation is incident upon the atom from all directions and with random polarization, T_{sn} must include equal contributions from x_{sn}, y_{sn}, and z_{sn}, so that

$$T_{sn} = \frac{\pi e^2 \rho(\omega_{sn}) t'}{3\hbar^2} |\langle \psi_s | \mathbf{r} | \psi_n \rangle|^2, \tag{8.64}$$

where the factor of 3 has been introduced into the denominator because each polarization direction is assumed to contribute $\frac{1}{3}$ of the intensity and $[|x_n|^2 + |y_{sn}|^2 + |z_{sn}|^2] = |\langle \psi_s | \mathbf{r} | \psi_n \rangle|^2 = |r_{sn}|^2 \cdot \langle \psi_s | \mathbf{r} | \psi_n \rangle$ is known as the dipole matrix element and Eq. (8.64) is the dipole approximation to the transition probability.

8.10.3 Spontaneous Transitions

Spontaneous transitions are those that occur in the absence of an external field. An example is the transition from an excited state to a ground state. To calculate such phenomena we rely on a treatment developed in 1917 by Einstein, which enables one to calculate the rate of spontaneous transitions from knowledge of the rate of induced transitions calculated in Section 8.10.2.

Consider a collection of atoms in thermal equilibrium; each atom must be emitting and absorbing radiation at the same rate. Let P_{ns} be the probability that a given atom will go from the nth state to the sth state in a short time dt. This *probability,* P_{ns} must be proportional to the *probability* P_n that the atom is in the nth state to begin with, multiplied by the transition probability to go from state n to state s, T_{ns}, [given by Eq. (8.64)]:

$$P_{ns} = T_{ns}P_n.$$

In view of Eq. (8.64), we may write this expression as

$$P_{ns} = A_{ns}\rho(\omega_{ns})P_n \, dt, \qquad (8.65)$$

where $\rho(\omega_{ns})$ is defined as before, dt takes the place of t' as the time interval, and A_{ns} is equal to all the other factors in Eq. (8.64).

The probability of a downward transition from state s to state n may be written as

$$P_{sn} = A_{sn}\rho(\omega_{sn})P_s \, dt,$$

but because of the symmetry of the equations leading to Eq. (8.64) we know that $A_{sn} = A_{ns}$ and $\omega_{sn} = \omega_{ns}$, so that

$$P_{sn} = A_{ns}\rho(\omega_{ns})P_s \, dt. \qquad (8.66)$$

Notice that P_{ns} does not equal P_{sn}, because $P_n > P_s$; the nth state, being lower in energy, is more heavily populated, according to the Boltzmann factor.

The system is in equilibrium so that the total number of transitions from n to s must equal the total number of transitions from s to n. Since the induced transition probabilities are unequal ($P_{ns} \neq P_{sn}$), there must be additional transitions from state s to state n which are spontaneous. The spontaneous transition probability by definition does not depend on the energy density of the externally applied field; this probability may be written $B_{sn}P_s \, dt$, where B_{sn} is the spontaneous transition probability. The coefficients A and B are referred to as the Einstein coefficients.

The total transition probability from s to n is therefore equal to $P_{sn} + B_{sn}P_s \, dt$, and this total should equal the transition probability P_{ns}. Thus, from Eqs. (8.65) and (8.66), we have

$$\rho(\omega_{ns})P_n A_{ns} = \rho(\omega_{ns})P_s A_{ns} + P_s B_{sn}$$

or

$$B_{sn} = \rho(\omega_{ns})A_{ns}\left[\frac{P_n}{P_s} - 1\right].$$

But the population of a state of energy E is proportional to the Boltzmann factor $e^{-E/kT}$, so that the ratio P_n/P_s may be written

$$\frac{P_n}{P_s} = e^{(E_s - E_n)/kT}$$

$$= e^{\hbar \omega_m / kt}.$$

Thus

$$B_{sn} = \rho(\omega_{ns}) A_{ns}(e^{\hbar \omega_m / kt} - 1). \tag{8.67}$$

The factor $\rho(\omega_{ns})$ in Eq. (8.67) is misleading because B_{sn} by definition does not depend on the energy density of the radiation field. We can eliminate this factor by using an expression for the energy density inside the cavity. Planck's law for thermal radiation gives

$$\rho(\omega) = \frac{\hbar \omega^3}{\pi^2 c^3 (e^{\hbar \omega / kT} - 1)}. \tag{8.68}$$

Substitution of this expression into Eq. (8.67) yields an expression for B_{sn} involving only A_{ns} and known constants:

$$B_{sn} = \frac{\hbar \omega_{ns}^3}{\pi^2 c^3} A_{ns}. \tag{8.69}$$

The transition rate for a spontaneous transition from a filled to an empty state is

$$W = \frac{4}{3} \frac{e^2}{\hbar} \left[\frac{\omega_{ns}}{c} \right]^3 |\langle \psi_s | \mathbf{r} | \psi_n \rangle|^2, \tag{8.70}$$

where $e^2 = 14.4$ eV Å. This can be expressed as

$$W = 0.38 \times 10^{16} (\hbar \omega)^3 \cdot |\langle \psi_s | \mathbf{r} | \psi_n \rangle|^2,$$

where $\hbar \omega$ is in units of keV and the matrix element has dimensions of (Å)². A typical value of W is 10^{15}/s for elements in the middle of the periodic table.

In summary,

$$W = \frac{2\pi}{\hbar} \cdot |H'_{sn}|^2 \rho(E_n); \text{ static perturbation [Eqs. (8.24) and 8.57)].}$$

$$W = \frac{\pi e^2 \rho(\omega_{sn})}{3\hbar^2} |\mathbf{r}_{sn}|^2; \text{ time-dependent, } H' = \mathcal{E}_0 \cos \omega t \text{ [Eq. (8.64)].}$$

$$W = \frac{4}{3} \frac{e^2}{\hbar} \left[\frac{\omega_{ns}}{c} \right]^3 |\mathbf{r}_{sn}|^2; \text{ spontaneous transitions [Eq. (8.70)].}$$

Problems

8.1. For Cu K_α radiation (E = 8.04 keV) incident on Al:
(a) Calculate the photoelectric cross section σ_{ph} for the K shell of Al and compare its value with the electron impact ionization cross section σ_e for 8.04 keV electrons.
(b) Calculate the mass absorption coefficient for Al based on only K shell absorption and compare with the values tabulated in Appendix 6B. The Al L shells have 8 electrons compared to 2 in the Al K shell.
(c) Estimate the contribution of L shell electrons to the mass absorption coefficient for 8.04 keV electrons.

8.2. Consider 5.41 keV radiation [photon (Cr K_α) and electron] incident on Si. Compare the values in microns of the linear absorption coefficient, the electron range and the electron mean free path.

8.3. You use an Ni filter to attenuate the Cu K_β radiation from a Cu X-ray source. If the Ni filter thickness is sufficient to a attenuate the K_β radiation by a factor of 1000, how much will the K_α radiation be attenuated. How thick the Ni filter? Estimate to zero order how much the Cu L_α radiation is attenuated.

8.4. Be is used a "window" material that allows X-rays to enter the X-ray detector with a minimum attenuation. The absorption of X-rays from elements with Z < 10 is one of the limitations of microprobe analysis. Estimate the absorption of carbon K_α X-rays by a 7-μm thick Be window.

8.5. Consider a photon of energy $\hbar\omega$ inducing a photoeffect process with an electron of binding energy E_B bound in an atom of mass M_n. Assume that the electron is emitted in the forward direction, i.e., along the same direction as the incident photon. In the approximation that $\hbar\omega \gg E_b$ show that the energy of the recoiling atom, E, is given by

$$E_n = \frac{(\hbar\omega)^2}{2M_n c^2} + \frac{M_e}{M_n}\hbar\omega - \frac{\hbar\omega}{M_n c^2}\sqrt{2M_e c^2 \cdot \hbar\omega}$$

where M_e is the electron mass. Evaluate this expression for $\hbar\omega = 1$ keV, 10 keV, 100 keV, and $M_n = 28$ (Silicon) and compare the result to 14 eV, the binding energy of Si in the Si lattice. Such considerations are important in determining the "destructiveness" of a given analysis technique.

8.6. Using conservation of momentum and energy, show that the photoeffect cannot occur with a free electron. Consider the nonrelativistic case, $\hbar\omega < m_e c^2$. The photoeffect occurs with photon irradiation of solids since all electrons in solids are bound to some degree.

References

1. E. E. Anderson, *Modern Physics and Quantum Mechanics* (W. B. Saunders, Philadelphia, 1971).
2. B. D. Cullity, *Elements of X-ray Diffraction*, 2nd Edition (Addison-Wesley, Reading, MA, 1978).
3. P. A. Lee, P. H. Citrin, P. Eisenberger, and P. M. Kincaid, Extended X-ray Absorption Fine Structure, *Rev. Mod. Phys.* **53**, 769 (1981).
4. R. Saxon, *Elementary Quantum Electrodynamics* (Holden-Day, San Francisco, 1968).
5. L. I. Schiff, *Quantum Mechanics*, 3rd Edition (McGraw-Hill Book Co., New York, 1968).
6. R. L. Sproull and W. A. Phillips, *Modern Physics*, 3rd Edition (John Wiley and Sons, New York, 1980).
7. P. A. Tipler, *Modern Physics* (Worth Publishers, New York, 1978).

9

X-RAY PHOTOELECTRON
SPECTROSCOPY (XPS)

9.1 Introduction

In this chapter electronic structure is the dominant theme. Photons
with energies of up to 10 keV interact with the atomic electrons
primarily via the photon absorption process (Chapter 8). The photoelectric process is a direct signature of the photon interaction with the
atom and is the basis of one of the major analytical tools—
photoelectron spectroscopy. This is referred to as UPS when ultraviolet light is incident on the sample and XPS when X-rays are used.
Another acronym is ESCA (for Electron-Spectroscopy for Chemical
Analysis); in this case the main concern is the chemical bonding.

The energy spectrum of electromagnetic radiation along with the
common nomenclature is shown in Figure 9.1. In material analysis,
the photon energy range of interest corresponds to the ultraviolet
(UV) and X-ray region. In practice it extends from 10 eV, close to the
binding energy (13.6 eV) of the electron in the hydrogen atom to
energies of around 0.1 MeV. At these energies photons can penetrate
within the solid and interact with the inner shell electrons. Lower
energy photons are used to establish the visible spectra associated
with the outermost, less tightly bound electrons. These outermost
electrons are involved in chemical bonding and are not associated
with specific atoms and hence are not useful for elemental identification. Photon-induced spectroscopies have undergone a major ad-

213

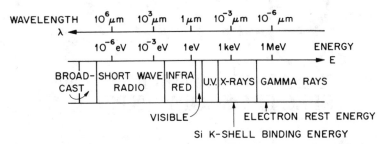

Figure 9.1 Electromagnetic spectrum indicating the region used for photoelectron spectroscopy. Ultraviolet photoelectron spectroscopy corresponds to incident photons in the UV region; X-ray electron spectroscopy corresponds to incident photons in the X-ray region.

vance due to the advent of electron synchrotrons which produce an intense source of monochromatic photons over a broad range of energies for materials science. Most laboratory instruments produce X-rays in the 1–10 keV region, which is the main region discussed in this chapter.

9.2 Experimental Considerations

The basic processes of interest in photoelectron spectroscopy are the absorption of a quantum of energy $\hbar\omega$ and the ejection of an electron, the photoelectron, whose kinetic energy, referenced to an appropri-

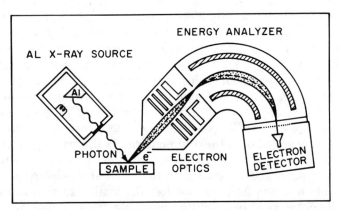

Figure 9.2 Schematic of the basic apparatus used in X-ray photoelectron spectroscopy. X-rays are produced at the Al anode by bombardment of electrons created at the filament. The X-rays impinge on a sample producing photoelectrons which are detected after analysis in the electron energy analyzer.

ate zero of energy, is related to the binding energy of an electron in the target atom. In this process, an incident photon transfers its entire energy to the bound electron, and element identification is provided by the measurement of the energy of the electrons that escape from the sample without energy loss. As indicated in Figure 9.2, photoelectron spectroscopy requires both a source of monochromatic radiation and an electron spectrometer. As is common to all the electron spectroscopies where the escape depth is 10–20 Å, careful sample preparation and clean vacuum systems are required.

9.2.1 Radiation Sources

A convenient source of characteristic X-rays is provided by electron bombardment of Mg or Al targets. The relative intensity of bremsstrahlung or X-ray continuum to characteristic X-rays is less important in the production of these soft X-rays (~ 1 keV) than for hard X-rays, i.e., from Cu bombardment. For Mg and Al about one-half of the X-rays produced by electron bombardment are the K_α X-rays. The contribution from the continuous spectrum is hardly noticeable since the bremsstrahlung spectrum is distributed over several keV while the K X-rays are concentrated in a peak of ~ 1 eV FWHM. In addition to the two K_α lines (K_{α_2} corresponding to $2p_{1/2} \rightarrow 1s$ and K_{α_1} corresponding to $2p_{3/2} \rightarrow 1s$, indicated in Figure 9.3a), there are lower intensities of higher energy characteristic lines that correspond to two electron excitations (a 1s ionization plus a 2p ionization) in the Al target. For most applications, however, the spectrum is sufficiently

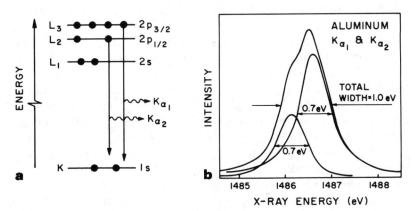

Figure 9.3 The two components of the K_α spectrum ($K_{\alpha_1} + K_{\alpha_2}$) which comprise the K_α spectrum of Al. [Spectrum from Siegbahn et al., 1967).]

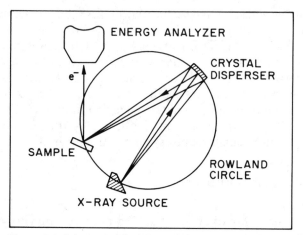

Figure 9.4 Schematic of an X-ray monochromation system.

clean for analysis purposes. If higher energy resolution is required in the photon source, a monochromator (Figure 9.4) must be used with a corresponding decrease in efficiency. X-ray monochromators usually make use of crystal diffraction to energy select the beam.

As shown in Figure 9.3, the Al $K_{\alpha_{1,2}}$ lines consist of two components separated by the 0.4 eV spin-orbit splitting of the 2p state. The K_{α_1} line which arises from the four electrons in the $2p_{3/2} \rightarrow$ 1s transition has about twice the intensity as that from the two electrons in the $2p_{1/2}$ state. For Mg K_α X-rays a somewhat better resolution (≈ 0.8 eV) can be obtained. The K_α lines from Cr (~ 5 keV) and Cu (~ 8 keV) have energy widths $\gtrsim 2.0$ eV, Mo (~ 17 keV) has an energy width of about 6 eV and all are not suitable for high resolution studies without further energy selection.

Ultraviolet photoemission spectroscopy (UPS) generally uses resonance light source such as He discharge lamp with energies in the 16–41 eV range. The energies are sufficient to allow analysis of the valence band density of states of most solids. The intensity of the light sources is high and the energy widths are sharp. The energy resolution in these experiments is generally limited by the electron analyzer in contrast to the case with X-ray sources. UPS studies are primarily directed at examining electron configurations in the valence shells or bonding orbits of a solid rather than determining elemental composition, the thrust of this book.

The use of synchrotron radiation from electron storage rings provides a continuous spectrum with intensities far in excess of the characteristic X-ray lines or resonance light sources. The use of polarized, tunable radiation from the synchrotron is a distinct advantage in experimental investigations. However the limited access to synchrotron facilities restricts its applicability in routine sample analysis.

9.2.2 Electron Spectrometers

The energy of photoelectrons is determined by their deflection in electrostatic or magnetic fields. Magnetic deflection analyzers such as those used in β-ray spectroscopy or early (Siegbahn et al., 1967) XPS measurements are difficult to use in routine analysis, and electrostatic analyzers are the instruments found in most laboratory systems. There are two general operating modes for analyzers: deflection and reflection (mirror). In deflectors, electrons travel along equipotential lines and in mirror–type analyzers the electrons travel across equipotentials. In the deflection type, shown in Figure 9.2, a potential is applied across two concentric sectors and the electrons pass through the analyzer without a change in energy. In the mirror–type analyzer, the electrons travel across potential lines and are reflected away from the reflecting electrode into the analyzer exit.

A common type of mirror analyzer is the cylindrical mirror analyzer (CMA) with angular entrance and exit slits so the entire spectrometer has cylindrical symmetry (Figure 9.5). The deflection is caused by the potential difference (set by the analyzer control) between the inner and outer cylinder. The CMA shown in Figure 9.5 is double pass with essentially two CMAs in series. The spherical retarding grids are used to scan the spectrum while the CMA is operated

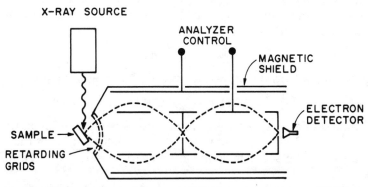

Figure 9.5 Schematic of a double pass cylindrical mirror analyzer (CMA) used in photoelectron spectroscopy.

at a constant pass energy in order to maintain a constant energy resolution.

The detection system is based upon the gain provided by electron multipliers, commonly a channel electron multiplier or channeltron. These channeltrons have a conelike opening and a continuous tube of high resistivity, semiconducting glass with a high secondary emission coefficient. A high electric field is applied along the tube, and incident electrons create a shower of secondary electrons which in turn hit the tube walls and create further secondaries. A gain of 10^8 can be achieved with the output taken through an amplifier–rate meter system.

9.3 Kinetic Energy of Photoelectrons

In photoelectron spectroscopy of solids, one analyzes the kinetic energy of electrons ejected when a solid is irradiated with monoenergetic photons of energy $\hbar\omega$. The relevant energy conservation equation is

$$\hbar\omega + E^i_{tot} = E_{kin} + E^f_{tot}(k), \tag{9.1}$$

where E^i_{tot} is the total energy of the initial state, E_{kin} is the kinetic energy of the photoelectron, and $E^f_{tot}(k)$ is the total final energy of the system after ejection of the photoelectron from the kth level. Contributions from a recoil energy E_r can be neglected (see problem 8.5). Only for the lightest atoms (H, He, Li) is E_r significant when compared to the instrumental linewidths in XPS spectra. The binding energy of the photoelectron is defined as the energy required to remove it to infinity with a zero kinetic energy. In XPS measurements $E^V_B(k)$, the binding energy of an electron in the kth level referred to the local vacuum level, is defined as

$$E^V_B(k) = E^f_{tot} - E^i_{tot}. \tag{9.2}$$

Substituting (9.1) into (9.2) results in the photoelectric equation

$$\hbar\omega = E_{kin} + E^V_B(k). \tag{9.3}$$

Binding energies are expressed relative to a reference level. In gas phase photoemission, binding energies are measured from the vacuum level. In the study of solids, the Fermi level is used as a reference.

In the case of a solid specimen, an electrical contact is made to the spectrometer. For metallic samples, the resulting energy levels are

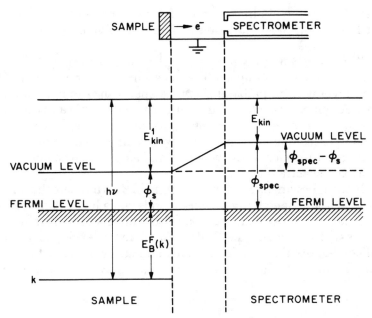

Figure 9.6 Schematic of the relevant energy levels for binding energy measurements. Note that the conducting specimen and spectrometer housing are in electrical contact and thus have common Fermi levels. The incoming photons, energy $h\nu$, create an electron of kinetic energy E_{kin}^1 relative to the vacuum level of the sample. The electron is detected by a spectrometer with a work function ϕ_{spec} so that the measured energy $E_{kin} = E_{kin}^1 - (\phi_{spec} - \phi_s)$.

shown in Figure 9.6. Because the sample and spectrometer are in thermodynamic equilibrium, their electrochemical potentials or Fermi levels are equal. In passing from the sample surface into the spectrometer, the photoelectron will feel a potential equal to the difference between the spectrometer work function ϕ_{spec} and the sample work function ϕ_s. Thus, the electron kinetic energy E_{kin}^1 at the sample surface is measured as E_{kin} inside the spectrometer analyzer

$$E_{kin} = E_{kin}^1 + (\phi_s - \phi_{spec}). \tag{9.4}$$

From Figure 9.6 it can be seen that the binding energy in a metallic specimen may be determined relative to the common Fermi level as follows:

$$\hbar\omega = E_B^F(k) + E_{kin} + \phi_{spec}, \tag{9.5}$$

where $E_B^F(k)$ is the binding energy referred to the Fermi level. Notice that the sample work function ϕ_s is not involved, but that of the spectrometer is.

When analyzing insulating samples, more care is required because of sample charging and the uncertainty in the location of the Fermi level within the band gap. One approach is to deposit a thin film of Au (or other metal) on the surface of the sample and use one of the known Au core levels to define the energy scale. Alternatively, the energies can be referenced to a well-defined feature of the electronic structure such as the valence band edge which can be located in the XPS spectra.

In the following, we will use the symbol E_B to indicate the binding energy without specifying the energy reference level. Although the Fermi level is most commonly used for metals and metallic substances such as silicides, a well-defined reference level is not found in semiconductors and insulators. This ambiguity along with sample charging indicates that care must be taken in evaluating spectra.

9.4 Photoelectron Energy Spectrum

The major features of the energy spectrum of X-ray excited photoelectrons are illustrated in Figure 9.7 for Mg K_α (E = 1.25 keV) irradiation of Ni. The spectrum exhibits the typical appearance of sharp peaks and extended tails through the allowed energy range. The peaks correspond to the energies of characteristic electrons which escape from the solid without undergoing energy loss. The higher energy tails correspond to electrons which have undergone inelastic scattering and energy loss on their outward path, thus emerging with lower kinetic energy (apparently higher binding energy).

The energy of the Mg K_α line is not sufficient to eject K shell electrons from Ni but can create vacancies in the L and M shells. The 2s and 2p as well as the 3s and 3p lines are clearly seen. The most prominent lines are the $2p_{1/2}$ and $2p_{3/2}$. Photoemission from p, d, and f electronic states with nonzero orbital angular momentum produces a spin-orbit doublet such as the $2p_{1/2} - 2p_{3/2}$ lines shown in the insert of Figure 9.7. The two lines correspond to final states with $j_+ = l + m_s = \frac{3}{2}$ and $j_- = l - m_s = \frac{1}{2}$. The intensity ratio of the lines is given by the ratio $(2j_- + 1)/(2j_+ + 1)$, which gives a ratio of line intensities of 1:2 for $p_{1/2}$ to $p_{3/2}$, 2:3 for $d_{3/2}$ to $d_{5/2}$, and 3:4 for $f_{5/2}$ to $f_{7/2}$.

After emission of a core electron such as the 2s or 2p electrons from the L shell, a hole is left in the core shell. The hole can be filled

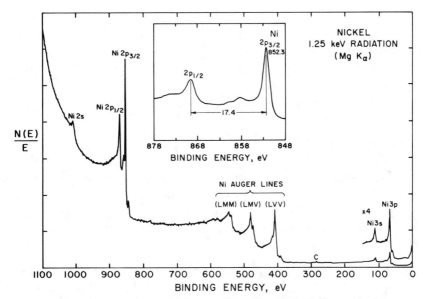

Figure 9.7 The energy spectrum of electrons from the 1.25 keV photon irradiation — Mg (K_α) — of nickel. The energy axis is in terms of binding energy, i.e., $h\nu - E_{kin}$. The vertical axis is denoted as N(E)/E to denote that the admittance of the spectrometer decreases as $1/E$. [From *Phi Handbook*, Eden Prairie, MN.]

by an electron from the M shell or valence band (V) with another M or V electron carrying away the energy. This Auger process is the dominant deexcitation process for elements lighter than $Z \approx 35$ and is described in more detail in Chapter 11. The Auger lines LMM, LMV, and LVV are clearly visible in the XPS spectrum of Figure 9.7. Since the Auger lines are element specific, they can also be used in element identification. As is the case for photoelectron lines, each Auger line is accompanied by a low energy tail which corresponds to electrons which have lost energy on the outward path. The energy of an Auger line is independent of the incident photon energy, while the energy of the photoelectron line varies linearly with the incident photon energy.

9.5 Binding Energy and Final State Effects

X-ray photoelectron spectroscopy is a straightforward and useful technique for the identification of atomic species at the surface of a solid. Adjacent elements throughout the periodic chart can easily be

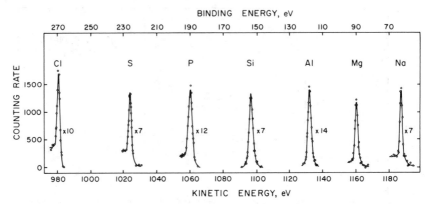

Figure 9.8 Electron lines from the L_1 subshells of the third period elements (sodium to chlorine) excited with magnesium K_α radiation (1.25 keV). [From Siegbahn et al., 1967.]

distinguished. The binding energies of adjacent elements are shown in Figure 9.8 for the $2s(L_1)$ lines of elements in the third period of the periodic table. Electron spectra for a variety of elements in the energy range of 600 – 20 eV are shown in Figure 9.9. The spin – orbit splittings are apparent for each group of elements. An overview of the variation in binding energy with atomic numbers is shown in Figure 9.10. The binding energies increase as the square of atomic number. For photon energies around 1 keV, only the outer M or N shells can be ionized for $Z > 30$. A compilation of binding energies is given in Appendix 5.

As indicated by Eq. (9.2), the binding energy as measured in XPS is the difference in the total energy between the initial and final state of the system in which one electron has been removed. This binding energy is not the same as the eigenvalue that one would calculate from an atom in its initial state with all orbitals occupied. In photoemission, the outer shells of the atom readjust when an inner electron is removed, because the Coulomb attraction of the positive nucleus is then less effectively screened. The difference between an initial state calculation with occupied orbitals and the experiment can be viewed as the following sequence: After the bound electron absorbs an energy $\hbar\omega$ from the photon, it loses some of its kinetic energy in overcoming the Coulomb attraction of the nucleus — that is, it loses kinetic energy. The outer orbitals readjust lowering the energy of the final state and giving this additional energy to the outgoing electron. As a result of the formation of a vacancy in an inner shell by photoelectron ejection, the readjustment in the orbital of the outer shell electron may not necessarily proceed to the ground state of the hole –

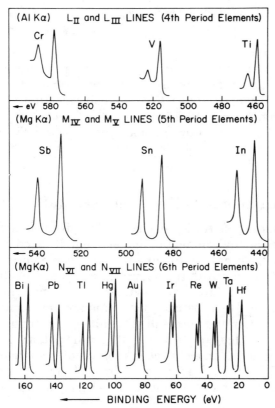

Figure 9.9 Electron spectra from a large variety of elements in the electron binding energy range of ~ 600 eV to ~ 20 eV. The incident photon radiations are Mg and Al K_α. [From Siegbahn et al., 1967.]

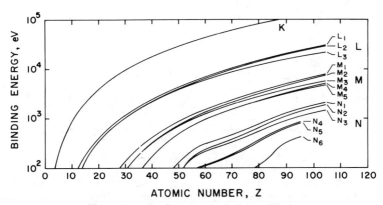

Figure 9.10 Binding energies of the elements.

state atom. The outer electron may go into an excited state (electron shakeup) or into a continuum state (electron shakeoff), and hence less additional energy is given to the outgoing electron. These transitions that produce excited final states result in satellite structure at the high binding energy (lower kinetic energy) side of the photoemission line.

9.6 Binding Energy Shifts; Chemical Shifts

The exact binding energy for an electron in a given element depends on the chemical environment of that element. If we consider a core level, the energy of an electron in this core state is determined by the Coulomb interaction with the other electrons and the attractive potential of the nuclei. Any change in the chemical environment of the element will involve a spatial redistribution of the valence electron charges of this atom and the creation of a different potential as seen by a core electron. This redistribution affects the potential of the core electrons and results in a change in their binding energies.

The shift in the binding energies of core electrons as a function of the chemical environment is demonstrated in Figure 9.11 for the Si 2p

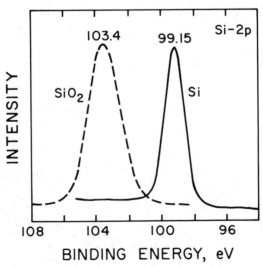

Figure 9.11 The chemical shift in binding energy of the Si 2p line for elemental Si and SiO$_2$. The spectra are taken with Al K$_\alpha$ radiation. [From *Phi Handbook*, Eden Prairie, MN.]

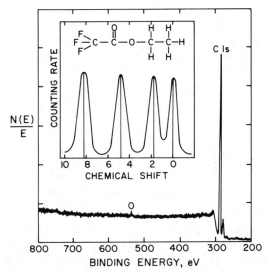

Figure 9.12 Carbon 1s chemical shifts in ethyl trifluoroacetate. The four carbon lines correspond to the four carbon atoms within the molecule. [Adapted from Ghosh, 1983.]

line. The measured binding energy of the Si 2p level shifts by more than 4 eV when the matrix is changed from Si to SiO_2. The existence of chemical shifts in XPS has led directly to analytical applications. The early work of the Uppsala group (Siegbahn et al., 1967) showed that core electron binding energies in molecular systems exhibit chemical shifts that are simply related to the covalency.

The concept of chemical shifts is based on the idea that the inner electrons feel an alteration in energy due to a change in the valence shell contribution to the potential based on the outer electron chemical binding. In the simplest picture valence electrons are drawn either from or toward the nucleus depending on the type of bond. The greater the electronegativity of the surrounding atoms, the more the displacement of electronic charge from the atom and the higher the observed binding energies of the core electrons. For example Figure 9.12 shows the binding energy shifts of the carbon atoms in ethyl trifluoroacetate, $C_4F_3O_2H_5$. Each carbon atom is in a different chemical environment and yields a slightly different XPS line. The binding energy shifts cover a change of about 8 eV.

An example of the shift in the Ni 2p XPS spectra as a result of the formation of Ni_2Si and NiSi is shown in Figure 9.13. Here the shift in the Ni signal represents 1.1 eV for the transition from Ni to NiSi. The decrease in the peak intensity is due to the decrease in the amount of

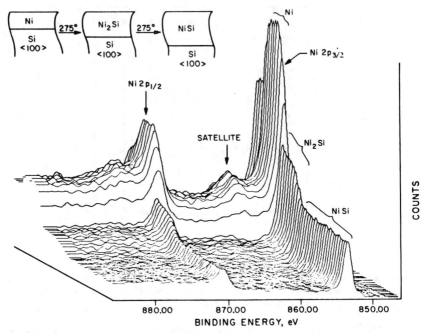

Figure 9.13 Three-dimensional plot of the Ni 2p XPS spectra with the Z axis representing time during heat treatment. The spectra illustrate the planar growth of different forms of nickel silicide as shown in the inset. Here chemical shifts are relatively small but the change in intensity indicates a change in composition. [From P. J. Grunthaner, Ph.D. thesis, Caltech, 1980.]

Ni atoms/cm² contained within the escape depth for the Ni 2p electrons as the compound becomes richer in Si. This is an example in which information on stoichiometry change comes primarily from intensity variations with only minor chemical shift changes.

9.7 Quantitative Analysis

Line intensities or the area of the photoelectron peaks are of interest for quantitative analysis. The intensity of a given line depends on a number of factors including the photoelectric cross section σ, the electron escape depth λ, the spectrometer transmission, surface roughness or inhomogeneities, as well as the presence of satellite structure (which results in a decrease in the main peak intensities).

The flux of X-rays is essentially unattenuated over the depths from which XPS signal peaks originate because the absorption

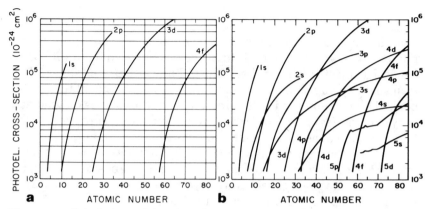

Figure 9.14 Calculations of the photoelectric cross section for different subshells throughout the periodic chart. The incident radiation is 1.5 keV. (a) The dominant shells most used in XPS; (b) the complete set of subshells. [From J. H. Scofield, *J. Electron Spectrosc.* **8**, 129 (1976).]

lengths of X-rays are orders of magnitude larger than the escape depth of the electrons. The probability P_{pe} per incident photon for creating a photoelectron in a subshell k is

$$P_{pe} = \sigma^k Nt, \tag{9.6}$$

where Nt is the number of atoms/cm² in a layer of thickness t and σ^k is the cross section for ejecting a photoelectron from a given orbital k. Some of the basic concepts underlying the determination of the photoelectric cross section are given in Chapter 8. The calculations of Scofield for the photoelectric cross section at 1.5 keV for different subshells are presented in Figure 9.14 in units of barns (one barn is 10^{-24} cm²). These calculations give an overview of the large variation in cross sections that can be found in the analysis of a given material. Empirical studies[1] indicate that these cross sections follow the strong Z dependence shown in Figure 9.14, but the values may be in error by a factor of 2 or greater.

As discussed in Chapter 6, the number of electrons that can escape from a solid without undergoing an elastic collision decreases with depth as $\exp(-x/\lambda)$, where λ is the mean free path. The "universal" curve for escape depth versus energy is given in Figure 6.4. The number of atoms per cm² that can produce a detectable photoelectron is then $N\lambda$ so that the probability P_d per incident photon of creating a detectable photoelectron from a subshell k is given by

---

[1] D. D. Wagner et al., *Surface Interface Anal.* **3**, 211 (1981).

$$P_d = \sigma^k N \lambda. \tag{9.7}$$

Not all the photoelectrons from a given subshell contribute to the photopeak which corresponds to the ground state configuration for a single vacancy in an inner shell. The influence of excited states (electron shakeup and shakeoff) is to decrease the intensity of the photopeak. The efficiency y in the production of a photopeak signal can vary over values of 0.7–0.8 for free atoms and, more importantly, can have a strong dependence on the chemical environment.

Finally the instrumental efficiency T is a function for the kinetic energy E of the electron and usually varies as E^{-1}. For example Figure 9.7 presents the energy spectrum $N(E)$ of the photoelectrons as $N(E)/E$ to compensate for the transmission efficiency.

In chemical analysis, one is generally interested in the relative concentrations, n_A/n_B, of elements A and B in a sample, so that only the ratio of the areas of lines (the intensity ratio I_A/I_B) is required. The composition ratio is then

$$\frac{n_A}{n_B} = \frac{I_A}{I_B} \cdot \frac{\sigma_B \cdot \lambda_B \cdot y_B T_B}{\sigma_A \cdot \lambda_A \cdot y_A T_A}. \tag{9.8}$$

If the photopeaks have about the same energy so that $\lambda_A \simeq \lambda_B$ and $T_A \simeq T_B$ and the photopeak efficiencies are about equal, then the composition ratio can be approximated by

$$\frac{n_A}{n_B} = \frac{I_A}{I_B} \cdot \frac{\sigma_B}{\sigma_A}. \tag{9.9}$$

This approach assumes that the sample is flat and homogeneous, that the photoelectrons are emitted isotropically, and that the sample surface is clean without a layer of surface contamination.

The sensitivity to the detection of trace elements depends upon the cross section of the element and the background of the signals from the other elements. Under favorable conditions, elemental analysis in bulk samples can reach sensitivities of 1 part in 1000. XPS measurements are extremely sensitive to the presence of surface layers. One often can detect as little as 0.01 monolayers of an element. The main use of XPS in materials analysis is determination of the chemical binding of atoms in the surface region of a solid.

Problems

9.1. (a) Compare the tabulated values of the binding energy, E_B (Appendix 5), for 1s and 2s levels with the prediction of the Bohr theory

[Eq. (1.15)] for atoms with $Z = 10, 20, 30$, and 40. Does the Bohr theory serve as a useful approximation (to within 10%) of the binding energies?

(b) Using the Pauli exclusion principle and quantum numbers with spin-orbit splitting, sketch an energy level diagram for copper based on Bohr values of the binding energy. Indicate the number of electrons in each level and verify that the ratio of electrons in the d levels, $d_{3/2}$ to $d_{5/2}$, is $2/3$.

9.2. For Al K_α X-rays incident on the compound NiSi calculate the ratio of yields using Eq. (9.8), assuming that the ratios of photopeak and instrument efficiencies are equal. Use Figures 9.14 and 6.4 to estimate the cross sections and escape depths:

(a) Ni 2s to Ni 2p.

(b) Ni 2s to Ni 3s.

(c) Ni 2p to Si 2p.

9.3. Consider incident radiation on Al at the same energy: 5.41 keV electrons and Cr K_α X-rays.

(a) Calculate the cross section for creating Al K shell vacancies for electrons [Eq. (6.11)] and photons [Eq. (8.37)].

(b) Calculate the electron range [Eq. (6.25)] and linear absorption coefficient [Eq. (8.35)]. Compare the calculated value of the mass absorption coefficient with the value listed in Appendix 6B.

(c) What is the energy of the K shell photoelectrons? Calculate the value of the escape depth λ [Eq. (6.20)] and compare it with the value estimated from Figure 6.4.

9.4. For Al K_α radiation incident on Cu, calculate the ratio of 2s to 3s photoelectron yields based only on cross sections (Figure 9.15) and escape depths.

9.5. Estimate the binding energy shift between the Li $(Z = 3)$ atom and LiF in the following way (the Li atom has an electron configuration of $1s^2 2s^1$):

(a) Calculate the probability of finding the 2s electron within the 1s orbit; i.e., calculate

$$\Delta q = \int_{r=0}^{r=a_0/Z} \psi_{2s}^2 r^2 dr \, \sin\theta d\theta d\phi,$$

where ψ_{2s} is a hydrogenic wave function (Chapter 8) and a_0/Z is the Bohr radius of the 1s shell. (Note that $0 < \Delta q < 1$).

(b) Estimate the binding energy of the 1s shell in the atom using the Bohr model with $Z_{eff}^{Li} = Z - \Delta q$.

(c) In LiF the outer electron is essentially on the F atom. Assume there is no contribution to the 1s screening from the 2s electron in LiF and estimate the binding energy shift using the Bohr model.

References

Atomic Physics and Quantum Mechanics

1. E. E. Anderson, *Modern Physics and Quantum Mechanics* (W. B. Saunders, Philadelphia, 1971).
2. J. D. McGervey, *Introduction to Modern Physics* (Academic Press, New York, 1971).
3. F. K. Richtmyer, E. H. Kennard, and J. N. Cooper, *Introduction to Modern Physics*, 6th Edition (McGraw-Hill Book Co., New York, 1969).
4. L. I. Schiff, *Quantum Mechanics*, 3rd Edition (McGraw-Hill Book Co., New York, 1968).
5. R. L. Sproull and W. A. Phillips, *Modern Physics*, 3rd Edition (John Wiley and Sons, New York, 1980).
6. P. A. Tipler, *Modern Physics* (Worth Publishers, New York, 1978).
7. R. T. Weidner and R. L. Sells, *Elementary Modern Physics*, 3rd Edition (Allyn and Bacon, Boston, MA, 1980).

X-ray Photoelectron Spectroscopy

1. D. Briggs, Ed., *Handbook of X-ray and Ultraviolet Photoelectron Spectroscopy* (Heydon and Son, London, 1977, 1978).
2. D. Briggs and M. P. Seah, *Practical Surface Analysis by Auger and X-ray Photoelectron Spectroscopy* (John Wiley and Sons, New York, 1983).
3. M. Cardona and L. Ley, Eds., *Photoemission in Solids I and II*, Topics in Applied Physics, Vols. 26 and 27 (Springer-Verlag, New York, 1978 and 1979).
4. T. A. Carlson, *Photoelectron and Auger Spectroscopy* (Plenum Press, New York, 1975).
5. G. Ertl and J. Kuppers, *Low Energy Electrons and Surface Chemistry* (Verlag Chemie International, Weinheim, 1974).
6. P. K. Ghosh, *Introduction to Photoelectron Spectroscopy* (Wiley-Interscience Publishers, New York, 1983).
7. H. Ibach, Ed., *Electron Spectroscopy for Surface Analysis*, Topics in Current Physics, Vol. 4 (Springer-Verlag, New York, 1977).
8. K. D. Sevier, *Low Energy Electron Spectrometry* (Wiley-Interscience Publishers, New York, 1972).

9. K. Siegbahn, C. N. Nordling, A. Fahlman, R. Nordberg, K. Hamrin, J. Hedman, G. Johansson, T. Bergmark, S. E. Karlsson, I. Lindgren, and B. Lindberg, *ESCA, Atomic, Molecular, and Solid State Structure Studied by Means of Electron Spectroscopy* (Almqvist and Wiksells, Uppsala, Sweden, 1967).

10

RADIATIVE TRANSITIONS AND THE ELECTRON MICROPROBE

10.1 Introduction

In previous chapters, we have calculated the cross section for creating an inner shell vacancy by irradiation with X-rays (photoelectric cross section, Chapter 8) and energetic electrons (impact ionization cross section, Chapter 6). After a vacancy is created, an electron can make a transition from an outer shell to fill the vacancy with the emission of a photon. This process is referred to as the spontaneous emission of radiation. In this chapter we consider the energies of X-ray transitions and calculate radiative transition rates. We will use the formula for the radiative transition rate from an initial state i to a final state f (derived in Chapter 8):

$$W = \frac{4}{3} \frac{(\hbar \omega_{fi})^3}{(\hbar c)^3} \cdot \frac{(e^2)}{\hbar} |\langle \psi_f | \mathbf{r} | \psi_i \rangle|^2, \tag{10.1}$$

where $\hbar \omega_{fi}$ is the energy of the emitted radiation, $\hbar \omega_{fi} = E_B^i - E_B^f$, where E_B is the binding energy of the initial or final state. The transition rate increases strongly with photon energy or for a given transition increases rapidly with Z. Evaluation of the matrix element indicates that $W = 0$ for some of the transitions, and hence selection rules for allowed transitions can be derived. These considerations are used in the description of the electron microprobe where bombardment of a solid by a beam of electrons leads to the emission of X-rays with

233

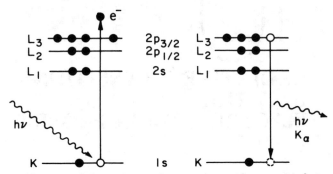

Figure 10.1 Schematic of photon interaction processes with atoms: (a) photon absorption in which an electron is emitted with energy, E, given by $E = \hbar\omega - E_B$, where E_B is the electron binding energy of the shell; (b) photon (X-ray) emission where the L shell electron makes a transition to fill the vacancy in the K shell.

energies characteristic of the atomic species. The relative advantages of proton induced X-ray excitation will be discussed in Section 10.9.

Consider an excited atom with a hole in the K or L shell. A straightforward way of deexcitation is the transition of an electron from an occupied state in a shell to the empty state (hole) with the emission of X-rays as shown by the L_3 to K transition (K_{α_1} X-ray) in Figure 10.1. X-ray emission is mainly due to dipole radiation and electron transition selection rules are obeyed ($\Delta l = \pm 1, \Delta j = 0, \pm 1$); the energy is given by the difference in binding energies,

$$\hbar\omega(K_\alpha) = E_B^K - E_B^{L_3}. \tag{10.2}$$

10.2 Nomenclature in X-Ray Spectroscopy

The arrows in Figure 10.2 show the allowed transitions of the atom and indicates their commonly used designation. For example, a K $\rightarrow$ L_3 transition, K_{α_1} X-ray line, is one in which there is an initial vacancy in the K shell and a final vacancy in the L shell. In spite of the fact that the energy levels belong to the atom as a whole, the one-electron quantum numbers n, l, j are used in descriptions of atomic levels as shown in Figure 10.2. States with the same values of n and l, but different values of $j(= l \pm \tfrac{1}{2})$, are well separated.

The main X-ray transitions for Pb are also shown in Figure 10.2. The intensity ratios are indicated at the top of the diagram and are referenced to $K_{\alpha_1} = 100$ for K X-rays, $L_{\alpha_1} = 100$ for L X-rays, etc. The strong transitions obey the dipole selection rules. The intensity ratios

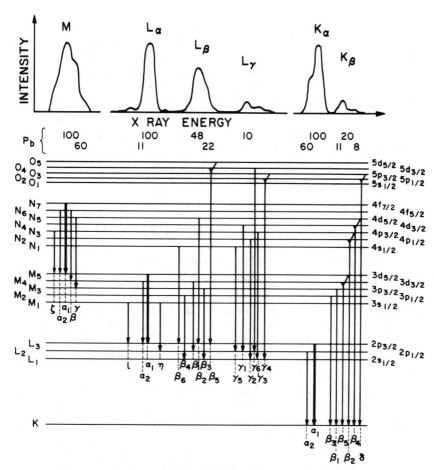

Figure 10.2 Principal X-ray transitions including their common designations. The upper portion of the figure indicates the relative intensities of the various lines which form the characteristic X-ray spectrum and indicates the spectral distribution for Pb. [From F. Folkmann in Thomas and Cachard, 1978.]

and energy positions serve to establish the fingerprint pattern which identifies the element.

10.3 Dipole Selection Rules

The formula for the transition probability Eq. (10.1) contains the matrix element $\langle \psi_f | \mathbf{r} | \psi_i \rangle$ known as the dipole matrix element. This quantity is identically zero—that is, transitions are forbidden—

between certain initial, ψ_i, and final, ψ_f, states. Basically a photon can be considered as a particle with unit angular momentum. Therefore, in order to conserve angular momentum as well as energy, transitions in which the angular momentum changes by one unit, $\Delta l = 1$, are allowed. These selection rules ($\Delta l = \pm 1$, $\Delta j = 0, \pm 1$) result in a simplification of the observed X-ray spectrum. The allowed transitions are illustrated in Figure 10.2 along with a few weak lines that arise from transitions which are dipole-forbidden but allowed for magnetic and electric quadrupole transitions.

10.4 Electron Microprobe

The detection and measurement of the characteristic X-rays for materials excited by energetic electrons is the basis of electron microprobe analysis. The essential feature of the electron microprobe (Figure 10.3) is the localized excitation of a small area of the sample surface with a finely focused electron beam. The volume of the sample material excited by the electrons has dimensions of the order of a micron, and hence the analytical technique is often referred to as electron probe microanalysis or electron microprobe analysis (EMA). The electron beam can be scanned across the surface to give an image of the lateral distribution of the material composition.

In materials analysis, there are only a few lines that are important, mainly K_α and K_β and the L series (L_α, L_β, and L_γ lines). The energies of K series and L series are tabulated in the Appendices. The energies of the more important characteristic lines are shown in Figure 10.4. In the analysis of X-ray spectra it is common to refer to either energy or wavelength, depending on the type of detection system,

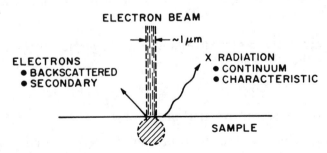

Figure 10.3 Electron beam interaction with a solid. Incident electrons create inner shell vacancies within the first micron of the solid.

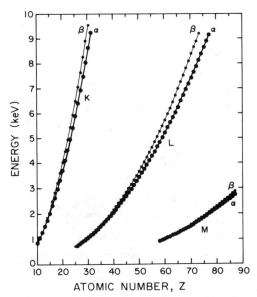

Figure 10.4 Energy of the K_α, K_β, L_α, L_β, M_α, and M_β lines of the elements as a function of atomic number.

either energy or wavelength dispersive. The most convenient form of analysis is the energy dispersive spectroscopy (EDS) mode using a Si(Li) detector whose basic operation is similar to that of a charged particle solid state detector described in Chapter 3. An incoming X-ray creates a photoelectron which eventually dissipates its energy through the formation of electron–hole pairs. The number of pairs is proportional to the incident photon energy. Under bias an electrical pulse is then formed with a magnitude proportional to the number of pairs or the X-ray energy. Such detectors provide a straightforward means of measuring the X-ray spectrum over a broad range of energies with an energy resolution of approximately 150 eV. The spectrum from Mn K X-rays taken with a Si(Li) detector is shown in Figure 10.5. The K_α and K_β lines are clearly resolvable. The full width at half maximum of the K_α line (148 eV) is determined by the resolution of the detector. The resolution is set by the statistical variations associated with the electron–hole pair creation process. Higher resolution is obtainable with wavelength dispersive techniques albeit at the expense of efficiency.

Wavelength dispersive spectroscopy (WDS) involves X-ray diffraction from an analyzer crystal; only those X-rays that satisfy the Bragg relation ($n\lambda = 2d \sin \theta$) are constructively reflected into the

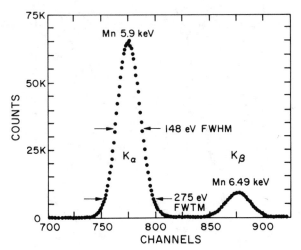

Figure 10.5 K X-ray spectra from Mn measured with an energy dispersive Si(Li) solid state detector. The K_α line is at 5.89 keV, the K_β line at 6.49 keV, and the resolution of the detector, full width half maximum, is 148 eV.

detector. Higher order reflections of a given wavelength can also be diffracted into the detector. The reflected wavelengths are λ, $\lambda/2$, $\lambda/3$, . . . corresponding to first-, second-, third-, . . . reflections. In Figures 10.6a and b, X-ray spectra are shown from a materials analysis of a nickel base alloy, using an energy dispersive system and a wavelength dispersive system. The energy resolution is sufficiently good in wavelength dispersive systems (~ 5 eV) that closely spaced lines can easily be resolved.

With energy dispersive systems it is possible to have signal interference from different elements when the energies of X-rays are close together. The K_α line of element Z falls close to the K_β line of element $Z - 1$ or $Z - 2$; i.e., Br and Rb. As can be seen in Figure 10.6b, an L transition from Ta falls close to the K transition of Ni; the Ta component of the material is, however, readily apparent from the wavelength dispersive system. These interferences can complicate analysis of multielement samples.

The chemical binding energy shifts found in photoemission (XPS) are not as readily detectable in X-ray analysis since the X-rays are due to a transition between two levels, both of which shift in the same direction due to bonding effects. Wavelength shifts to both longer and shorter wavelength are observed in very high resolution X-ray spectra for various elements upon chemical combination.

In X-ray spectroscopy as in most atomic transitions there is an energy (line width) width or breadth associated with the lifetime of

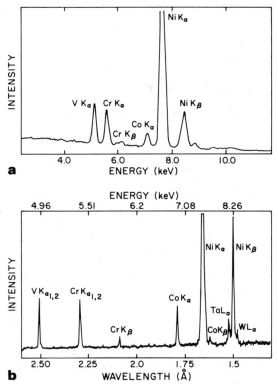

Figure 10.6 (a) Energy dispersive and (b) wavelength dispersive X-ray spectra from a nickel base alloy. The energy dispersive spectrum uses a Si(Li) solid state detector and the wavelength dispersive spectrum is with a LiF diffracting crystal. [From Goldstein et al., 1981.]

the core hole. This is a fundamental width set by the atomic processes involved and is the minimum energy width that could be observed in a high resolution detector. The uncertainty principle states that

$$\Delta E \, \Delta t \geq \hbar. \tag{10.3}$$

For atomic energy levels, we may take $\Delta t = \tau$, where τ is equal to the mean life of the excited state. The mean life is determined by radiative and Auger transitions. Then,

$$\Delta E \cong \frac{\hbar}{\tau}. \tag{10.4}$$

The lifetime broadening can be as large as 10 eV for the L shells and more than 100 eV for the K shell of high Z elements. These widths are a consequence of the filling of an inner electron shell vacancy by outer electrons within a time interval of the order of 10^{-16} s.

10.5 Transition Rate for Spontaneous Emission

The rate for a spontaneous transition from an electron in an initial state i to a final state f is

$$W = \frac{4}{3} \frac{e^2}{\hbar} \cdot \left(\frac{\omega_{fi}}{c}\right)^3 |\langle \psi_f | \mathbf{r} | \psi_i \rangle|^2 \tag{10.5}$$

where $e^2 = 14.4$ eV Å, c is the speed of light, $\hbar\omega_{fi}$ is the energy of the emitted photon, and ψ_i and ψ_f are the wave functions of the initial and final states, respectively. This can be expressed as

$$W = 0.38 \times 10^{16} (\hbar\omega)^3 \cdot |\langle \psi_f | \mathbf{r} | \psi_i \rangle|^2$$

in units of inverse seconds, where $\hbar\omega$ is in units of keV and the matrix element $(|\langle \psi_f | \mathbf{r} | \psi_i \rangle|^2)$ has dimensions of $(\text{Å})^2$. A typical value of W is 10^{15}/s for K shell transitions of elements in the middle of the periodic table. This formula is derived by considering the absorption transition probability in an electromagnetic field and the Einstein coefficients which describe the balance between photon absorption and induced and spontaneous transitions (Section 8.10).

10.6 Transition Rate for K_α Emission in Ni

We illustrate the use of formula (10.5) by explicitly calculating the transition rate for a particular case. Our starting point is the transition rate formula

$$W = \frac{4}{3} \frac{\omega^3 e^2}{\hbar c^3} |\langle \psi_f | \mathbf{r} | \psi_i \rangle|^2.$$

As a first estimate we note that $\langle \psi_f | \mathbf{r} | \psi_i \rangle \sim a_0/Z$. The wave functions $\psi_f(\mathbf{r})$ and $\psi_i(\mathbf{r})$ have a finite intensity only for values of $|\mathbf{r}| \lesssim a_0/Z$, where a_0 is the Bohr radius, $a_0 = 0.53$ Å. Then

$$W = \frac{4}{3} \frac{\omega^3 e^2}{\hbar c^3} \frac{a_0^2}{Z^2}. \tag{10.6}$$

The energy of the transition, $\hbar\omega$, for $2p \rightarrow 1s$ transitions in hydrogenic atoms is $10.2 Z^2$ (eV) [i.e., $\hbar\omega = 13.6 (1 - 1/n^2)Z^2$, where $n = 2$].

Therefore, the transition probability is proportional to Z^4,

$$W \propto Z^4; \tag{10.7}$$

this dependence will hold in more sophisticated treatments as well.

We can give a simple order of magnitude estimate of W. The units of W are often given in terms of $eV/\hbar$. Thus we rewrite W [Eq. (10.6)] as

$$W = \frac{4}{3} \frac{1}{\hbar} (\hbar\omega)^3 \cdot \left(\frac{e^2}{\hbar c}\right)^3 \cdot \left(\frac{a_0}{e^2}\right)^2 \cdot \frac{1}{Z^2} \tag{10.8}$$

(note that $e^2/\hbar c = 1/137$ and $e^2/a_0 = 27.2$ eV). Consider the K_α transition in Ni ($Z = 28$, $\hbar\omega = 7.5 \times 10^3$ eV). Then

$$W \simeq 0.38 \text{ eV}/\hbar.$$

Actually the six 2p electrons in Ni can contribute to the process so that

$$W \cong 2.3 \text{ eV}/\hbar.$$

The accepted result is $W = 0.551$ eV$/\hbar$. A factor of 4 is close for a crude estimate! In units of time and using the accepted value $W = 8.3 \times 10^{14}$ s^{-1}, or

$$\frac{1}{W} = 1.2 \times 10^{-15} \text{ s}.$$

A detailed evaluation of the transition probability is given in Section 10.9.

The energy width Γ of an atomic state is related to the mean life τ of the state through the Heisenberg uncertainty principle:

$$\Gamma\tau = \hbar. \tag{10.9}$$

In the previous section we estimated the radiative contribution to the width or more precisely the decay probability (per unit time) for a 2p to 1s transition, i.e.,

$$W_{rad} = \Gamma_{rad}/\hbar. \tag{10.10}$$

The width of the state is made up of all the processes which contribute to its finite lifetime:

$$\Gamma = \Gamma_{rad} + \Gamma_{nonrad} \tag{10.11}$$

where Γ_{rad} represents all the radiative processes which contribute to the lifetime (i.e., X-ray emission) and Γ_{nonrad} contributes to all nonradiative contributions (i.e., Auger emission). The probability of a ra-

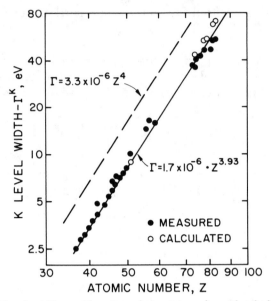

Figure 10.7 K level width as a function of atomic number. The dashed line is the calculated result [Eq. (10.14)]. The solid line is an empirical fit to the measured points. [From Bambynek et al., 1972.]

diative decay is $\Gamma_{rad}/(\Gamma_{rad} + \Gamma_{nonrad})$ and is known as the fluorescence yield ω_X (see Section 11.3). For a state consisting of a K vacancy with atomic $Z > 40$, radiative processes dominate; $\Gamma_{nonrad} \ll \Gamma_{rad}$. In a hydrogenic approximation

$$\Gamma = \frac{4}{3}(\hbar\omega)^3 \left(\frac{e^2}{\hbar c}\right)^3 \left(\frac{a_0}{e^2}\right)^2 \frac{0.74}{Z^2}(6), \tag{10.12}$$

where the extra factor of 6 allows for six 2p electrons. In the spirit of the hydrogenlike model we note that the energy of the 2p–1s transition is given by

$$\hbar\omega = 13.6(1 - \tfrac{1}{4})Z^2 \tag{10.13}$$

so that

$$\Gamma = 3.3 \times 10^{-6} Z^4 \text{ (eV)}. \tag{10.14}$$

This calculated result is only a factor of ~2 greater than the fit to experiment shown in Figure 10.7. Note that the Z^4 dependence is close to that observed experimentally.

10.7 Electron Microprobe: Quantitative Analysis

The electron microprobe is used for identification of elements and the quantitative analysis of elemental composition. All elements with atomic number greater than that of Be can be analyzed in principle, but in practice the technique is mostly applied to $Z \gtrsim 10$. The detection limit for elements is 50–100 ppm except for low Z elements (below Mg) which are detectable in concentrations greater than about 0.1 atomic percent.

Quantitative analysis of the concentration of a given element can be carried out with an accuracy of about 1% if suitable standards are available. The simplest procedure is to measure the yield Y_p of a given element at a wavelength λ_p, subtract the background yield Y_b, and determine the ratio K of the corrected yield in the sample to that in the standard $Y_p^s - Y_b^s$, where the background is due to Bremsstrah-

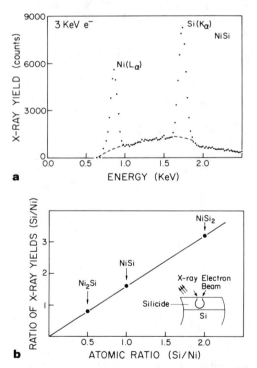

Figure 10.8 (a) Energy spectra from 3 keV electron bombardment of nickel silicide. (b) Ratio of X-ray yields as a function of known atomic ratio for the different nickel silicides. [From Zheng et al., *J. Vac. Sci. Technol. A* **1**, 758 (1983).]

lung (see Section 6.11). Using this definition

$$K = \frac{Y_p - Y_b}{Y_p^s - Y_b^s}. \tag{10.15}$$

The concentration c_A of element A can be determined from the concentration c_A^s in the standard by

$$c_A = c_A^s \cdot K, \tag{10.16}$$

where the standard and sample are subject to identical electron beam impingement and X-ray detection conditions and the standard has a composition near that of the sample to allow for equivalent X-ray absorption effects.

Often it is the concentration ratio c_A/c_B of two elements in a bulk or thin film sample that is desired. In this case it is useful to establish a calibration curve which relates peak intensity ratios to atomic ratios. An example of this procedure is illustrated in Figure 10.8, which shows the Ni_{L_α} and Si_{K_α} lines and the ratio of yields versus the measured Si/Ni atomic ratio in nickel silicide standards whose composition was determined by Rutherford backscattering analysis. The electron beam energy was sufficiently low (see electron ranges, Chapter 6) so that the penetration of the electrons was confined to the silicide layer and X-ray generation in the Si substrate was minimal.

10.7.1 Quantitative Analysis

The determination of an absolute concentration of an element in an unknown matrix represents a complicated problem. Consider first the yield of X-rays, Y_X, produced from a thin layer of width Δt at depth t into the sample:

$$Y_X(t) = N \, \Delta t \sigma_e(t) \omega_X e^{-\mu t/\cos\theta} I(t) \epsilon \, \frac{d\Omega}{4\pi} \tag{10.17}$$

where

N	is the number of atoms/vol;
$\sigma_e(t)$	is the ionization cross section at depth t where the particle has energy E_t;
μ	is the X-ray absorption coefficient;
ω_X	is the fluorescence yield;
θ	is the detector angle;
$\epsilon \, d\Omega$	is the efficiency and solid angle of the detector;
$I(t)$	is the intensity of electron beam at depth t.

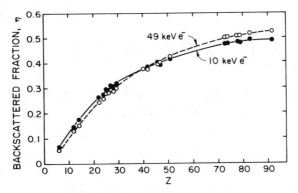

Figure 10.9 Variation of the backscattered fraction as a function of atomic number at $E_0 = 10$ keV and $E_0 = 49$ keV. [From Goldstein et al., 1981.]

The total observed yield Y is given by

$$Y = \int_{t=0}^{R} Y(t)\, dt + \text{Secondary fluorescence}, \qquad (10.18)$$

where R is the range of the electron and the second term includes the effect of secondary fluorescence due to absorption of high energy X-rays (generated by other heavy atoms within the matrix) and re-emission of the X-rays of interest.

Equation (10.17) takes into account (1) the change in cross section as a function of depth into the sample due to the change in electron energy as the beam penetrates the sample and (2) the attenuation of the beam, $I(t)$, as it penetrates the sample due to backscattering of the electrons.

The attenuation of the beam can be a surprisingly large factor. Figure 10.9 shows the backscattered fraction as a function of Z for two different incident energies. The backscattered fraction is almost energy independent.

10.7.2 Correction Factors

In practice, Eqs. (10.17) and (10.18) are usually not used explicitly in the evaluation of composition; rather comparisons of X-ray yields from the unknown sample and a standard are used. Even under these conditions, however, corrections must be made since many of the factors in (10.17) are matrix-dependent. Extensive investigations in the field of microprobe analysis have generated an approach based on empirical correction factors which represent those matrix-dependent effects.

Some insight into the correction factors involved in quantitative analysis can be gained by considering the procedure for determining the concentration c_A of an element A in an alloy from the ratio K of X-ray intensities from the sample and from a standard composed of element A. An expression of the form

$$c_A = KZAF \qquad (10.19)$$

is used (Birks, 1979; Goldstein et al., 1981), where Z is the atomic number correction factor, A is the absorption correction factor, and F is the fluorescence correction factor. These corrections relate to the three major effects which arise from the differing characteristics of the sample and the standard with respect to electron and X-ray interactions. The atomic number correction Z allows for the fact that the generation of primary X-rays in the samples does not increase linearly with concentration. The proportion of incident electrons that are backscattered and the volume of the sample in which X-rays are generated depend on sample composition. The absorption correction A is required since the absorption coefficients — i.e., the attenuation

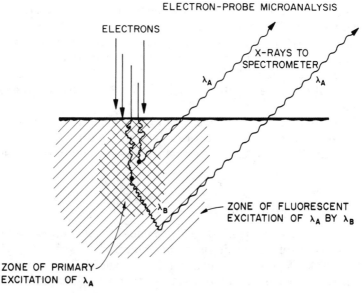

Figure 10.10 Schematic illustration of the generation of secondary radiation as a result of fluorescence excitation by primary radiation.

of the emerging X-radiation—will be different in the sample and standard. The fluorescent correction F accounts for the generation of secondary X-rays from element A due to fluorescent excitation by X-rays emitted by another element (Figure 10.10). The effect is strongest when the exciting lines have an energy slightly greater than the binding energy associated with the line that is being measured, i.e., near the maximum of the absorption cross section. For example, in a sample containing Fe and Ni, the Ni K X-rays could excite Fe K X-rays and in a Cu–Au sample, the Au L lines can excite Cu K_α. The correction for fluorescence by characteristic lines depends on the atomic fluorescence yield ω_X, and the fraction of the exciting element in the sample. For X-ray lines with energies below 3 keV, the values of ω_X are small and the fluorescence becomes negligible. The magnitude of each of the correction factors in Eq. (10.19) is between 2 and 10% in most cases.

10.8 Proton- (and Helium-) Induced X-ray Emission (PIXE)

Inner shell ionization is caused by the time-dependent electric field created by the passage of a charge near an atom. In a classical sense the field from a proton is precisely the same (aside from a sign) as that of an electron at the same velocity. In terms of kinetic energy this occurs at an ion energy of $(M/m)E$, where M and m are the masses of the ion and electron respectively and E is the electron kinetic energy. Thus the velocity matching criteria requires proton kinetic energies of 1836 E, which corresponds to the MeV ion range to match keV electrons.

Proton induced K shell ionization cross sections are shown in Figure 10.11 in a reduced plot. The horizontal axis is E/E_B^K, the ratio of the ion energy to the K shell binding energy. The vertical axis is $(E_B^K/Z_1)^2\sigma$, the product of the square of the K shell binding energy and the reduced ionization cross section σ/Z_1^2. This scaling allows the formulation of a universal plot for the cross section. In Figure 10.11 the ratio M_P/M_1 allows scaling for heavier projectiles. For protons $M_P/M_1 = 1$ and $Z_1 = 1$, for He ions $M_P/M_1 = \frac{1}{4}$ and $Z_1 = 2$, etc. X-ray production cross sections are shown explicitly in Figure 10.12. The X-ray production cross section σ_{K_α} is related to the ionization cross section σ (Figure 10.11) through the fluorescence yield ω_X; $\sigma_{K_\alpha} = \omega_X\sigma$. The fluorescence yield is the probability of a radiative transition relative to all possible transitions (radiative and nonradiative). The maximum value of the cross section decreases with increasing atomic

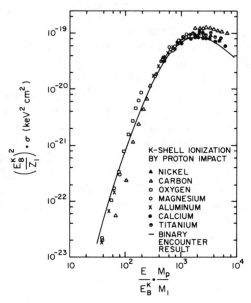

Figure 10.11   Scaled K shell ionization cross sections by proton impact. E/E_B^K is the ratio of the incident energy to the K shell binding energy. M_1/M_P is the ratio of the projectile mass to that of the proton. [From Garcia, 1970.]

number or binding energy. Further, the maximum of the cross section is at higher energy for the heavier elements, corresponding to the idea of velocity matching, similar to that for electrons where the maximum occurs at energies of three to four times the binding energy E_B.

Particle-induced X-rays (PIXE) have been used for a variety of analytical problems with MeV accelerators. The major advantage of ion-induced X-ray analysis is a reduction in background relative to that of electrons, yielding a better sensitivity for trace element analysis. The background in the electron-μ-probe-induced X-ray spectrum is due to electron bremsstrahlung (see Section 6.11). This is a continuous distribution associated with the deceleration of the electron as it traverses a solid. A quantum treatment of the bremsstrahlung process shows that the probability of photon emission decreases strongly with increasing mass of the charged particle projectile. Thus at the same velocity protons and electrons have approximately the same characteristic X-ray production probability but a large difference in bremsstrahlung background. Particle induced X-rays can be combined with the formation of μ-beams (of order 10 μm) to do high sensitivity-lateral mapping of trace element distributions. Such beams can also

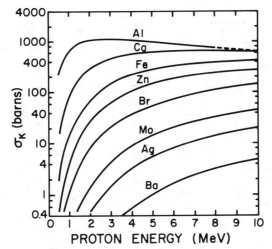

Figure 10.12 K_α X-ray production cross sections ($\sigma\omega_x$) versus proton energy. [From Cahill, 1980.]

be brought into air for analysis of biological and vacuum-degradable samples (Cahill, 1980).

10.9 Evaluation of the Transition Probability for Radiative Transitions

In this section we make a detailed evaluation of radiative transition probabilities and first concentrate on the dipole moment $\langle \psi_f | \mathbf{r} | \psi_i \rangle$. For ψ_i and ψ_f we will use hydrogenic wave functions (see Table 8.1):

$$\psi_f = \psi_{1s} \tag{10.20}$$

$$\psi_{1s} = R_{10}(r)Y_{0,0}(\theta,\phi), \tag{10.21}$$

where

$$R_{10} = \frac{2}{a^{3/2}} e^{-\rho} \tag{10.22}$$

and $\rho = rZ/a_0 = r/a$;

$$\psi_i = \psi_{2p} = \frac{1}{\sqrt{3}} R_{21}(r)\{Y_{1,0}(\theta,\phi) + Y_{1,-1}(\theta,\phi) + Y_{1,1}(\theta,\phi)\}, \tag{10.23}$$

where $R_{21} = [1/4(2\pi)^{1/2}] \cdot (1/a^{3/2})\rho e^{-\rho/2}$ and the spherical harmonics, the Ys, are defined in Table 10.1. The factor $1/\sqrt{3}$ assures proper normalization, i.e., $\langle \psi_{2p} | \psi_{2p} \rangle = 1$.

Table 10.1 Spherical Harmonics[a]

l	m	Y_{lm}
0	0	$Y_{00} = \dfrac{1}{\sqrt{4\pi}}$
1	1	$Y_{11} = -\left(\dfrac{3}{8\pi}\right)^{1/2} \sin\theta \exp(i\phi)$
1	0	$Y_{10} = \left(\dfrac{3}{4\pi}\right)^{1/2} \cos\theta$
1	-1	$Y_{1-1} = \left(\dfrac{3}{8\pi}\right)^{1/2} \sin\theta \exp(-i\phi)$
2	2	$Y_{22} = \left(\dfrac{15}{32\pi}\right)^{1/2} \sin^2\theta \exp(2i\phi)$
2	1	$Y_{21} = -\left(\dfrac{15}{8\pi}\right)^{1/2} \cos\theta \sin\theta \exp(i\phi)$
2	0	$Y_{20} = \left(\dfrac{5}{16\pi}\right)^{1/2} (3\cos^2\theta - 1)$
2	-1	$Y_{2-1} = \left(\dfrac{15}{8\pi}\right)^{1/2} \cos\theta \sin\theta \exp(-i\phi)$
2	-2	$Y_{2-2} = \left(\dfrac{15}{32\pi}\right)^{1/2} \sin^2\theta \exp(-2i\phi)$

[a] $\int\int Y_{lm}^* Y_{lm} \sin\theta \, d\theta \, d\phi = 1$.

The matrix element of $\mathbf{r}$ is given by $\langle \psi_f | x | \psi_i \rangle + \langle \psi_f | y | \psi_i \rangle + \langle \psi_f | z | \psi_i \rangle$ or in spherical coordinates as

$$\langle \psi_f | x | \psi_i \rangle = \langle \psi_f | r \sin\theta \cos\phi | \psi_i \rangle, \qquad (10.24)$$

$$\langle \psi_f | y | \psi_i \rangle = \langle \psi_f | r \sin\theta \sin\phi | \psi_i \rangle, \qquad (10.25)$$

and

$$\langle \psi_f | z | \psi_i \rangle = \langle \psi_f | r \cos\theta | \psi_i \rangle. \qquad (10.26)$$

All the terms can be separated in the following sense:

$$\langle \psi_f | z | \psi_i \rangle = \tfrac{1}{3}\Bigg\{ \int R_{21}(r) \cdot r \cdot R_{10}(r) r^2 \, dr$$
$$\times \int [(Y_{1,1} + Y_{1,-1} + Y_{1,0}) \cdot Y_{0,0}] \qquad (10.27)$$
$$\times \cos\theta \sin\theta \, d\theta \, d\phi \Bigg\}.$$

Consider the ϕ dependence of integrals such as

$$\int Y_{1,-1}(\theta,\phi)Y_{0,0}(\theta,\phi) \sin\theta \, d\theta \, d\phi = \frac{1}{\sqrt{4\pi}}\left(\frac{3}{8\pi}\right)^{1/2}$$
$$\times \int \sin^2\theta \, d\theta \int e^{-i\phi} \, d\phi,$$

where $Y_{1,-1} = (3/8\pi)^{1/2} \sin\theta \, e^{-i\phi}$.

In general any integral of the form

$$\int Y_{l,m}(\theta,\phi) \, Y_{l',m'}(\theta,\phi) \sin\theta \, d\theta \, d\phi$$

will result in a factor of the form

$$\int_0^{2\pi} e^{-i(m-m')\phi} \, d\phi.$$

This second integral is identically zero unless $m = m'$, thus yielding a selection rule. That is, the matrix element and hence the transition probability is identically zero unless $m = m'$. For the z component the selection rule reduces to $m_f = m_i$. For the x and y components the integral is of the form

$$\int \sin\phi \, e^{-i(m-m')\phi} \, d\phi$$

or

$$\cos\phi \, e^{-i(m-m')\phi} \, d\phi,$$

which yield a selection rule $m' = m \pm 1$ or $m_f = m_i \pm 1$. These "dipole selection rules" govern the observed spectra.

The z matrix element for 2p to 1s hydrogenic wave functions reduces to:

$$\tfrac{1}{3}\langle \psi_{100}|r\cos\theta|\psi_{210}\rangle = 0.248a. \tag{10.28}$$

In a similar manner one can show that

$$\langle \psi_{1s}|x|\psi_{2p}\rangle = \langle R_{10}|r|R_{21}\rangle \delta_{l',l\pm1}\delta_{m',m\pm1}, \tag{10.29}$$

where

$$\begin{aligned}\delta_{l',l\pm1} &= 1 \quad \text{if } l' = l\pm1 \\ &= 0 \quad \text{otherwise.}\end{aligned} \tag{10.30}$$

Then $\langle \psi_{1s}|x|\psi_{2p}\rangle$ reduces to

$$= \frac{2}{3\sqrt{24}} \frac{1}{a^3} \int e^{-3r/2a} \frac{r}{a} r^3 \, dr \qquad (10.31)$$

or

$$\langle |x| \rangle = \left(\frac{81}{54}\right)^{1/2} \cdot \langle |z| \rangle \qquad (10.32)$$

and

$$\langle |y| \rangle = \left(\frac{81}{54}\right)^{1/2} \langle |z| \rangle .$$

Finally

$$\langle |x| \rangle + \langle |y| \rangle + \langle |z| \rangle = \langle |r| \rangle \qquad (10.33)$$

and

$$\langle |r| \rangle = 0.248a \left[1 + 2\left(\frac{81}{54}\right)^{1/2} \right],$$

$$\langle |r| \rangle^2 = 0.74a^2 \qquad (10.34)$$

for $2p \to 1s$. Using Eq. (10.5), we find a value for the transition probability for Ni of

$$W = 1.2 \text{ eV}/\hbar,$$

a factor of 2 greater than the accepted value of $0.55 \text{ eV}/\hbar$.

10.10 Calculation of K_β/K_α Ratio

As a further example of the use of the transition probability formula we calculate the K_β/K_α ratio for hydrogenic atoms. We use the following form for the radiative transition rate W,

$$W = \frac{4\omega^3 e^2}{3\hbar c^3} |\langle \psi_f|\mathbf{r}|\psi_i \rangle|^2,$$

and the ratio R is

$$R = \frac{W_{K_\beta}}{W_{K_\alpha}} = \frac{\omega_\beta^3}{\omega_\alpha^3} \frac{|\langle \psi_{100}|\mathbf{r}|\psi_{31} \rangle|^2}{|\langle \psi_{100}|\mathbf{r}|\psi_{21} \rangle|^2}, \qquad (10.35)$$

where

$$\psi_{3,1} = \frac{1}{\sqrt{3}} (\psi_{310} + \psi_{311} + \psi_{31-1})$$

and

$$\psi_{2,1} = \frac{1}{\sqrt{3}} (\psi_{210} + \psi_{211} + \psi_{21-1}).$$

These represent wave functions which are properly normalized and reflect the statistical weights of the degenerate sublevels, i.e.,

$$\langle \psi_{3,1} | \psi_{3,1} \rangle = \tfrac{1}{3}(1 + 1 + 1) = 1.$$

We have also used the $\Delta l = \pm 1$ selection rule by considering p $\rightarrow$ s transitions only.

Consider the z component, $z = r \cos \theta$:

$$\langle \psi_{100} | r \cos \theta | \psi_{31} \rangle = \tfrac{1}{3} \{ \langle \psi_{100} | r \cos \theta | \psi_{310} \rangle$$
$$+ \langle \psi_{100} | r \cos \theta | \psi_{311} \rangle \qquad (10.36)$$
$$+ \langle \psi_{100} | r \cos \theta | \psi_{31-1} \rangle \},$$

which reduces to $\sim 0.1a$.

For the $n = 2 \rightarrow 1$ transition: $\langle \psi_{100} | r \cos \theta | \psi_{21} \rangle$ reduces to $\tfrac{1}{3} \langle \psi_{100} | r \cos \theta | \psi_{210} \rangle$, which is the matrix element in the K_α transition ($= 0.25a$) [Eq. (10.28)] given in the previous section. The ratio of K_β to K_α intensities is

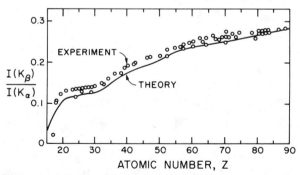

Figure 10.13 Measured K_β/K_α ratios as a function of atomic number. The theoretical curve represents a more sophisticated treatment which includes all electronic wave functions. The hydrogenic atom model result for $I(K_\beta)/I(K_\alpha)$ is $\simeq 0.25$. [From Bambynek et al., 1972.]

$$R = \frac{\omega_{K_\beta}^3}{\omega_{K_\alpha}^3} \times \frac{(0.1)^2}{(0.25)^2}.$$

In hydrogenic atoms,

$$\omega_{K_\alpha} = \tfrac{3}{4} E_K,$$

$$\omega_{K_\beta} = \tfrac{8}{9} E_K$$

so that

$$R = \left(\frac{\tfrac{8}{9}}{\tfrac{3}{4}}\right)^3 \cdot \left(\frac{0.1}{0.25}\right)^2 = 0.25. \tag{10.37}$$

The measured values of the K_β/K_α intensity ratio are given in Figure 10.13 and shown for a Mn spectrum in Figure 10.5. Except for low Z values, the ratio is nearly independent of Z and the calculation above is a reasonable estimate for $Z \gtrsim 40$. At low Z the calculation is influenced simply by the number of electrons available for the transitions (i.e., $R \cong 0$ for $Z \sim 10$, where there are no M shell electrons).

Problems

10.1. Compare the values of L_α radiation given in Figure 10.4 with values you would predict from the Bohr theory for elements with $Z = 30$, 40, 50, and 60. Does the Bohr theory give a reasonable estimate (within 10%)?

10.2. Consider X-radiation emitted from a sample of CuTa irradiated by 15 keV electrons. List the appropriate bindings energies associated with $K_{\alpha 1}$, $K_{\beta 1}$, $L_{\alpha 1}$, $L_{\beta 1}$. What energy resolution would be required to separate the signals for Cu K and Ta L radiation?

10.3. The energy width for the K vacancy state of Ca is about 1 eV. Estimate the ratio of radiative to nonradiative transitions.

10.4. Calculate the cross section for Ni K shell production by 20 keV electrons and for 6.5 MeV protons. Compare the proton velocity with that of the Bohr velocity of Ni K shell electrons.

10.5. You have thin films of $Hg_x Cd_{1-x} Te$ whose composition is to be evaluated for a program on infrared detectors. You want to determine the Hg to Cd ratio and suspect that the sample may contain 1 at.% of Cu as an impurity. You mount a 1000 Å thick film in an electron microscope (125 keV electrons) that has an energy dispersive spectrometer for measurement of X-ray intensities with an energy resolution of 200 eV.

(a) For a given composition $x = 0.2$ of $Hg_x Cd_{1-x}$ estimate the ratio of Hg to Cd K_α X-ray intensities from the ratio of cross sections and fluorescence yields (Appendix 9).

(b) Would you expect interference from K, L_α, or M X-ray lines from Hg, Cd, or Te?

(c) Estimate from Figure 8.6 the mass absorption coefficient for Hg and Cd X-rays in Cd Te (assume values for Sn). Would X-ray absorption have a major influence in the intensity ratio?

(d) Estimate the Cu/Cd K X-ray intensity for 1 at.% Cu. Would interference or absorption have an influence?

10.6. Compare the Rutherford scattering cross section with the K_α X-ray production cross section for 4 MeV protons incident on Ag. Compare the two techniques on the basis of mass resolution and depth resolution.

10.7. Compare electron and X-ray emission processes for ^{4}keV X-rays and electrons incident on Al.

(a) What is the range, R_x, for the electrons in K shell X-ray production and what is the absorption depth where the incident X-ray flux decreases by $1/e$ (use Figure 8.6)?

(b) What is the cross section for K shell ionization by electrons and X-rays?

(c) What is the escape depth of K shell photoelectrons and the absorption length ($1/e$ attenuation) for the Al K_α X-rays?

(d) What is the ratio of photoelectron to electron–induced X-ray emission yields for a 30 Å film and a 3000 Å film?

References

1. W. Bambynek, B. Craseman, R. W. Fink, H.-U. Freund, H. Mark, C. D. Swift, R. E. Price, and Venugopala Rao, X-ray Fluorescence Yields, Auger and Coster-Kronig Transition Probabilities, *Rev. Mod. Phys.* **44,** 716 (1972).

2. L. S. Birks, *Electron Probe Microanalysis* (R. E. Krieger, New York, 1979).

3. T. A. Cahill, *Annu. Rev. Nucl. Part. Sci.* **30,** (1980).

4. J. D. Garcia, *Phys. Rev. A* **1,** 1402 (1970).

5. J. I. Goldstein, D. E. Newbury, P. Echlin, D. C. Joy, C. Fiori, and E. Lifshin, *Scanning Electron Microscopy and X-ray Microanalysis* (Plenum Press, New York, 1981).

6. S. A. E. Johansson and T. B. Johansson, *Nucl. Instrum. Meth.* **137,** 473 (1976).

7. J. W. Mayer and E. Rimini, Eds., *Ion Beam Handbook for Material Analysis* (Academic Press, New York, 1977).

8. J. McGervey, *Introduction to Modern Physics* (Academic Press, New York, 1971).

9. F. H. Read, *Electromagnetic Radiation* (John Wiley and Sons, New York, 1980).

10. P. K. Richtmyer, E. H. Kennard, and J. N. Cooper, *Introduction to Modern Physics* (McGraw-Hill Book Co., New York, 1969).

11. L. Schiff, *Quantum Mechanics* (McGraw-Hill Book Co., New York, 1968).

12. J. P. Thomas and A. Cachard, Eds., *Material Characterization Using Ion Beams* (Plenum Press, New York, 1978).

13. D. B. Williams, *Practical Analytical Electron Microscopy in Materials Science* (Verlag Chemie International, Weinheim, 1984).

14. J. C. Willmott, *Atomic Physics* (John Wiley and Sons, New York, 1975).

15. R. Woldseth, *X-ray Energy Spectrometry* (Kevex Corp., 1973).

11

NONRADIATIVE TRANSITIONS AND AUGER ELECTRON SPECTROSCOPY

11.1 Introduction

In previous chapters, we have discussed inner shell vacancy formation by photon irradiation (the basis of X-ray photoelectron spectroscopy) or energetic electron and proton irradiation. The excited atoms can release their energy in radiative transitions (Chapter 10) with the emission of X-rays or in nonradiative transitions with the emission of electrons. The latter process forms the basis for Auger electron spectroscopy (AES) in which one determines composition by measuring the energy distribution of electrons emitted during irradiation with a beam of energetic electrons. As with other electron spectroscopies, the observation depth is about 10–30 Å and is determined by the escape depth (Chapter 6). The identification of atoms by core level spectroscopies is based upon the values of the binding energies of the electrons. With Auger electron spectroscopy, the energy of the emergent electron is determined by the differences in binding energies associated with the deexcitation of an atom as it rearranges its electron shells and emits electrons (Auger electrons) with characteristic energies. Figure 11.1 shows the Auger radiationless deexcitation processes, in which the atom is left in the final state with two vacancies (or holes). If one of the final state vacancies lies in the same shell as the primary vacancy (although not in the same subshell), the radiationless transition is referred to as a Coster–Kronig transition. This transition is significant because the Coster–Kronig transition rates are

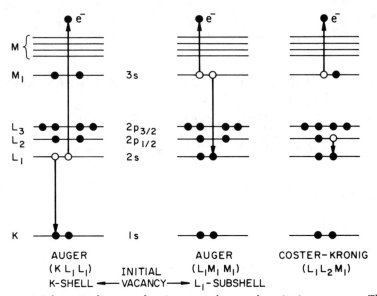

Figure 11.1 Schematic diagram of various two-electron deexcitation processes. The KL_1L_1 Auger transition corresponds to an initial K hole which is filled with an L_1 electron and simultaneously the other L_1 electron is ejected to the vacuum. The LM_1M_1 Auger transition is the corresponding process with an initial 2s vacancy. The Coster–Kronig $L_1L_2M_1$ transition contains an initial L_1 hole which is filled with an electron from the same shell (but different subshell, L_2).

much higher than the normal Auger transitions and influence the relative intensities of the Auger lines. For example in Figure 11.1, if an L_1 shell has a vacancy, the L_2 to L_1 transition will be rapid (Coster–Kronig), therefore reducing M electron to L_1 vacancy transitions.

11.2 Auger Transitions

11.2.1 Nomenclature

The nomenclature used to describe the Auger processes is shown in Figure 11.1. For vacancies in the K shell, the Auger process is initiated when an outer electron such as an L_1 electron (dipole selection rules are not followed) fills the hole. The energy released can be given to another electron such as another L_1 or an L_3 electron which is then ejected from the atom. The energy of the outgoing electron is $E_K - E_{L_1} - E_{L_1}$. The process described is called a KLL Auger transition in general terms and, more specifically, denoted as KL_1L_1 or KL_1L_3. If there are vacancies in the L shell, one can have Auger processes in which an electron from the M shell (M_1 electron) fills the L hole and another M shell electron (for example, an M_1 electron) is ejected, an $L_1M_1M_1$ Auger transition. Since electron–electron interactions are strongest between electrons whose orbitals are closest together, the strongest Auger transitions are of the type KLL or LMM. For Coster–Kronig transitions the vacancy is filled by electrons which come from the same shell, i.e., LLM. Auger transitions involving the outermost orbitals, the valence band, have an energy width of about twice that of the valence band. In Figure 11.2, the Si $KL_1L_{2,3}$ and $L_{2,3}V_1V_2$ (or LVV) Auger transitions are indicated with V_1 and V_2 located at positions of maxima in the density of states in the valence band.

A complete nomenclature describing Auger transitions indicates the shells involved and the final state of the atom. The final state is usually described using the spectroscopic notation describing the orbitals. For example, a KL_1L_1 transition would leave the 2s shell empty (two vacancies) and the 2p shell with six electrons; the transition is KL_1L_1 ($2s^02p^6$). A KL_2L_3 would leave the vacancies in the 2p shell and would be indicated KL_2L_3 ($2s^22p^4$). Even in the relatively simple KLL transition, there are a large variety of final states which can have slightly different energies and hence correspond to slightly different Auger lines. In the following we discuss this in detail.

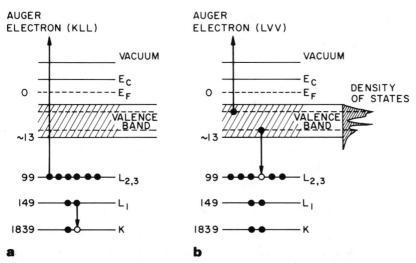

a   **b**

Figure 11.2 Schematic of the (a) $KL_1L_{2,3}$ and (b) LVV Auger deexcitation processes in Si. Binding energies are indicated on the left. The energy of Auger electron in the $KL_1L_{2,3}$ process is approximately 1591 eV and the $L_{2,3}$VV Auger electron has an energy of approximately 90 eV.

11.2.1.1 KL_1L_1

In the usual X-ray notation this transition corresponds to an initial state of a single 1s hole and a final state of two 2s holes. We can consider electron holes as electrons to find the possible final configurations of the final states. The n = 2 shell is now considered as $2s^0 2p^6$ (where the filled shell is $2s^2 2p^6$) and has states given by the possible allowed quantum numbers consistent with the Pauli exclusion principle: $m_s = \pm 1$, $M_L = 0$, $M_s = 0$, where M_L and M_s are the total orbital and spin angular momenta, respectively. The notation ^1S indicates a state of total orbital momentum zero (S). This transition is properly written $KL_1L_1(^1S)$, although the final state (^1S) is the only one allowed and, in this case, the notation is slightly redundant.

11.2.1.2 KL_1L_2 or KL_1L_3

In this case the final state electron configuration is written $2s^1 2p^5$. The possible quantum states are ^1P and ^3P, where P denotes the total orbital angular momentum. This corresponds to two states coupling to total angular momentum L = 1, i.e., a P state with the electron spins aligned ^3P and antialigned ^1P.

11.2.1.3 KL_2L_2; KL_2L_3; KL_3L_3 Transitions

Here the final states can couple to total angular momenta states of D(L = 2), P(L = 1), and S(L = 0) with different possible spin alignments to yield states ^1D, ^3P, and ^1S.

Thus in KLL type of transitions there are a total of six final states possible:

$$KL_1L_1 - 2s^0 2p^6 (^1S),$$
$$KL_1,L_{2,3} - 2s^1 2p^5 (^1P, {}^3P),$$
$$KL_{2,3},L_{2,3} - 2s^2 2p^4 (^1D, {}^3P, {}^1S).$$

These final states are shown experimentally in Figure 11.3 for the case of magnesium. [Actually the $2s^2 2p^4 (^3P)$ state is not observed due to lack of intensity.]

11.2.2 Energies

The energy of the Auger electrons can in principle be determined in the same way as that of X-rays: by the difference of the total energies before and after the transition. An empirical way of doing that, for example, is by

$$E_{\alpha\beta\gamma}^Z = E_\alpha^Z - E_\beta^Z - E_\gamma^Z - \tfrac{1}{2}(E_\gamma^{Z+1} - E_\gamma^Z + E_\beta^{Z+1} - E_\beta^Z), \qquad (11.1)$$

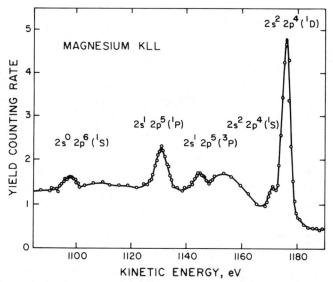

Figure 11.3 KLL Auger spectrum of magnesium containing 5 of the 6 lines predicted in L–S coupling. [From Siegbahn et al., 1967.]

where $E^Z_{\alpha\beta\gamma}$ is the Auger energy of the transition $\alpha\beta\gamma$ of the element Z. The first three terms correspond to the difference in the binding energies of shells α, β, γ of the element Z. The correction term is small and involves the average of the increase in binding energy of the γ-electron when a β-electron is removed and of the β-electron when a γ-electron is removed. Measured values of Auger KLL transitions are given in the Appendices along with values of the binding energies. A

Table 11.1 Tabulation of Values Used to Calculate the Energy of the KL_1L_2 Auger Transition in Ni

(1) $E^{Ni}_{KL_1L_2} = E^{Ni}_K - E^{Ni}_{L_1} - E^{Ni}_{L_2} - \frac{1}{2}(E^{Cu}_{L_2} - E^{Ni}_{L_2} + E^{Cu}_{L_1} - E^{Ni}_{L_1})$

(2) Electron binding energies in keV from Appendix 5

$\begin{aligned} E^{Ni}_K &= 8.333 \\ -E^{Ni}_{L_1} &= 1.008 \\ -E^{Ni}_{L_2} &= \underline{0.872} \\ &\quad 6.453 \end{aligned}$ $\qquad \begin{aligned} E^{Cu}_{L_2} &= 0.951 \\ -E^{Ni}_{L_2} &= \underline{0.872} \\ &\quad 0.079 \\ \frac{1}{2}(0.079 &+ 0.088) = 0.084 \end{aligned}$ $\qquad \begin{aligned} E^{Cu}_{L_1} &= 1.096 \\ -E^{Ni}_{L_1} &= \underline{1.008} \\ &\quad 0.088 \end{aligned}$

$E^{Ni}_{KL_1L_2} = 6.453 - 0.084 = 6.369$ keV

(3) Auger transition energy from Appendix 7

$E^{Ni}_{KL_1L_2} = 6.384$ keV

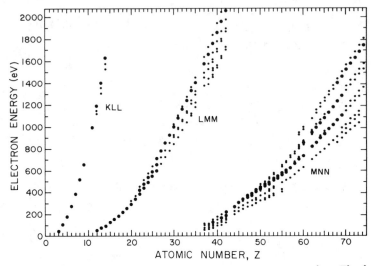

Figure 11.4 Principal Auger electron energies versus atomic number. The heavy points indicate the strong transitions for each element. [From Davis et al., 1976.]

numerical test of the approximation of Eq. (11.1) is given in Table 11.1 for KLL transitions. The agreement is good. Figure 11.4 shows the dominant Auger energies versus atomic number. The strong Z dependence of the binding energies leads to a straightforward elemental identification using this technique.

11.2.3 Chemical Shifts

The chemical environment of an atom is reflected in changes in the valence shell orbitals which in turn influence the atomic potential and the binding energy of the core electrons. The binding energies of the inner core K and L shells shift in unison with changes in the chemical environment. For this reason, the K_α X-ray emission lines, which are transitions between K and L shells, have only small shifts. For KLL Auger electron lines both the K and L shells are involved, but unlike the K_α X-ray emission lines, the L shell is involved twice in the transition. The inner shell electron that is ejected in KLL Auger processes therefore will display a chemical shift. Thus, one would expect chemical shifts in both AES and XPS spectra.

Chemical shifts are evident in both AES and XPS spectra. However, the chemical shifts are more difficult to interpret in the two electron Auger process than in the one electron photoelectric process. Further, Auger linewidths are broader than XPS lines. Consequently, the latter technique is typically used to explore changes in chemical binding.

11.2.4 Estimate of the Auger Transition Probability in a Hydrogenlike Atom: KLL Transition

In the usual manner the transition probability for the Auger effect, W_A, can be written as

$$W_A = \frac{2\pi}{\hbar} \rho(k) |\phi_f(\mathbf{r}_1)\psi_f(\mathbf{r}_2) \frac{e^2}{|\mathbf{r}_1 - \mathbf{r}_2|} \phi_i(\mathbf{r}_1)\psi_i(\mathbf{r}_2) \, d\mathbf{r}_1 \, d\mathbf{r}_2|^2, \quad (11.2)$$

where $\rho(k) = m \, (V/8\pi^3 \, \hbar^2) \, k \sin\theta \, d\theta d\phi$ is the density of states associated with normalization in a box of volume V and for a KLL Auger transition from a hydrogenlike atom:

$$\phi_i(\mathbf{r}_1) = \frac{1}{\sqrt{6a^3}} \frac{\mathbf{r}_1}{a} e^{-\mathbf{r}_1/2a} Y_1^m(\theta_1,\phi_1), \quad (11.3)$$

$$\phi_f(\mathbf{r}_1) = \frac{2}{\sqrt{a^3}} e^{-\mathbf{r}_1/a}, \quad (11.4)$$

$$\psi_i(\mathbf{r}_2) = \frac{1}{\sqrt{6a^3}} \frac{\mathbf{r}_2}{a} e^{-\mathbf{r}_2/2a} Y_1^m(\theta_2,\phi_2), \quad (11.5)$$

$$\psi_f(\mathbf{r}_2) = \frac{1}{\sqrt{V}} e^{i\mathbf{k}\cdot\mathbf{r}_2}. \quad (11.6)$$

These wave functions represent electrons in the 2p state, the 1s state, the 2p state, and a free electron, respectively, with $a = a_0/Z$ and a_0 is the Bohr radius. For convenience it is useful to write these equations as functions of $\mathbf{r}/a$:

$$\phi_i(\mathbf{r}_1) = \frac{1}{\sqrt{6a^3}} \phi_i'\left(\frac{\mathbf{r}_1}{a}\right), \quad (11.3')$$

$$\phi_f(\mathbf{r}_1) = \frac{2}{\sqrt{a^3}} \phi_f'\left(\frac{\mathbf{r}_1}{a}\right), \quad (11.4')$$

$$\psi_i(\mathbf{r}_2) = \frac{1}{\sqrt{6a^3}} \psi_i'\left(\frac{\mathbf{r}_2}{a}\right), \quad (11.5')$$

$$\psi_f(\mathbf{r}_2) = \frac{1}{\sqrt{V}} e^{ia\mathbf{k}\cdot\mathbf{r}_2/a} \quad (11.6')$$

and the potential as

$$\frac{e^2}{a}\left(\frac{1}{|\mathbf{r}_1/a - \mathbf{r}_2/a|}\right) = \frac{e^2}{a}V'(\mathbf{r}_1/a, \mathbf{r}_2/a). \tag{11.7}$$

In this calculation we are considering a KL_2L_2 Auger transition, the transition of a 2p electron (L_2) to the 1s state (K) and the subsequent emission of another 2p electron (L_2) to a free electron. In this hydrogenic model the energy of the Auger electron, E_A, is

$$E_A = E_K - E_L - E_L = E_K/2,$$

where E_K and E_L are the binding energies of the K and L shells, respectively, and $E_L = \frac{1}{4}E_K$ in the Bohr model. In the hydrogenic model

$$E_K = e^2 Z^2/2a_0$$

and

$$a = a_0/Z$$

so that

$$ak = (a_0/Z)\sqrt{mE_K/\hbar^2},$$
$$= 1/\sqrt{2},$$

where we have used the Bohr relation $a_0 = \hbar^2/me^2$. Then

$$\psi_f(\mathbf{r}_2) = \frac{1}{\sqrt{V}} e^{i r_2 \cos \theta_2/a\sqrt{2}}.$$

In this form we can extract the basic dependence of the transition probability on atomic parameters such as the atomic number Z. Substituting (11.3)–(11.7) (in modified form) into the formula for the transition probability, we find

$$W_A = \frac{2\pi}{\hbar} \frac{mk\,d\Omega}{\hbar^2 8\pi^3} \frac{e^4 a}{9} F, \tag{11.8}$$

where $d\Omega = \sin\theta\,d\theta\,d\phi$ and

$$F = \left| \iint \int \frac{d\mathbf{r}_1}{a^3} \cdot \frac{d\mathbf{r}_2}{a^3} \phi_f'\left(\frac{\mathbf{r}_1}{a}\right) \psi_f'\left(\frac{\mathbf{r}_2}{a}\right) \right.$$
$$\left. \cdot V'\left(\frac{\mathbf{r}_1}{a}, \frac{\mathbf{r}_2}{a}\right) \phi_i'\left(\frac{\mathbf{r}_1}{a}\right) \psi_i'\left(\frac{\mathbf{r}_2}{a}\right) \right|^2.$$

F is a definite integration over all space for $\mathbf{r}_1/a$ and $\mathbf{r}_2/a$, resulting in a definite number which represents a matrix element of the potential factor $(|\mathbf{r}_2/a - \mathbf{r}_1/a|)^{-1}$.

Remembering that $ak = 1/\sqrt{2}$ and taking $d\Omega = 4\pi$ we can write the transition probability as

$$W_A = C\,\frac{e^4 m}{\hbar^3},$$

where C is a numerical constant dependent on the various factors in W_A and F. Noting that $a_0 = \hbar^2/me^2$ and $v_0 = e^2/\hbar$, we have the simple relationship

$$W_A = Cv_0/a_0, \qquad (11.9)$$

where v_0 is the Bohr velocity, 2.2×10^8 cm/s, a_0 is the Bohr radius, 0.53 Å, and a_0/v_0 is a characteristic atomic time, 2.4×10^{-17} s. The integral F can be evaluated in the crude approximation $r_2 > r_1$ so that $1/|\mathbf{r}_2 - \mathbf{r}_1| \cong [1 + (r_1/r_2)\cos\theta_{1,2}]/r_2$, where $\theta_{1,2} = \theta_1 - \theta_2$. This approximation is based on the fact that the radial extension of the 1s wave function is small compared to the 2p function. The calculation

Table 11.2 Comparison of Auger Transition Rates and K level X-ray Emission Rates (Scofield) in eV/$\hbar$ (from Bambynek et al., 1972)

Atomic no.	Element	Auger	K X-ray
10	Ne	0.23	0.005
11	Na	0.29	0.007
12	Mg	0.36	0.010
13	Al	0.40	0.014
14	Si	0.44	0.02
15	P	0.48	0.03
16	S	0.51	0.04
17	Cl	0.54	0.05
18	Ar	0.58	0.07
20	Ca	0.65	0.12
22	Ti	0.69	0.19
24	Cr	0.72	0.28
26	Fe	0.75	0.40
28	Ni	0.78	0.55
32	Ge	0.83	1.0
36	Kr	0.89	1.69
40	Zr	0.94	2.69
46	Pd	0.99	4.94
52	Te	1.04	8.40
58	Ce	1.07	11.6
65	Tb	1.10	21.8
70	Yb	1.13	29.6

of C is tedious but straightforward; resulting in $C = 7 \times 10^{-3}$. The result for W_A [Eq. (11.9)] is independent of Z as suggested by Table 11.2. In a more complete calculation one must properly account for all the different equivalent pairs of electrons available for Auger decay. The calculations can be done more precisely via numerical techniques, which include more sophisticated wave functions and a better description of the interaction potential (Bambynek et al., 1972). The main feature is that the Auger transition probability is roughly independent of Z in contrast to the strong Z dependence of radiative transitions.

11.3 Yield of Auger Electrons and Fluorescence Yield

The lifetime of an excited state, τ, (a hole in a shell) is determined by the sum of all possible decay processes. Radiative transitions occur with probability W_X. Auger transitions have a probability W_A and Coster–Kronig (where the hole is filled by an electron of the same shell) W_K. There are no other deexcitation mechanisms, so that

$$1/\tau = W_X + W_A + W_K. \qquad (11.10)$$

For transitions to vacancies in the K shell (as well as for holes in the L_3 and M_5 shells), Coster–Kronig transitions do not occur and the probability for X-ray emission, ω_X, is given by

$$\omega_X = \frac{W_X}{W_A + W_X}; \qquad (11.11)$$

ω_X is commonly called the fluorescence yield. For transitions to K shell vacancies, the probability for radiative decay is proportional to Z^4 (Chapter 10), and the Auger probability is essentially independent of Z. Burhop[1] has suggested a semiempirical relation for ω_X of the form

$$\omega_X = \frac{W_X/W_A}{1 + W_X/W_A}, \qquad (11.12)$$

where

$$\frac{W_X}{W_A} = (-a + bZ - cZ^3)^4, \qquad (11.13)$$

[1] E. H. S. Burhop, *J. Phys. Radium* **16**, 625 (1955).

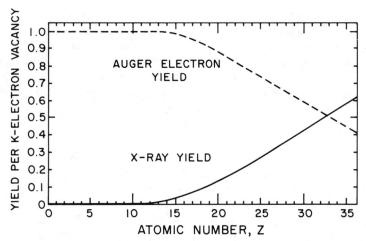

Figure 11.5 Auger electron and X-ray yields per K vacancy as a function of atomic number. The curves are from Eq. (11.12). [From Siegbahn et al., 1967.]

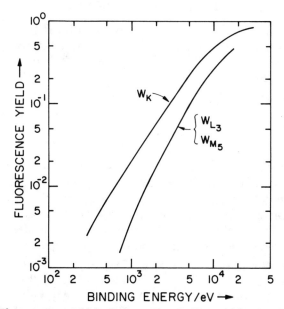

Figure 11.6 Fluorescence yield for K, L_3, and M_5 shells versus binding energy. [From J. Kirschner, in Ibach, 1977.]

with the numerical values $a = 6.4 \times 10^{-2}$, $b = 3.4 \times 10^{-2}$, and $c = 1.03 \times 10^{-6}$. This relationship yields the solid curve shown in Figure 11.5, the "Auger electron yield" is $1 - \omega_X$. This figure shows the dominance of Auger transitions for low Z elements; in these cases Auger emission is the important mechanism for relaxation of K vacancies. This curve does not imply that the Auger rate decreases at high Z, but emphasizes that the X-ray transition becomes the preferred method of deexcitation at high Z.

The fluorescence yield for K, L_3, and M_5 shells versus binding energy is shown in Figure 11.6. The point of the figure is that the fluorescence yield is approximately the same for comparable transition energies independent of the electronic shell, in those cases where Coster–Kronig transitions do not occur. For K shell transitions, the fluorescence yield is less than 0.1 for binding energies less than 2 keV, and the total Auger yield is larger than 90% for low Z elements (Z < 15). Similarly, for L_3 transitions (Coster–Kronig transitions not allowed), Auger transitions dominate for Z < 50 where L shell binding energies are less than 5 keV.

11.4 Atomic Level Width and Lifetimes

As pointed out in Chapter 10, the energy width ΔE or more conventionally Γ is related to the mean life τ of the state through the uncertainty principle, $\Gamma\tau = \hbar$. The decay probability per unit time is equal to the sum of the transition probabilities so that the total energy width of the state is given by

$$\Gamma = \Gamma_{\text{radiative}} + \Gamma_{\text{nonradiative}}. \qquad (11.14)$$

There is a decay probability for each atomic process, but there is only a single lifetime for the hole. The natural linewidth for each process is given by the total lifetime. In the Z < 30 regime where Auger emission dominates, Table 11.2 shows that Auger rates vary from 0.23 to 0.80 eV/$\hbar$. The total width of the atomic transition then is 0.23–0.8 eV. For Z > 30, the K X-ray emission rates range up to 30 eV/$\hbar$ with a corresponding increase in atomic level width. The total lifetime $\tau = \hbar/\Gamma$, where $\hbar = 6.6 \times 10^{-16}$ s, will vary from about 10^{-17} to 10^{-15} s. Consequently, the measured X-ray spectrum will exhibit more line broadening at higher Z than at low Z, hence Al or Mg are used as X-ray sources for XPS.

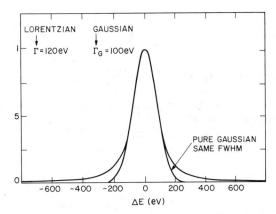

Figure 11.7 Comparison of a Lorentzian line shape [Eq. (11.15)] and a Gaussian line shape with the same full width and half maximum. The Lorentzian is characterized by extended tails.

An X-ray spectrum measured with high resolution would be in the form of a Lorentzian centered about an energy E_X (Figure 11.7):

$$Y(E) = \frac{A}{(E - E_X)^2 + \Gamma^2/4}. \qquad (11.15)$$

11.5 Auger Electron Spectroscopy

As with the other electron spectroscopies, Auger analysis is carried out under high vacuum conditions. Figure 11.8 shows schematically an experimental apparatus. The cylindrical mirror analyzer (CMA) has an internal electron gun whose beam is focused to a point on the specimen sample at the source point of the CMA. Electrons ejected from the sample pass through an aperture and then are directed through the exit aperture on the CMA to the electron multiplier. The pass energy E is proportional to the potential applied to the outer cylinder, and the range ΔE of transmitted electrons is determined by the resolution $R = \Delta E/E$, where R is typically 0.2–0.5%.

A schematic overall spectrum of electrons emitted from a solid irradiated by a 2 keV electron beam is shown in Figure 11.9. The narrow peak on the right side is made up of elastically scattered electrons (no energy loss). Features at slightly lower energy correspond to electrons with characteristic energy losses due to electronic and plasma excitations. Auger electron transitions generally appear as small features superimposed on the large background of secondary electrons. The usual practice is to use derivative techniques and

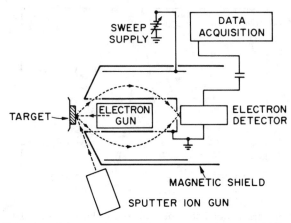

Figure 11.8 Experimental apparatus used in Auger spectroscopy. [After Palmberg in Czanderna et al., 1975.]

generate a $dN(E)/dE$ function (insert on Figure 11.9). Differential analysis of a hypothetical spectrum is shown in Figure 11.10. The contribution from the slowly varying background is minimized by the derivative technique. The total backscattered background current with energy greater than 50 eV is typically 30% of the primary beam current. The noise level due to this current and the ratio of the analyzer ΔE to Auger line width generally establishes the signal-to-noise ratio and hence the detection limit for impurities in the sample. A typical value for the detection limit is 1000 ppm, $\cong 0.1$ atomic %.

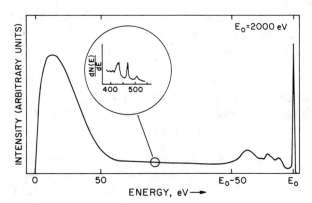

Figure 11.9 Spectrum of 2 keV electrons scattered from a solid. The inset shows the regime for Auger spectroscopy. The energy scale is nonlinear. [From Ibach, 1977.]

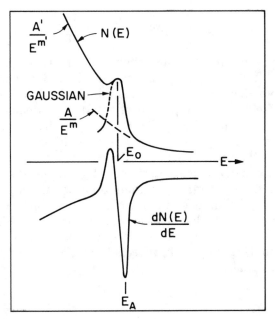

Figure 11.10 A hypothetical spectrum $N(E)$ containing a continuous background AE^{-m}, a Gaussian peak, and a low energy step of form $A'E^{-m'}$. The lower spectrum is the derivative spectrum. Note that the energy E_A of the most negative excursion of the derivative corresponds to the steepest slope of $N(E)$.

In practice, Auger spectroscopy is usually carried out in the derivative mode because of the small signal. The "differentiation" is conveniently done electronically by superimposing a small ac voltage on the outer cylinder voltage and synchronously detecting the in-phase signal from the electron multiplier with a lock-in amplifier. The y axis of the recorder is then proportional to $dN(E)/dE$ and the x axis to the kinetic energy E of the electrons. The derivative spectrum is extracted directly. In this scheme a perturbing voltage

$$\Delta V = k \sin \omega t \qquad (11.16)$$

is superposed on the analyzer energy so that the collected electron current $I(V)$ is modulated. $I(V + \Delta V)$ can be written in a Taylor expansion:

$$I(V + k \sin \omega t) = I_0 + I'k \sin \omega t + k^2 \frac{\sin^2 \omega t}{2!} I'' + \cdots , \qquad (11.17)$$

where the prime denotes differentiation with respect to V. Including higher order terms in the expansion

$$I = I_0 + \left[kI' + \frac{k^3}{8} I''' \right] \sin \omega t - \left[\frac{k^2}{4} I'' + \frac{k^4}{48} I'''' \right] \cos 2\omega t, \quad (11.18)$$

where I_0 contains all non-time-dependent terms. In this calculation we assumed $k \ll V$ so that terms of order k^3 and higher can be neglected in practice. Using a lock-in amplifier, for phase sensitive detection, we select the component of the signal associated with the frequency ω, which is simply the desired quantity I' or dN/dE for a cylindrical mirror Auger analyzer. To satisfy this criteria, we require that k be less than the Auger width of ~ 5 eV.

An example of the use of derivative techniques is shown in Figure 11.11 for 2 keV electrons incident on a Co sample. In the direct spectrum, n(E), the main features are the peak of elastically scattered electrons and a nearly flat background. The arrows in Figure 11.11a indicate the energies of oxygen and Co Auger transitions. The derivative spectrum (Figure 11.11b) reveals the LMM Co and KLL carbon and oxygen signals.

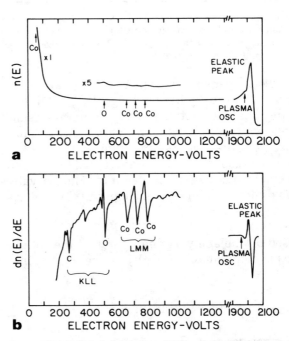

Figure 11.11 Comparison of (a) the spectrum N(E) and (b) the derivative dN(E) for 2 keV electrons incident on a Co sample. [From Tousset, in Thomas and Cachard, 1978.]

For a free atom, the Auger yield Y_A is determined by the product of the electron impact ionization cross section (Chapter 6) and the probability for the emission of an Auger electron $(1 - \omega_X)$:

$$Y_A \propto \sigma_e \cdot (1 - \omega_X). \tag{11.19}$$

In a solid, the situation is more complicated even when considering the yield from a layer of the thickness of the electron escape depth λ. For example, primary electrons that penetrate the surface layer and then are backscattered can contribute to the Auger yield when the energy E_p of the primary electron is much greater than the binding energy. The yield is also strongly affected by the angles of incidence (diffraction effects influence the number of elastically scattered primaries) and of emission (geometric projection of the escape depth). Consequently, surface roughness plays a role; the escape probability of electrons from a rough surface is less than that from a smooth surface. In analyzing solids, then, one must consider the modification of both the incident beam and the Auger electrons on passing through the solid.

Auger electron spectroscopy is a surface sensitive technique. Figure 11.12 shows the oxygen signal corresponding to the absorption of 0.5 monolayers of oxygen atoms. In general, small amounts of the typical contaminants, C, N, and O, are easily detected. Hydrogen cannot be detected in Auger measurements since 3 electrons are needed in an Auger transition.

The Auger signal from a substrate is sensitive to the presence of surface layers. In Chapter 6, we noted that the substrate signal de-

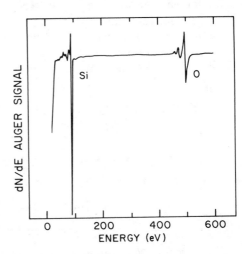

Figure 11.12 Auger spectra from single crystal Si (111) after adsorption of ~0.5 monolayers of oxygen atoms.

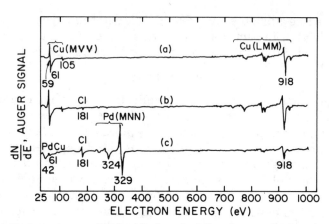

Figure 11.13 Auger traces of (a) freshly deposited Cu substrate, (b) Cu substrate just prior to Pd deposition, and (c) Pd/Cu bilayer, 13.5 Å of Pd.

creased as $e^{-x/\lambda}$, where $\lambda = 5$ Å for Si electrons penetrating Ge. Figure 11.13 shows Auger spectra from a Cu substrate before and after the deposition of 13.5 Å of Pd. It is clear from the figure that the Cu signal is strongly attenuated by the Pd coverage. In particular, the low energy Cu(MVV) line is completely attenuated due to the small escape length for 60 eV electrons; the high energy line at 918 eV is only partly attenuated.

11.6 Quantitative Analysis

The determination of an absolute concentration of an element x in a matrix from the yield Y_A of Auger electrons is complicated by the influence of the matrix on the backscattered electrons and escape depth. For simplicity let us consider $Y_A(t)$ the yield of KLL Auger electrons produced from a thin layer of width Δt at a depth t in the sample,

$$Y_A(t) = N_x \Delta t \cdot \sigma_e(t)[1 - \omega_x]e^{-(t\cos\theta/\lambda)} \cdot I(t) \cdot T \cdot d\Omega/4\pi, \quad (11.20)$$

where

$N_x =$ the number of x atoms/unit vol;

$\sigma_e(t) =$ the ionization cross section at depth t;

$\omega_x =$ the fluorescence yield;

$\lambda =$ the escape depth;

$\theta =$ the analyzer angle;

T = the transmission of the analyzer;

$d\Omega$ = the solid angle of the analyzer;

$I(t)$ = the electron excitation flux at depth t.

It is convenient to separate the excitation flux density into two components,

$$I(t) = I_P + I_B(t) = I_P(t)[1 + R_B(t)],$$

where I_P is the flux of primary electrons at depth t and I_B is the flux due to backscattered primary electrons and R_B is the backscattering factor (Section 10.7).

When external standards are used with a known concentration N_x^S of element x in the standard, the concentration N_x^T in the test sample can be found from the ratio of Auger yields:

$$\frac{N_x^S}{N_x^T} = \frac{Y_x^S}{Y_x^T}\left(\frac{\lambda^T}{\lambda^S}\right)\left[\frac{(1 + R_B^T)}{(1 + R_B^S)}\right].$$

In this approach the ionization cross section and the fluorescence yield are not required because the Auger yields from the same atom are measured. In addition, if the composition of the standard is close to that of the test sample, the element composition can be determined directly from the ratio of Auger yields if the measurements are made under identical experimental conditions. When the composition of the standard differs substantially from that of the test specimen, the influence of the matrix on electron backscattering and escape depth must also be considered.

Elemental sensitivities are acquired using pure element standards and are applied to unknown determinations in multielemental matrices. One must correct for the highly matrix-dependent parameters, which include the inelastic mean free path λ.

Even with corrections for escape depth and backscattering, the measured surface composition may not be related to the bulk composition of the sample because of the ion bombardment used in sputtering for sample cleaning and depth profiling (see Chapter 4).

11.7 Auger Depth Profiles

A major use of Auger electron spectroscopy is determining the composition as a function of depth in thin films and layered structures. The conventional apparatus is illustrated in Figure 11.8, which consists of an electron gun and CMA assembly as well as a sputter ion gun. The Auger signal is generated in the near surface region of the sample (~ 30 Å) and ion sputtering provides the layer sectioning technique required for depth analysis. In routine laboratory use, the depth profiles are shown as Auger signal height versus sputter time. Further calibrations are required to convert sputter time to depth and signal height to atomic concentration. The combination of Rutherford backscattering spectrometry (RBS) and Auger electron spectroscopy (AES) is quite useful in such depth profile analyses because RBS gives quantitative information on depths and heavy mass constituents without the complications introduced by the intermixing due to sputtering. As discussed in Chapter 4, ion sputtering causes a change in the composition of the surface layers due to surface segregation and preferential sputtering. As compared to RBS, Auger depth profiling provides better depth resolution and is sensitive to both heavy and light elements.

In Figure 11.14 we illustrate the data obtained from RBS and AES measurements on a sample prepared by depositing 1000 Å of Ni on $\langle 100 \rangle$ InP (Figure 11.14a) and annealing at 250°C for 30 min (Figure 11.14b). In the RBS spectrum for the as-deposited case, the Ni signal is superimposed on the signal from the InP substrate. In the AES spectrum, both the In and P signals have comparable heights and can be clearly resolved. The long tail on the Ni signal which extends well beyond the interface region is clearly an artifact of the sputtering process because the Ni/InP interface is sharp, as can be inferred from the rear edge of Ni signal in the RBS spectrum. After annealing, the layer is partially reacted with an outer layer of Ni on a layer of $In_xP_yN_z$. The Ni layer and the reacted InPNi layer can be clearly seen in the AES spectrum, which has a P/In yield ratio of $\simeq 2/1$. In the RBS spectrum, the heights of the Ni and In signals are nearly equal, which indicates that the ratio of Ni to In is about 3 $[\sigma_{In}/\sigma_{Ni} \simeq 3.08]$. Analysis of the RBS spectra yields a P/In ratio of 0.5, a value quite different from the P-rich composition deduced from the AES data. The origin of the discrepancy possibly is due to preferential sputtering and segregation. The region of pure Ni in the reacted film is better resolved with AES due to its superior depth resolution. Further AES allowed a determination of the carbon and oxygen at the interface region (not shown), which is not possible with RBS.

One of the advantages of Auger electron spectroscopy is its sensitivity to low mass impurities, such as carbon or oxygen, which are common contaminants at surfaces and interfaces. The presence of these interfacial contaminants plays a disruptive role in thin film reactions by retarding interdiffusion. The degradation of the planar-

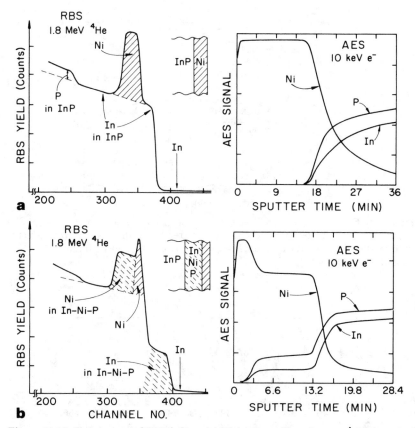

Figure 11.14 Comparison of RBS (left) and AES depth profiles of a 1000 Å Ni deposited on InP: (a) as deposited; (b) annealed at 250°C for 30 min. [From A. Appelbaum, private communication.]

ity of thin film structures following thermal processing is often directly correlated with these contaminants. The presence of a native oxide of about 15 Å thickness is readily apparent in the AES depth profile shown in Figure 11.15. The removal of this native oxygen layer is crucial for the formation of thin, uniform oxide layers on top of the Ta-silicide layers during thermal oxidation. The presence of the native oxide layer retards the release of Si from the poly-Si layer and leads to the oxidation of the whole Ta-silicide layer rather than the formation of a SiO_2 layer on the surface. Auger electron spectroscopy in conjunction with sputter depth profiling has the prerequisite sensitivity to detect contaminant layers that impede thin film reactions.

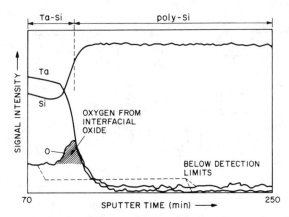

Figure 11.15 Sputter depth profiling with AES of the interface region of a Ta–Si film deposited on polycrystalline Si. The shaded area represents the oxygen signal from the native oxide at the interface. [From D. Pawlik, H. Oppolzer, and T. Hillmer, *J. Vac. Sci. Technol.* B **3**, 492 (1985).]

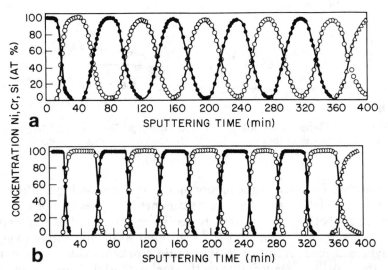

Figure 11.16 Auger electron spectroscopy sputter depth profiles of multilayer Cr/Ni thin film structures deposited on a Si substrate. The top Ni layer is about 250 Å thick and the other films are about 500 Å thick. Sputtering was carried out with a rastered beam of 5 keV Ar ions with a stationary sample in the upper portion (a), and a rotating sample in the lower portion (b). The symbols represent: ● nickel; ○ chromium; △ silicon. [From A. Zalar, *Thin Solid Films* **124**, 223 (1985).]

Multilayer films are used in integrated circuits and optical structures as well as in many other aspects of solid-state science. Auger electron spectroscopy with sputter depth profiling has a natural application to the analysis of these structures.

Figure 11.16 shows sputter depth profiles of multilayer Cr/Ni thin film structures deposited in a Si substrate. This impressive figure demonstrates the ability of Auger spectroscopy, combined with sputtering, to profile a multilayer film of nearby elements in the periodic chart in a semiquantitative manner. The rounding in the traces in the upper portion of the figure reflects the irregularity in the surface topology that developed during sputtering with a rastered beam of 5 keV Ar ions (see Chapter 4 and Carter et al., 1983). In this example, the surface roughness could be minimized by rotating the sample (lower portion of Figure 11.16) during sputtering.

Modern analytical laboratories are now equipped with a variety of systems for depth profiling of samples. When confronted with a layered or thin film sample containing unknown impurities or contaminants, the analyst will use all techniques at hand. Sputter depth profiling with Auger analysis often represents the starting point for initial analyses.

Problems

11.1. You irradiate an AlP sample with 5 keV electrons and measure the KLL Auger electrons. Calculate the ratio of Al to P ionization cross sections; fluorescence yields ω_X; escape depths λ; and Auger yields.

11.2. Compare the Auger yields in Problem 11.1 with the electron microprobe K X-ray yields ignoring X-ray absorption or electron backscattering corrections.

11.3. A Mg K_α X-ray creates a vacancy in the Cu $L_1(2s)$ subshell. Estimate the energies of photoelectrons, LM_1M_1 Auger electrons, $L_1L_2M_1$ Auger electrons (Coster-Kronig transitions), and L X-rays. Would this L shell vacancy preferentially be filled by radiative or nonradiative transitions and make an estimate of the upper value of the fluorescence yield? Which Cu L level would not deexcite by Coster–Kronig transitions?

11.4. A beam of 10 keV electrons irradiates a 1000 Å thick film of Ni on a Si substrate. Calculate the ratio of K X-ray and KLL Auger yields.

11.5. You are given a 200 Å thick layer of $Ga_x Al_{1-x} As$ on an InP substrate about 1 mm thick and are asked to determine the Ga to Al ratio. You can carry out XPS, AES or EMA analysis using 20 keV electrons and an Al K_α X-ray source. In order to compare the different techniques, you carry out the following calculations or comparisons.

(a) What is the cross section ratio σ_{Ga}/σ_{Al} for the K shell electron impact ionization and L shell photoeffect?

(b) What is the fluorescence yield ratio $\omega_X(Ga)/\omega_X(Al)$ for the K shell hole?

(c) You measure the intensity of the K_α X-ray emission from Ga and Al with a detector system with 200 eV resolution. Would you expect interference from K, L_α or M X-rays from As atoms or from the InP substrate? Would you expect electron backscattering from the InP substrate to influence the total or the ratio of X-ray yields from Ga and Al?

(d) In XPS measurements (neglecting work functions) what are the Ga and Al photoelectron energies and associated escape depths (λ)? What is the intensity ratio assuming the same detector efficiency for both electron energies?

(e) In measurements of KL_1L_1 Ga and Al Auger electrons, what are the energies and associated escape depths? What is the ratio of Ga to Al transition rates?

(f) Compare the three techniques in terms of analysis depth, corrections or interferences, and yield ratios for values of x near 0.9.

11.6. Compare transitions for K shell holes in $Z = 20$ and $Z = 36$ elements.

(a) What are the W_X/W_A ratios [Eq. (11.13)]? Compare these values with the curve in Figure 11.5.

(b) What are the atomic level widths and lifetimes? Compare the lifetime values for the two elements with the time for an electron to make a circular orbit in the Bohr model of the atom.

11.7. In an XPS analysis system with an Al K_α X-ray source, Auger electrons as well as photoelectrons are detected (see, for example, Figure 9.7). For a vanadium target, what would be the energies and escape depths of the 2s photoelectrons and L_1MM Auger electrons? In comparison with 1.5 keV electrons, what is the ratio of electron to photon cross sections, σ_e/σ_{ph}, to form a 2s hole? The L shell fluorescence yield is small (Appendix 8 or Figure 11.6), so estimate the ratio of photoelectron to Auger electrons assuming Coster–Kronig transitions can be neglected. Is this a good assumption?

References

1. W. Bambynek et al., X-ray Fluorescence Yields, Auger and Coster–Kronig Transitions, *Rev. Mod. Phys.* **44**, 716 (1972).

2. T. A. Carlson, *Photoelectron and Auger Spectroscopy* (Plenum Press, New York, 1975).

3. G. Carter, B. Navinšek, and J. L. Whitton, "Heavy Ion Surface Topography Development" in *Sputtering by Particle Bombardment II*, R. Behrisch, Ed. (Springer-Verlag, Berlin, 1983).

4. C. C. Chang, "Analytical Auger Electron Spectroscopy" in *Characterization of Solid Surfaces*, P. F. Kane and G. R. Larrabee, Eds. (Plenum Press, New York, 1974), Chap. 20.

5. A. W. Czanderna, Ed., *Methods of Surface Analysis* (Elsevier, Amsterdam, 1975).

6. L. E. Davis, N. C. MacDonald, P. W. Palmberg, G. E. Riach, and R. E. Weber, *Handbook of Auger Electron Spectroscopy* (Physical Electronics Industries, Inc., Eden Prairie, MN, 1976).

7. G. Ertl and J. Kuppers, *Low Energy Electrons and Surface Chemistry* (Verlag Chemie International, Weinheim, 1974).

8. G. Herzberg, *Atomic Spectra and Atomic Structure* (Dover, New York, 1944).

9. H. Ibach, Ed., *Electron Spectroscopy for Surface Analysis*, Topics in Current Physics, Vol. 4 (Springer-Verlag, New York, 1977).

10. A. Joshi, L. E. Davis, and P. W. Palmberg, "Auger Electron Spectroscopy" in *Methods of Surface Analysis*, A. W. Czanderna, Ed. (Elsevier Science Publishing Co., New York, 1975), Chap. 5.

11. G. E. McGuire, *Auger Electron Spectroscopy Reference Manual* (Plenum Press, New York, 1979).

12. K. D. Sevier, *Low Energy Electron Spectroscopy* (Wiley-Interscience, New York, 1972).

13. K. Siegbahn, C. N. Nordling, A. Fahlman, R. Nordberg, K. Hamrin, J. Hedman, G. Johansson, T. Bergmark, S. E. Karlsson, I. Lindgren, and B. Lindberg, *ESCA, Atomic, Molecular and Solid State Structure Studied by Means of Electron Spectroscopy* (Almqvist and Wiksells, Uppsala, Sweden, 1967).

14. J. P. Thomas and A. Cachard, Eds., *Material Characterization Using Ion Beams* (Plenum Press, New York, 1978).

CHAPTER

12

NUCLEAR TECHNIQUES:
ACTIVATION ANALYSIS
AND PROMPT
RADIATION ANALYSIS

12.1 Introduction

Analytical methods based on electronic interactions have long been used in the materials analysis laboratory, but determinations based on nuclear spectroscopy are much more recent. If radioactivity is produced by the irradiation and detected afterward, the method is called activation analysis; if radiations emitted instantaneously are detected, it is termed prompt radiation analysis. These two categories will be used for labelling major sections of the discussion which follows. For example, when a material containing carbon is irradiated with a beam of deuterons, one of the nuclear reactions with the carbon, ^{12}C, in the sample is the transformation to radioactive nitrogen, ^{13}N, by the prompt emission of a neutron, n. The carbon content of the sample can be determined either by measurement of the radiation emitted from the radioactive product nuclide, ^{13}N, activation analysis, or by measuring the yield of neutrons, prompt radiation analysis. Radioactive nuclides that are used in analysis decay with half-lives ranging between milliseconds and thousands of years (half-life of ^{13}N is 9.96 min), whereas the prompt radiation from a nuclear reaction is emitted within times less than 10^{-12} s after the nuclear reaction is initiated.

Several different kinds of nuclear interactions can be used for analysis:

283

1. A charged particle can elastically scatter from the charged target nucleus as in Rutherford scattering (Chapter 2) or nuclear elastic scattering.
2. Particles can excite the nucleus to a higher energy state (analogous to promoting an electron to a higher energy state in atomic spectroscopy); the nucleus can then deexcite by γ-ray emission.
3. A different nucleus may be formed as a result of the nuclear reaction.

In most nuclear reactions we have two particles or nuclei interacting to form two different nuclei. Thus

$$a + b \longrightarrow c + d .$$
$$\text{Reactants} \qquad \text{Products}$$

Any reaction must meet the requirement that the sum of the atomic numbers and mass numbers of reactants and products must balance. In other words

$$Z_a + Z_b = Z_c + Z_d ,$$
$$A_a + A_b = A_c + A_d .$$

Mass, however, *is* changed.

Although there is no theoretical limitation on what the nuclides a, b, c, and d can be, as a practical matter each side of the equation usually includes a very light nuclide. These are frequently termed "particles." If we designate the "particle" (keep in mind that the distinction between particle and nucleus is a bit ambiguous) by the lower case, we can write a nuclear reaction,

$$a + X \longrightarrow b + Y.$$

In a common shorthand notation, one would write

$$X(a,b)Y.$$

A nucleus with four neutrons and three protons is designated

$$^7_3 X,$$

where the subscript is Z, the number of protons, and the superscript is the total nucleons, which we term the mass number, A. More generally, therefore, a nucleus is designated

$$^A_Z X.$$

A nuclear reaction of interest in determining boron depth profiles is

$$^1_0n + ^{10}_5B \longrightarrow ^4_2He + ^7_3Li.$$

In shorthand,

$$^{10}B(n,\alpha)^7Li.$$

The reactant and product light particles are placed in parentheses and separated by a comma.

As in any specialized field, a certain nomenclature has developed based on convenience and tradition. The terms most frequently used are given below.

Nucleon	Either a proton or a neutron.
Nuclide	A specific nuclear species with a given proton number Z and neutron number N.
Isotopes	Nuclides of same Z and different N.
Isobars	Nuclides of same mass number A, where $A = Z + N$.
Isotones	Nuclides with same N but different Z.
Isomer	Nuclide in an excited state with a measurable half-life.
Deuteron (d)	^2_1H, one proton and one neutron.
Tritron (t)	^3_1H, one proton and two neutrons.
Alpha (α)	^4_2He, two protons and two neutrons.

Let us consider the irradiation of nuclei in a flux of protons, specifically proton irradiation of ^{12}C (Figure 12.1). Some of the incident protons can be scattered due to long-range Coulomb interaction with the nuclei, elastic scattering (as described in Chapter 2). Charged particles cannot effectively react through the nuclear force

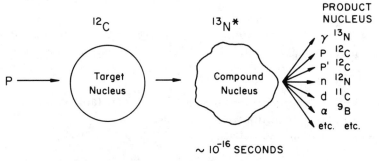

Figure 12.1 Schematic representation of the formation and decay of the compound nucleus ^{13}N during a nuclear reaction between protons and ^{12}C.

unless they have an energy comparable to the Coulomb barrier $Z_1Z_2e^2/R \cong Z_1Z_2/A^{1/3}$ (in MeV) of the target atoms, which sets the lower limit of usable energy. If the protons have sufficient energy to overcome the Coulomb barrier, they may actually be captured by the nucleus to form a "compound nucleus." The compound nucleus is now in a highly excited state and the kinetic energy of the incident particle also adds to the excitation energy. In the compound nucleus model, it is assumed that the excitation energy is randomly distributed among all the nucleons in the resultant nucleus so that none of them has enough energy to escape immediately, and thus the compound nucleus has a lifetime which is long $(10^{-14}-10^{-18}$ s) compared to the time it takes for a nucleon to traverse a nucleus $(10^{-21}-10^{-22}$ s). The highly excited compound nucleus can now deexcite in many different ways by emitting γ-rays, protons, neutrons, alpha particles, etc. The incident protons can, however, also transfer sufficient energy to single nucleons or groupings of nucleons (such as deuterons and alphas), so that they may be directly ejected from the nucleus. Examples of such *direct interactions* are (p,n) (p,α), (α,p), and (α,n) reactions. Compound nucleus reactions are more likely at relatively low energies, whereas the probability for a direct interaction increases with energy. Some of the nuclear reactions which can occur during proton irradiation are:

(p,p)   Elastic scattering (Rutherford).
(p,p)   Compound nucleus elastic scattering.
(p,p')   Inelastic scattering.
(p,γ)   Prompt γ-ray emission.
(p,n)   Prompt neutron emission.
(p,α)   Prompt alpha emission.

The probability of reaction between an incident particle and a target nucleus can be approximated by the geometrical cross section presented by the target nucleus to a point-size projectile. The radius of a nucleus is given rather accurately by the empirical formula

$$R = R_0 A^{1/3},$$

where A is the mass number and R_0 is a constant equal to 1.4×10^{-13} cm. For a medium weight nucleus such as ^{66}Zn we can therefore calculate the geometrical cross section as follows:

$$\sigma_{\text{geo}} = \pi(1.4 \times 10^{-13} \times 66^{1/3})^2$$
$$= 1.006 \times 10^{-24} \text{ cm}^2.$$

Since most cross sections are of the order of 10^{-24} cm², it has become convenient to express cross sections in units of the barn, where

$$1 \text{ barn} = 10^{-24} \text{ cm}^2.$$

The reaction cross sections in general cannot be given by simple analytical functions. For example, Figure 12.2 shows that the $^{18}O(p,\alpha)^{15}N$ reaction has a cross section which varies smoothly with energy below and above the resonance at 0.629 MeV. For specific purposes, values of the cross section are found in various compilations (see, for example, Feldman and Picraux, 1977).

The Breit–Wigner treatment deals with resonant cross sections in a quantitative way. The probability of the reaction

$$X(a,b)Y$$

may be denoted by the cross section $\sigma(a,b)$. According to the two-step compound nucleus view of nuclear reactions,

$$\sigma(a,b) = \sigma_c(a) \times \text{(relative probability of emission b)},$$

where $\sigma_c(a)$ is the cross section for the formation of the compound nucleus. The relative probability for the emission of b is just Γ_b/Γ, where Γ_b is the transition rate for emission of b, also called the partial level width for b, and Γ is the total level width $\Gamma = \hbar/\tau$, where τ is the mean life for a state so that

$$\sigma(a,b) = \sigma_c(a) \cdot \Gamma_b/\Gamma.$$

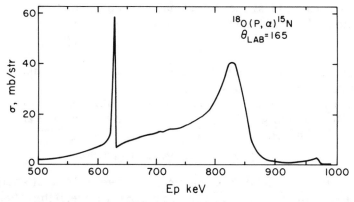

Figure 12.2 Cross section σ in millibars per steradian versus incident proton energy E_p at a detection angle $\theta = 165°$ for $^{18}O(p,\alpha)$ ^{15}N reaction. [From Amsel and Samuel, *Anal. Chem.* **39**, 1689 (1967).]

In general, the values of the cross sections and level widths depend on the energy of the incident particle, and on the charge and mass of the target nucleus. In its simplest form, the Breit–Wigner formula gives the value of the cross section in the neighborhood of a single resonance level in the compound nucleus, formed by an incident particle with zero angular momentum. Under these conditions, the formula is

$$\sigma(a,b) = \frac{\lambda^2}{4\pi} \frac{\Gamma_a \Gamma_b}{(E - E_0)^2 + (\Gamma/2)^2},\tag{12.1}$$

where λ is the de Broglie wavelength of the incident particle ($\lambda = h/mv$), E_0 the energy at the peak of the resonance, E the energy of the incident particle, and Γ_a is the partial level width for the emission of a in the inverse reaction. It is clear from this equation that the cross section will be at a maximum for $E = E_0$.

12.2 *Q* Values and Kinetic Energies

Nuclear reactions obey the following conservation laws:

1. Conservation of nucleons (A).
2. Conservation of charge (Z).
3. Conservation of mass–energy (E).
4. Conservation of momentum (p).

If the exact rest masses of the reactants and of the products of a nuclear reaction are totaled, there is likely to be a difference between the two because mass and energy may be exchanged according to the equation

$$E = mc^2,$$

where E is the energy, m is the mass, and c is the speed of light with 1 mass unit equal to 931.4 MeV.

The mass difference will correspond to either an emission or an absorption of energy. Thus a complete nuclear reaction should be written in the form

$$X + a \longrightarrow Y + b \pm Q,$$

where Q is the energy balance, usually given in MeV. If energy is released by the reaction, the Q value will be positive. If the Q value is negative, energy must be supplied and there will be a definite threshold below which these endoergic reactions will not occur.

Once a nuclear reaction has occurred, the radiations which are emitted are characteristic of the excited nuclei in much the same way that optical radiation is characteristic of an excited emitting atom. It is the existence of a unique set of well-defined energy levels in the atom or the nucleus that permits the use of the emitted radiation as an identification of the source.

Symbolically, a nuclear reaction may be written

$$M_1 + M_2 \longrightarrow M_3 + M_4 + Q$$

where M_1 is the incident nucleus, M_2 the target, M_3 is the emitted radiation, which may be either a nuclear particle or a gamma ray, M_4 is the residual nucleus, and Q is the energy released (absorbed) in the reaction. Q is simply the difference between the total energy, at rest, of the interacting system before the reaction takes place and after the reaction has occurred. If the M's are taken to be masses,

$$Q = (M_1 + M_2)c^2 - (M_3 + M_4)c^2. \qquad (12.2)$$

Let us consider the reaction

$$^{35}_{17}\text{Cl} + ^1_0\text{n} \longrightarrow ^{32}_{15}\text{P} + ^4_2\text{He} + Q,$$

which can be written in an abbreviated form as

$$^{35}\text{Cl}(n,\alpha)^{32}\text{P}.$$

The reaction is balanced with respect to nucleons and charge in that the reactants and products have the same total number of nucleons (36) and protons (17). The reaction energy Q is equivalent to the difference in mass between the reactants and the products.

$$\Delta M = M(^{35}\text{Cl}) + M(n) - M(^{32}\text{P}) - M(^4\text{He})$$

$$= 34.96885 + 1.00867 - 31.97391 - 4.00260$$

$$= +0.00101 \text{ amu}$$

$$Q = 0.00101 \times 931.4 = +0.94 \text{ MeV}.$$

Similarly the Q value can be calculated for the following reaction:

$$^{14}\text{N}(p,n)^{14}\text{O}, \quad Q = -5.931 \text{ MeV}.$$

Q values can thus be positive or negative.

Positive Q value — exoergic nuclear reaction.

Negative Q value — endoergic nuclear reaction.

If the residual nucleus M_4 is left in an excited energy state, the Q for the reaction will be reduced, relative to the value which would be obtained if the residual nucleus were left in its ground state. The reduction is just the amount of the excitation energy. For a well-defined beam energy, the energy spectrum of M_3 will be "characteristic" of the Q values possible in the reaction or, equivalently, to the excited states of the residual nucleus. Even if the emitted particles M_3 are not observed, the prompt gamma rays emitted in the decay of the excited M_4 nucleus will be characteristic of that nucleus.

In the case of activation analysis, the identifying characteristics for the radioactive M_4 nucleus can be the half-life for decay, the types of radiation emitted, and the characteristic gamma rays emitted from the daughter nuclei of M_4.

For the prompt radiation analysis (PRA) case, if M_3 is a gamma ray, the nuclear reaction is called a direct capture reaction. The case $M_1 = M_3$ and $Q = 0$ is just the elastic scattering reaction. When $M_1 = M_3$ but $Q \neq 0$, the reaction is called inelastic scattering, and, finally, when $M_1 \neq M_3$, it is commonly termed a rearrangement collision.

In contrast to the case of atoms, nuclear characteristics usually differ markedly between two isotopes of the same chemical element. The emitted radiations or reaction products are specific not only to the chemical element but also to a particular isotope of that element. It is this property that provides the basis for the many important applications of stable and radioisotope tracers. Nuclear reactions with only a few exceptions are not affected by the state of atomic electrons and so do not give information directly about the chemical bonds or the chemical compound form of the elements in a sample.

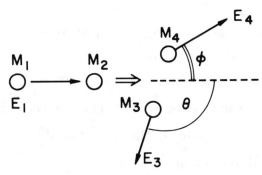

Figure 12.3 Notation used in nuclear reactions where the incident particle is denoted by mass M_1 and energy E_1 and the emitted particle M_3 is detected at an angle θ relative to the direction of the incident particle.

For a reaction induced by an incident particle M_1 of energy E_1 (Figure 12.3), the energy E_3 of the emitted particle M_3 in the direction θ (relative to the incident direction in the laboratory) is determined by the conservation of total energy and momentum and in the non-relativistic case is given by

$$E_3^{1/2} = A \pm (A^2 + B)^{1/2}, \tag{12.3}$$

where

$$A = \frac{(M_1 M_3 E_1)^{1/2}}{M_3 + M_4} \cos \theta \quad \text{and} \quad B = \frac{M_4 Q + E_1(M_4 - M_1)}{M_3 + M_4}, \tag{12.4}$$

using $M_1 + M_2 = M_3 + M_4$.

Equations (12.3) and (12.4) show that E_3 is characteristic of the reaction for a given E_1 and θ. In fact, the residual nucleus can be left in ground state or in excited states, each state corresponding to a different Q value for the same reaction, and hence to a different value of E_3. The energy spectrum of the emitted particles will exhibit a series of peaks which are specific to the reaction and lead to the detection of a given nucleus M_2. The energy of a peak allows the identification of

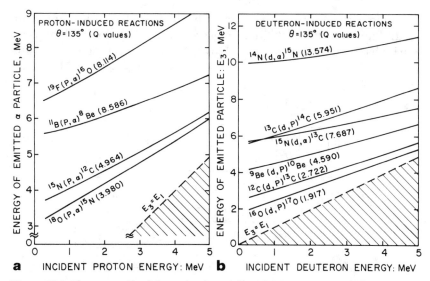

Figure 12.4 The energy E_3 of the emitted particle at $\theta = 135°$ versus incident particle energy E_1 for (a) proton–induced reactions and (b) deuteron–induced reactions. The Q values for the reaction is given in parentheses and the dashed lines $E_3 = E_1$ is the maximum energy of a particle scattered from a heavy mass element in the target. [From Feldman and Picraux, 1977.]

the reaction (and hence of the nucleus M_2) and from the intensity of the peak the amount of the M_2 species can be determined.

Equation (12.3) can be approximated within a wide energy range by

$$E_3 = \alpha E_1 + \beta, \qquad (12.5)$$

where α and β [as A and B in (12.3)] are specific to the reaction under study and depend on the detection angle θ.

The kinematics of deuteron induced reactions are shown in Figure 12.4 for some specific reactions. The relation between emitted particle energy E_3 and incident energy E_1 roughly follows Eq. (12.5) with different values of α and β for each reaction. The dashed line $E_3 = E_1$ denotes the maximum energy of elastically scattered particles and may be considered the high energy limit for the more abundant elastic scattering. The detection of light species in a heavy mass substrate can be carried out in many cases without interference from incident particles elastically scattered from the substrate.

12.3 Radioactive Decay

A nuclear reaction can take place through the formation of a compound nucleus in two distinct stages as follows: (a) the incident particle is absorbed by the target nucleus to form a compound nucleus and

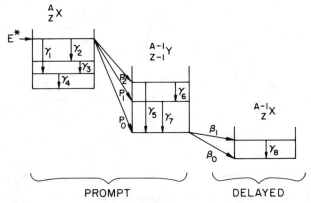

Figure 12.5 Energy level representation of an excited compound nucleus $^A_Z X$ in an excited state E^* which promptly decays by γ-ray emission to its ground state or by proton emission to the states of the residual nucleus $^{A-1}_{Z-1}Y$. This nucleus is unstable and decays at a longer time by beta emission to the nucleus $^{A-1}_Z X$.

(b) the compound nucleus disintegrates by ejecting a particle or emitting a γ-ray. We designate the compound nucleus $_Z^A X$ and assume that the nuclear reaction produces the compound nucleus in the excited state E^* as shown in the left side of Figure 12.5.

The level E^* can decay either by the emission of the "radiative capture γ-rays" γ_1, γ_2, γ_3, and γ_4 to reach the ground state of $_Z^A X$, or (as in this case) by ejection of the protons p_0, p_1, and p_2 of three distinct energies. The proton groups feed excited states of the residual nucleus $_{Z-1}^{A-1} Y$, which can deexcite by γ-ray emission (γ_5, γ_6, and γ_7) to yield the ground state of $_{Z-1}^{A-1} Y$. This nucleus is itself unstable and decays by ejection of a β-particle to the excited or ground state of $_Z^{A-1} X$. The transitions $\gamma_1 - \gamma_7$ and proton emission are most likely to occur very rapidly after the formation of the compound nucleus, i.e., within $< 10^{-12}$ s, but the half-life for the β-decay and hence for the emission of γ_8 will be very much longer. Thus two types of activation techniques can be distinguished: prompt techniques, where the samples are measured while the irradiation is in progress; and delayed methods, which depend on the measurement of a radionuclide with a half-life which is sufficiently long for the sample to be removed from the place of irradiation before the radioactivity is counted.

12.3.1 Beta Decay

As the atomic number increases there is an excess of neutrons needed for nuclear stability, and nuclides with neutron to proton ratios, N/Z ratios, that differ from the stability line undergo radioactive decay. This decay occurs by emission of beta particles, either electrons, β^-, or positrons, β^+. When the N/Z ratio of a radioactive nucleus is greater than that of a stable nucleus with the same mass number, a neutron is converted into a proton with the emission of an electron and an antineutrino ($\bar{\nu}$);

$$N/Z \text{ ratio too large: } \quad n \longrightarrow p^+ + \beta^- + \bar{\nu}.$$

If, however, the N/Z ratio is too small, the nucleus can become stable by converting protons into neutrons within the nucleus by the following processes: the emission of a positron, or the capture of an atomic orbital electron [electron capture (EC)];

$$N/Z \text{ ratio too small: } \begin{cases} p^+ \longrightarrow n + \beta^+ + \nu, \\ p^+ + e^- \longrightarrow n + \nu. \end{cases}$$

The neutrino in β-decay shares the decay energy with the β-particle. β-particles thus have a continuous energy spectrum, with an average

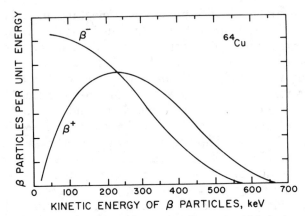

Figure 12.6 Energy spectra of the positive (β^+) and negative (β^-) electrons emitted by ^{64}Cu. The pronounced difference between the two spectral shapes results largely from the Coulomb effect.

energy of about one third of the maximum beta decay energy. Energy spectra for ^{64}Cu (half-life = 12.9 hr) emission, are shown in Figure 12.6.

If a nucleus decays by electron capture (EC), the resultant hole in the electron orbital (usually the K shell) can be filled by an electron from an outer shell. Decay by electron capture is thus associated with X-ray emission, which may also be measured for analytical purposes.

Coulomb instability of the nucleus becomes very large for the heavier nuclides. Since the helium nucleus is very stable, α-decay takes place for $Z > 83$. Delayed neutron and proton emission also occurs for nuclides far from the stability line. These modes of decay are, however, not very common and as such are not very important for nuclear analysis.

12.3.2 Gamma Decay

During β-decay, the product nucleus may be left in an excited state. Deexcitation usually occurs by the emission of a gamma ray. The decay schemes for ^{27}Mg and ^{64}Cu are shown in Figure 12.7. Because deexcitation by γ-ray emission is much more probable than β-decay, the γ-decay rate will be the same as the rate of the β-decay with which it is associated.

From the decay scheme for ^{64}Cu it can be seen that it decays by both β^+ (19%) and β^- (39%) emission, while the other 42% of its decay is by electron capture, which will lead to ^{64}Ni X-ray emission.

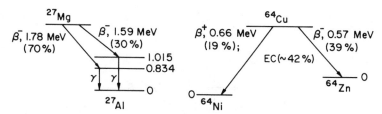

Figure 12.7 The principal decay schemes for ^{27}Mg (half-life = 9.5 min) and ^{64}Cu (half-life = 12.9 hr). The decay of ^{27}Mg shows the gamma rays associated with β decay and of ^{64}Cu the β^+-β^- branching.

When positrons pass through matter, they can annihilate with electrons (their antiparticle). If such a positron comes to rest and then annihilates with a free electron, conservation of linear momentum requires that two gamma rays be emitted, in opposite directions (180°), each having an energy equal to m_0c^2 (0.51 MeV). Positron decay is thus associated with 0.51 MeV gamma-rays, which is referred to as annihilation radiation. If the electron is bound in an atom, annihilation with the production of a single photon can occur, because the atom can take up some momentum. The probability for such a process is, however, extremely small.

12.4 Radioactive Decay Law

The rate of radioactive decay, A, is proportional to the number of radioactive nuclei, N, present:

$$A = \frac{dN}{dt} = -\lambda N, \tag{12.6}$$

where λ is the decay constant. If at some particular time there are N_0 radioactive nuclei in a sample, then we can find the number of radioactive nuclei N_t, remaining at a later time t, by integration of the above equation

$$\int_{N_0}^{N_t} \frac{dN}{N} = -\int_0^t \lambda \, dt, \tag{12.7}$$

$$N_t = N_0 e^{-\lambda t}.$$

It is convenient to express the decay constant λ as a half-life, which is defined as the time $T_{1/2}$ required for any number of radioactive nuclei to decay to half its initial value:

$$\frac{N_t}{N_0} = \frac{1}{2} = e^{-\lambda T_{1/2}} \qquad (12.8)$$

and

$$\lambda = \frac{\ln 2}{T_{1/2}} = 0.693/T_{1/2}.$$

From Eq. (12.6) we now have

$$A_t = -\lambda N_0 e^{-0.693t/T_{1/2}}$$
$$A_t = A_0 e^{-0.693t/T_{1/2}}. \qquad (12.9)$$

Equation (12.9) is known as the radioactive decay law. If the decay rate (A_t) is plotted as a function of decay time (t) on semilog paper, the decay rate will follow a straight line with intercept A_0. The half-life of the radionuclide may be obtained from the slope.

Information regarding the half-lives, types of decay, and decay energies can be obtained from the nuclide chart. More details about decay level schemes can be obtained from compilations such as the *Table of Isotopes* (Lederer et al., 1967).

12.5 Radionuclide Production

In activation analysis we are interested in the amount of radioactivity formed as a function of irradiation time. Let us consider the nuclear reaction

$$X(a,b)Y,$$

with Y the radioactive product nucleus, which we determine as a measure of X in the sample. The rate at which nuclide Y is formed may be expressed as

$$\left(\frac{dN_Y}{dt}\right)_{\text{growth}} = N_X \sigma \phi,$$

where N_X is the number of target nuclides and N_Y the number of product nuclides formed, σ the reaction cross section (cm^2), and ϕ the flux of incident particles (particles $cm^{-2} s^{-1}$). If the product nucleus is stable, the total number of nuclides Y formed during an irradiation time of t seconds is

$$N_Y = N_X \sigma \phi t. \qquad (12.10)$$

If Y is radioactive, however, and decays with a decay constant λ_Y, we have

$$\left(\frac{dN_Y}{dt}\right)_{\text{decay}} = -N_Y\lambda_Y.$$

The total rate for forming the radioactive product nuclide Y is then given by

$$\frac{dN_Y}{dt} = N_X\sigma\phi - N_Y\lambda_Y. \tag{12.11}$$

If the target number N_X is a constant, as it usually is, one obtains by integration

$$N_Y = \frac{N_X\sigma\phi}{\lambda_Y}(1 - e^{-\lambda_Y t}).$$

The activity A_t (in disintegrations per second) of the radioactive nuclide Y at time t is given by $A_t = N_Y\lambda_Y$ so that

$$A_t = N_X\sigma\phi(1 - e^{-\lambda_Y t}). \tag{12.12}$$

For $t \to \infty$ the term in brackets (growth factor) equals unity and $A_\infty = N_X\sigma\phi$. The activity A_∞ is referred to as the saturation A_{sat}, and Eq. (12.12) may be written as

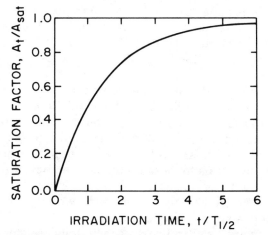

Figure 12.8 Growth of radioactive nuclide activity A_t/A_{sat} as a function of irradiation time $t/T_{1/2}$. After an irradiation time of 4 half-lives, about 94% of the saturation activity A_{sat} is formed.

$$A_t = A_{sat}(1 - e^{-\lambda t}). \qquad (12.13)$$

The saturation factor A_t/A_{sat} is plotted in Figure 12.8 as a function of irradiation time t expressed in number of half-lives, where $\lambda = 0.693/T_{1/2}$.

12.6 Activation Analysis

Activation analysis is a highly sensitive, nondestructive (if chemical separations are not used) technique for qualitative and quantitative determination of trace amounts of elements in a sample. It has been particularly useful for the simultaneous determination of many elements in complex samples because it provides a simple alternative to much more difficult, tedious, and destructive analytical techniques.

In activation analysis the element to be determined is uniquely identified by the half-life and energy of the radiation emitted by its radioactive product nuclide. For optimum analytical sensitivity it is desirable to form the maximum amount of activity in the sample from the elements to be determined and to measure such activities with maximum efficiency in the presence of other interfering activities. Fortunately, the irradiation and measurement parameters can be changed over wide ranges in order to select the optimum conditions for analysis.

It has been shown (in Figure 12.8) that 50% of the maximum activity is obtained after an irradiation time equal to one half-life and more than 90% after four half-lives. Irradiations longer than a few half-lives will thus only increase the amount of unwanted background activity from other longer-lived activities also found in the sample.

The measurement of activities with short half-lives should be carried out as soon as possible after the irradiation. For longer-lived activities, better sensitivity is usually obtained if measurement takes place after time is allowed for the shorter-lived activities in the sample to decay. Figure 12.9 shows the decay curve for two activities. It is clear that the best time to measure the 24-h activity would be after allowing 20 h of decay time. The different components contributing to a decay curve are usually determined by decay curve stripping using a graphic approach or a computer.

Activation analysis is primarily a technique used for detection of impurities in bulk materials. The activation processes, involving neutrons, are deeply penetrating and thus sample many atoms

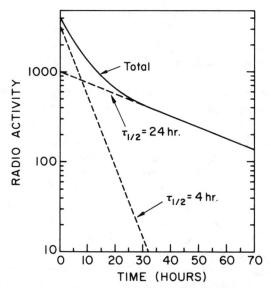

Figure 12.9 Decay diagram of a mixture of two independently decaying nuclides with half-lives of 4 and 24 h.

throughout the material. The detection limits for thermal neutron activation analysis are generally in the range of $10^{-8} - 10^{-10}$ grams. In terms of monolayers, 5×10^{-8} g of Ni would correspond to one monolayer on a 1 cm^2 sample. There is no intrinsic depth sensitivity in neutron activation analysis so that depth profiles must be generated by the use of sample thinning techniques such as sputtering or chemical etching.

12.7 Prompt Radiation Analysis

In prompt radiation analysis the presence of an element in a sample is detected through the nuclear radiations emitted instantaneously from nuclear reactions produced in the target by the irradiating beam. Detection limits can be quite good, but typically not as good as can be achieved under ideal conditions with activation analysis.

One of the important advantages of prompt analysis and the backscattering techniques discussed in Chapter 3 is that they can be used to measure the depth distribution of elements in the surface or near-surface regions of the sample. The dependence of the characteristics of the emitted radiations on depth is due to the energy loss suffered by the incident ions as they penetrate into the sample and also to the energy losses suffered by charged particles emitted from the reaction as they emerge from within the sample.

Since nuclear-reaction analysis can provide essentially back-ground-free detection of light elements ($Z \leq 15$), depth distributions of trace amounts within the near-surface region can be measured. The primary emphasis in our discussion of prompt radiation analysis is the determination of concentration depth profiles of trace element impurities. In the use of prompt analysis for depth profiles, two different methods are applied, namely, the energy-analysis method and the resonance method. The former is used when the nuclear-reaction cross section is a smoothly varying function of energy. During an analysis, the energy of the analyzing beam is kept fixed, and the energy spectrum of particles emitted by the reaction, from which the depth profile is derived, is recorded. The latter method is used when a sharp peak (resonance) (see Figure 12.2) in the cross section as a function of energy is present, and the depth profile is derived from a measurement of the nuclear-reaction yield as a function of the energy of the analyzing beam.

12.7.1 Energy-Analysis Method

12.7.1.1 Thermal Neutron-Induced Reactions

Elemental depth distributions of certain trace elements can be determined by a thermal neutron beam to produce reactions with certain elements that yield monoenergetic charged particles.[1] These isotropically emitted particles lose energy in passing through matter, and their residual energy on leaving the sample surface is primarily dependent on the amount of matter the particle has passed through. For a given sample with a known atomic density, the energy of a detectable particle is determined by the depth at which the initial reaction took place. A quantitative image of the elemental distribution with depth — a depth profile — is generated directly from the charged particle spectrum. For impurity distributions close to the surface so that the energy loss rate is nearly constant, the energy difference ΔE between detected energies of particles emitted from atoms at the surface or at depth t is determined by the rate of energy loss dE/dx along the outgoing path

[1] J. F. Ziegler, G. W. Cole, and J. E. E. Baglin, *J. Appl. Phys.* **43**, 3809 (1972); J. P. Biersack, D. Fink, R. Henkelmann, and K. Muller, *Nucl. Instr. Methods* **149**, 93 (1978).

Table 12.1 Energies and Cross Sections for Thermal Neutron Reactions

Element	Reaction	Energy of emitted particles in keV	Cross section (barns)
Li	^{6}Li(n,α)T	2056	940
B	^{10}B(n,α)^{7}Li	1472	3836
Be	^{7}Be(n,p)^{7}Li	1439	48,000
Na	^{22}Na(n,p)^{22}Ne	2248	29,000

$$\Delta E = t(dE/dx) \tag{12.14}$$

whereas, in Rutherford backscattering (Chapter 3), the energy to depth conversion is determined by the energy loss along both the inward and outward paths.

Thermal neutron reaction cross sections can be substantially greater than geometric cross sections (≈ 1 barn) as indicated in Table 12.1. For a total thermal neutron flux of 10^8 n/cm^2, the sensitivity for impurity detection is about 10^{14} atoms/cm^2 for boron.

Figure 12.10 shows the charged particle spectrum of a thin film of boron (100 Å) on Ni. The four peaks in the spectrum correspond to the secondary and primary α-particles and lithium ions from the ^{10}B(n,α)^{7}Li reaction. The measured energies of the four charged particles are highest when the reaction occurs at the surface of the sample. When the reaction occurs within the sample, the particles must pass through overlying matter, and the entire charged particle spectrum is shifted to lower energies. The maximum distance the particles can travel and still exit the surface — the range — varies with sample composition, but is typically 1–10 μm for solids. Figure 12.10b shows the spectrum shift of the primary alpha peak that occurs when the thin boron film is covered with 500 Å of Cu. Besides the spectral shift, the covered B film also shows indications of diffusion into the nickel substrate. The left side of the peak has broadened, and the height has decreased, indicating movement of the boron.

12.7.1.2 Charged-Particle-Induced Reactions

In charged-particle irradiation of targets at energies sufficiently high so that the incident particle can penetrate the Coulomb barrier, a variety of nuclear reactions can occur, as shown in Figure 12.11 for deuteron irradiation of a thin aluminum nitride target.

The yield depends, as for Rutherford backscattering, on the differential cross section of the reaction; but, unlike backscattering, there is no simple analytical formula for the cross section. The cross-

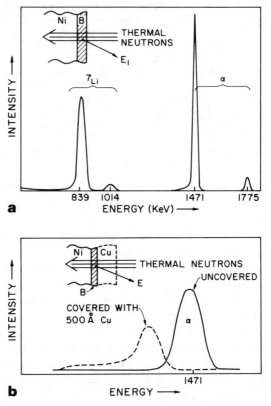

Figure 12.10 (a) Charged particle spectrum from ^{10}B for a thermal neutron irradiated sample of Ni covered with 100 Å of B. The portion of the spectra associated with 1471 keV α is expanded in (b) for samples covered (dashed line) and samples not covered (solid line) with a 500 Å thick Cu film. [From J. E. Riley, Jr. and R. F. Fleming, private communication.]

section curves can be obtained from the nuclear physics literature (see Feldman and Picraux, 1977). As a general feature, medium and high Z nuclei do not undergo nuclear reactions (in MeV range) owing to Coulomb barrier repulsion. This, combined with the fact that the emitted particles have energies well above that of the incident particle energy (due to the high Q values of most of the reactions) allows background-free detection of light elements on heavier substrates. The abundant elastically scattered particles are stopped in a thin absorber to prevent count-rate saturation of the detector and electronic systems.

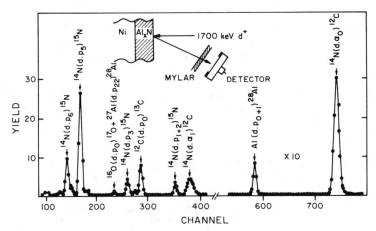

Figure 12.11 Energy spectrum at 8.23 keV/channel from deuteron bombardment of a 1700 Å thick aluminum nitride film on an Ni backing for a deuteron incident energy of 1700 keV and a detection angle of 160°. The mylar film is used to block the scattered deuterons from the detector and hence decrease the number of background particles which cause pulse pile-up in the detection system. [From Cachard and Thomas, 1978.]

The number of detected particles, Q_D, is proportional to the total number N_s of atoms/cm^2,

$$Q_D = N_s \sigma(\theta) Q \Omega, \qquad (12.15)$$

where $\sigma(\theta)$ is the differential cross section, Ω the detection solid angle, and Q the number of incident particles [Eq. (2.9)].

For incident energy E_0, the energy E_3 of the detected species for a surface reaction is $E_3 \cong \alpha E_0 + \beta$ as given in Eq. (12.5). Then the energy difference ΔE between the detected particles originating from the surface and from depth t depends both on the energy loss dE/dx of the incident particle on its inward path and the energy loss of the reaction particle on its outward path,

$$\Delta E = t \left[\alpha \left. \frac{dE}{dx} \right|_{in} + \left. \frac{dE}{dx} \right|_{out} \right]. \qquad (12.16)$$

Again we use the approximation that the energy losses are constant in the near surface region. The reaction factor α weights the energy loss in the inward path in the same fashion as the backscattering kinematic factor weights $dE/dx|_{in}$ [Eq. (3.20a)]. This formula (12.16) defines the correspondence between the depth scale and the energy scale. If the cross section is known, the concentration profile can be deduced from the shape of the experimental spectrum.

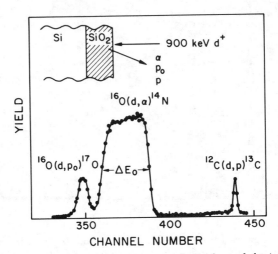

Figure 12.12 Portion of the energy spectrum at 6.75 keV/channel obtained at $\theta = 145°$ from the irradiation of a 6000 Å SiO_2 layer with 900 keV deutrons. The energy width ΔE_0 of the oxygen signal is directly related to the oxide thickness. [From A. Turos, L. Wielunski, and A. Barez, *Nucl. Inst. Meth.* **111**, 605 (1973).]

The $^{16}O(d,\alpha)^{14}N$ reaction can be used for oxygen depth profile measurements. This reaction, at low deuteron energy, emits only a ground state α group, α_0. For α-particles corresponding to the first excited state of the ^{14}N nucleus, the reaction has a negative Q value of 0.829 MeV (and therefore a threshold energy) and will not occur for a deuteron energy below 933 keV. At low deuteron bombarding energies the α_0 energy at large angles is low, and the stopping power or energy loss per unit length therefore is relatively high and provides improved depth resolution. Figure 12.12 shows the energy spectrum which was observed for a 6000 Å thick SiO_2 layer with the beam at normal incidence to the target ($\phi = 0°$) and the detector at 145°. In order to avoid interference from the $^{16}O(d,p_0)^{17}O$ reaction the detector depletion depth was not allowed to exceed 26 μm. While the α-particles were stopped in this thickness and deposited their full energy in the detector, the protons deposited only a portion of their energy and were displaced thereby to lower energies in the particle energy spectrum.

Figure 12.12 shows clearly the advantage of making measurements with particles having a high stopping power. Thus while the proton groups from the $^{16}O(d,p)^{17}O$ are quite narrow and cannot be used for depth profile measurements, the α_0 group is quite wide and can be so used.

12.7.2 The Resonance Method

Many nuclear reactions have the property that the reaction yield exhibits one or more sharp peaks or "resonances" as a function of bombarding energy. Such a resonance is measured experimentally by varying the incident beam energy in small increments and measuring the quantity of radiation emitted per unit beam fluence at each energy. The use of the resonance method in depth profiling of trace elements takes advantage of the sharp peak (see, for example, Fig. 12.2) in the nuclear reaction cross section as a function of energy. Consider the ideal case shown in Figure 12.13, where only one resonance exists in the cross-section curve and where off-resonance cross-section values can be neglected. The method consists of measuring the reaction yield (most often γ-rays) due to the interaction between the incident beam and the impurity atoms as a function of incident beam energy. Incident ions having an energy E_0 (i.e., larger than E_R, the resonance energy) are slowed down until E_R is reached at depth x, where the nuclear reaction will then occur at a rate propor-

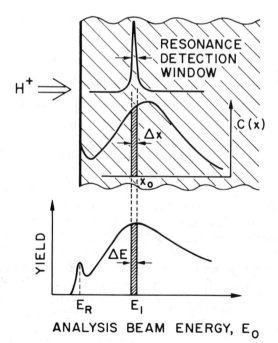

Figure 12.13 Principle of concentration profile measurements using resonant reactions.

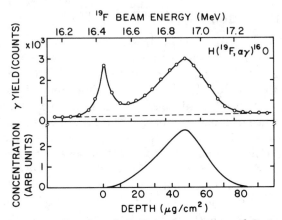

Figure 12.14 Range profile of 12 keV H implanted into Al_2O_3 to a fluence of 4×10^{16}/cm² measured by use of the nuclear reaction $^1H(^{19}F,\alpha\gamma)$ ^{16}O. The upper part of the figure is the raw experimental data and the lower part is the extracted depth profile. [From Figure 1 of J. Bottiger, S. T. Picraux, and N. Rud, in *Ion Beam Surface Layer Analysis, Vol. 2*, O. Meyer, G. Linker, and F. Käppeler, Eds. (Plenum Press, New York, 1976), p. 811.]

tional to the impurity concentration. The depth x and the incident-beam energy E_0 are related through the equation

$$E_0 = E_R + \left(\frac{dE}{dx}\right)_{in} \frac{x}{\cos \theta_1}, \tag{12.17}$$

where θ_1 is the angle between the incident beam and the surface normal. The stopping power $(dE/dx)_{in}$ for the incident beam is assumed to be a constant. A more elaborate analysis can be carried out by taking into account the detailed cross section function, energy straggling, and other factors.[2]

Neglecting the finite experimental depth resolution, it is seen that the yield curve in Figure 12.13 can be converted into the desired concentration profile by simply changing scales of yield and energy to corresponding scales of concentration and depth, respectively. An example of the use of nuclear resonance is shown in Figure 12.14, which gives the gamma yield as a function of beam energy for a hydrogen-implanted target. The reaction between fluorine and hydrogen has a strong resonance at about 16.4 MeV so that the hydrogen

[2] See, for example, G. Amsel and B. Maurel, *Nucl. Inst. Methods* **218**, 8 (1983).

Table 12.2 Most Used Charged Particle Reactions for Light-Atom Detection[a]

Nucleus	Reaction	Incident energy (E_0) (MeV)	Emitted energy[b] (MeV)	Approximately σ_{LAB} (E_0) (mb/sr)	Yield[c] (counts/μC)
^{2}H	^{2}H(d,p)^{3}H	1.0	2.3	5.2	30
^{2}H	^{2}H(^{3}He,p)^{4}He	0.7	13.0	61	380
^{3}He	^{3}He(d,p)^{4}He	0.45	13.6	64	400
^{6}Li	^{6}Li(d,α)^{4}He	0.7	9.7	6	35
^{7}Li	^{7}Li(p,α)^{4}He	1.5	7.7	1.5	9
^{9}Be	^{9}Be(d,α)^{7}Li	0.6	4.1	~1	6
^{11}B	^{11}B(p,α)^{8}Be	0.65	5.57(α_0)	0.12(α_0)	0.7
		0.65	3.70(α_1)	90(α_1)	550
^{12}C	^{12}C(d,p)^{13}C	1.20	3.1	35	210
^{13}C	^{13}C(d,p)^{14}C	0.64	5.8	0.4	2
^{14}N	^{14}N(d,α)^{12}C	1.5	9.9(α_0)	0.6(α_0)	3.6
		1.2	6.7(α_1)	1.3(α_1)	7.0
^{15}N	^{15}N(p,α)^{12}C	0.8	3.9	~15	90
^{16}O	^{16}O(d,p)^{17}O	0.90	2.4(p_0)	0.74(p_0)	5
		0.90	1.6(p_1)	4.5(p_1)	28
^{18}O	^{18}O(p,α)^{15}N	0.730	3.4	15	90
^{19}F	^{19}F(p,α)^{16}O	1.25	6.9	0.5	3
^{23}Na	^{23}Na(p,α)^{20}Ne	0.592	2.238	4	25
^{31}P	^{31}P(p,α)^{28}Si	1.514	2.734	16	100

[a] From Feldman and Picraux, 1977.
[b] Laboratory emission angle of 150° with recoil nucleus in ground state (excited state).
[c] For a 1×10^{16}/cm^2 surface layer and a solid angle of 0.1 sr at 150°.

concentration profile can be obtained directly. The extracted hydrogen concentration profile indicates only the hydrogen within the sample and does not include the surface hydrogen present due to contamination.

Nuclear reaction analysis (NRA) is a method of determining the absolute concentration (atoms/cm^2) of light impurities in and on a solid. It thus provides an absolute calibration for other surface sensitive techniques—particularly Auger analysis and secondary ion mass spectroscopy. In a typical application, a light particle of interest is implanted into a heavier substrate. SIMS provides a sensitive depth profile while NRA determines an absolute concentration. Reaction analysis is particularly useful for hydrogen detection and absolute hydrogen surface coverages. Table 12.2 lists the most used charged particle reactions for light atom detection.

Problems

12.1. A magnesium foil 0.1 mm thick is irradiated with a beam of 22 MeV deuterons (beam current = 100 μA), which has a cross-sectional area less than that of the foil. Sodium-24 (half-life = 15.0 h) is formed by the ^{26}Mg (d,α)^{24}Na reaction, which has an average cross section $\bar{\sigma} = 25$ mbarns, throughout the thickness of the foil. What is the activity of ^{24}Na (disintegrations/sec) in the foil during a 2-h irradiation?

12.2. A steel sample weighing 2.5 g is irradiated for 30 min in a reactor with a thermal neutron flux $\phi = 4.2 \times 10^{13}$ neutrons/cm^2/s. The ^{27}Mg activity (half-life = 9.5 min) formed by the ^{26}Mg (n,γ)^{27}Mg reaction is measured 10 min after the irradiation and the 0.834 MeV γ-ray gives a count rate of 625 cpm. For a counting efficiency of 3% and σ(n,γ) = 30 mbarns calculate the wt % Mg concentration in the sample.

12.3. How would you dope silicon with phosphorous using a nuclear reactor? Calculate the dopant concentration of a silicon sample irradiated in a thermal neutron flux of 2×10^{14} neutrons s^{-1} cm^{-2} for 6 h? (^{30}Si (n,γ)^{31}Si; $\sigma = 0.12$ barns.)

12.4. Find the threshold kinetic energy (MeV) for the following incident particles to disintegrate the deuteron into a proton and a neutron.
(a) Electrons
(b) Protons
(c) Alpha particles

12.5. A thin 8.0 mg foil of ^{113}Cd is exposed to a flux of thermal neutrons. How many ^{114}Cd nuclei are formed? $\phi_n = 1.6 \times 10^{13}$ n/s/cm^2, the irradiation time is 2 h and σ(n,γ) = 2×10^4 barns.

12.6. Compare proton backscattering and ^{15}N(p,α)$^{1/2}$C reactions from one monolayer of nitrogen atoms (10^{15} N/cm^2) on a carbon substrate.
(a) For backscattering with 0.8 MeV protons through 180° with a detector solid angle of 0.01 str, calculate the yield [assuming pure Rutherford scattering, Eq. (2.17)] and
(b) Compare the RBS yield with the nuclear reaction yield (Table 12.2) for the same detector.
(c) What are the relative advantages of the two techniques?

12.7. The geometrical cross section for a reaction is given by $\sigma_{geom} = \pi R^2$ where R is the nuclear radius and the distance of closest approach d equals $Z_1 Z_2 e^2 / E$.

(a) Calculate these values for 1 MeV p,d and α-particles incident on ^{14}N.

(b) Compare the values of the geometrical cross section with the Rutherford scattering cross sections for 1 MeV particles ($\theta = 180°$) and with the ^{15}N(p,α) and ^{14}N(d,α) cross sections deduced from Table 12.2.

(c) Do these values give support to the rule of thumb that the onset of nuclear reactions occurs when the particles penetrate the Coulomb barrier?

12.8. Compare the depth scales in eV/Å for detection of F in an Al thin film with RBS for 3 MeV ^{4}He ions incident ($\theta = 180°$) and for the ^{19}F(p,α) reaction with 1.25 MeV protons.

12.9. For $X(a,b)Y$ resonant nuclear reactions, there is an inverse reaction $a(X,Y)b$. If the energy of the resonance at E_a is known, what is E_X in terms of E_a and the masses of the reactants? Evaluate your answer using energy values for ^{19}F(p,α) on Table 12.2 compared with those for ^{1}H(F,α) on Figure 12.14.

References

1. A. Cachard and J. P. Thomas, "Microanalysis by Direct Observation of Nuclear Reactions," in *Material Characterization Using Ion Beams*, J. P. Thomas and A. Cachard, Eds. (Plenum Press, New York, 1978).

2. G. R. Choppin and J. Ryaberg, *Nuclear Chemistry* (Pergamon Press, Oxford, 1980).

3. R. D. Evans, *The Atomic Nucleus* (McGraw-Hill Book Co., New York, 1955).

4. F. Everling, L. A. Koenig, J. H. E. Mattauch, and A. H. Wapstra, "Consistent Set of Energies Liberated in Nuclear Reaction," 1960 Nuclear Data Tables (U.S. Government Printing Office, Washington, DC, 1961).

5. L. C. Feldman and S. T. Picraux, "Selected Low Energy Nuclear Reaction Data, in *Ion Beam Handbook for Material Analysis*, J. W. Mayer and E. Ŕimini, Eds. (Academic Press, New York, 1977).

6. G. Friedlander, J. W. Kennedy, and J. M. Miller, *Nuclear and Radiochemistry* (John Wiley and Sons, New York, 1964).

7. B. G. Harvey, *Nuclear Chemistry* (Prentice-Hall, Englewood Cliffs, NJ, 1965).

8. I. Kaplan, *Nuclear Physics* (Addison-Wesley, Reading, MA, 1964).

9. C. M. Lederer, J. M. Hollander, and I. Perlman, *Table of Isotopes*, 6th Edition, (John Wiley and Sons, New York, 1967).

10. W. E. Meyerhof, *Elements of Nuclear Physics* (McGraw-Hill Book Co., New York, 1967).

11. P. A. Tipler, *Modern Physics* (Worth Publishers, New York, 1978).

12. E. A. Wolicki, "References to Activation and Prompt Radiation Analysis: Material Analysis by Means of Nuclear Reactions," in *New Uses of Ion Accelerators*, J. F. Ziegler, Ed. (Plenum Press, New York, 1975), Chap. 3.

1

K_{M_2} FOR ^{4}He AS PROJECTILE AND INTEGER TARGET MASS, $M_2{}^a$

ATOMIC MASS (amu)	180°	170°	160°	145°	90°	60°	45°	30°
4					0.000	0.250	0.500	0.750
6	0.040	0.041	0.043	0.051	0.200	0.476	0.659	0.832
8	0.111	0.113	0.118	0.134	0.333	0.589	0.738	0.873
10	0.184	0.186	0.193	0.213	0.429	0.661	0.787	0.897
12	0.250	0.253	0.260	0.282	0.500	0.711	0.820	0.914
14	0.309	0.311	0.319	0.342	0.556	0.748	0.844	0.926
16	0.360	0.363	0.371	0.394	0.600	0.776	0.863	0.935
18	0.405	0.408	0.416	0.439	0.636	0.799	0.877	0.942
20	0.444	0.447	0.455	0.478	0.667	0.817	0.889	0.948
22	0.479	0.482	0.490	0.512	0.692	0.833	0.899	0.952
24	0.510	0.513	0.521	0.542	0.714	0.846	0.907	0.956
26	0.538	0.540	0.548	0.569	0.733	0.857	0.914	0.960
28	0.563	0.565	0.572	0.592	0.750	0.866	0.920	0.962
30	0.585	0.587	0.594	0.614	0.765	0.875	0.925	0.965
32	0.605	0.607	0.614	0.633	0.778	0.882	0.929	0.967
34	0.623	0.626	0.632	0.650	0.789	0.889	0.933	0.969
36	0.640	0.642	0.649	0.666	0.800	0.895	0.937	0.971
38	0.655	0.657	0.664	0.681	0.810	0.900	0.940	0.972
40	0.669	0.671	0.678	0.694	0.818	0.905	0.943	0.974
42	0.682	0.684	0.690	0.706	0.826	0.909	0.946	0.975
44	0.694	0.696	0.702	0.718	0.833	0.913	0.948	0.976
46	0.706	0.707	0.713	0.728	0.840	0.917	0.950	0.977
48	0.716	0.718	0.723	0.738	0.846	0.920	0.952	0.978
50	0.726	0.727	0.733	0.747	0.852	0.923	0.954	0.979
52	0.735	0.736	0.742	0.755	0.857	0.926	0.956	0.980
54	0.743	0.745	0.750	0.763	0.862	0.929	0.958	0.980

^{a} Kinematic factor, K_{M_2}, defined by Eq. (2.5) for a ^{4}He atom as projectile and integral atomic masses for the target atom M_2. The angle indicates the scattering angle as measured in the laboratory frame of reference.

ATOMIC MASS (amu)	180°	170°	160°	145°	90°	60°	45°	30°
56	0.751	0.753	0.758	0.771	0.867	0.931	0.959	0.981
58	0.759	0.760	0.765	0.778	0.871	0.933	0.960	0.982
60	0.766	0.767	0.772	0.784	0.875	0.935	0.962	0.982
62	0.772	0.774	0.778	0.791	0.879	0.937	0.963	0.983
64	0.779	0.780	0.784	0.796	0.882	0.939	0.964	0.983
66	0.784	0.786	0.790	0.802	0.886	0.941	0.965	0.984
68	0.790	0.792	0.796	0.807	0.889	0.943	0.966	0.984
70	0.795	0.797	0.801	0.812	0.892	0.944	0.967	0.985
72	0.801	0.802	0.806	0.817	0.895	0.946	0.968	0.985
74	0.805	0.807	0.811	0.821	0.897	0.947	0.969	0.986
76	0.810	0.811	0.815	0.826	0.900	0.949	0.970	0.986
78	0.814	0.816	0.819	0.830	0.902	0.950	0.970	0.986
80	0.819	0.820	0.824	0.834	0.905	0.951	0.971	0.987
82	0.823	0.824	0.827	0.837	0.907	0.952	0.972	0.987
84	0.826	0.828	0.831	0.841	0.909	0.953	0.972	0.987
86	0.830	0.831	0.835	0.844	0.911	0.955	0.973	0.988
88	0.834	0.835	0.838	0.847	0.913	0.956	0.974	0.988
90	0.837	0.838	0.842	0.851	0.915	0.957	0.974	0.988
92	0.840	0.841	0.845	0.854	0.917	0.957	0.975	0.988
94	0.843	0.844	0.848	0.856	0.918	0.958	0.975	0.989
96	0.846	0.847	0.851	0.859	0.920	0.959	0.976	0.989
98	0.849	0.850	0.853	0.862	0.922	0.960	0.976	0.989
100	0.852	0.853	0.856	0.864	0.923	0.961	0.977	0.989
102	0.855	0.856	0.859	0.867	0.925	0.962	0.977	0.990
104	0.857	0.858	0.861	0.869	0.926	0.962	0.978	0.990
106	0.860	0.861	0.864	0.872	0.927	0.963	0.978	0.990
108	0.862	0.863	0.866	0.874	0.929	0.964	0.979	0.990
110	0.865	0.866	0.868	0.876	0.930	0.964	0.979	0.990
112	0.867	0.868	0.871	0.878	0.931	0.965	0.979	0.990
114	0.869	0.870	0.873	0.880	0.932	0.966	0.980	0.991
116	0.871	0.872	0.875	0.882	0.933	0.966	0.980	0.991
118	0.873	0.874	0.877	0.884	0.934	0.967	0.980	0.991
120	0.875	0.876	0.879	0.886	0.935	0.967	0.981	0.991
122	0.877	0.878	0.881	0.888	0.937	0.968	0.981	0.991
124	0.879	0.880	0.882	0.889	0.938	0.968	0.981	0.991
126	0.881	0.882	0.884	0.891	0.938	0.969	0.982	0.992
128	0.882	0.883	0.886	0.892	0.939	0.969	0.982	0.992
130	0.884	0.885	0.887	0.894	0.940	0.970	0.982	0.992
132	0.886	0.887	0.889	0.896	0.941	0.970	0.982	0.992
134	0.887	0.888	0.891	0.897	0.942	0.971	0.983	0.992
136	0.889	0.890	0.892	0.898	0.943	0.971	0.983	0.992
138	0.890	0.891	0.894	0.900	0.944	0.971	0.983	0.992
140	0.892	0.893	0.895	0.901	0.944	0.972	0.983	0.992
142	0.893	0.894	0.896	0.903	0.945	0.972	0.984	0.992
144	0.895	0.896	0.898	0.904	0.946	0.973	0.984	0.993
146	0.896	0.897	0.899	0.905	0.947	0.973	0.984	0.993
148	0.898	0.898	0.900	0.906	0.947	0.973	0.984	0.993
150	0.899	0.900	0.902	0.908	0.948	0.974	0.984	0.993
152	0.900	0.901	0.903	0.909	0.949	0.974	0.985	0.993
154	0.901	0.902	0.904	0.910	0.949	0.974	0.985	0.993
156	0.902	0.903	0.905	0.911	0.950	0.975	0.985	0.993
158	0.904	0.904	0.906	0.912	0.951	0.975	0.985	0.993
160	0.905	0.906	0.908	0.913	0.951	0.975	0.985	0.993
162	0.906	0.907	0.909	0.914	0.952	0.976	0.986	0.993
164	0.907	0.908	0.910	0.915	0.952	0.976	0.986	0.993
166	0.908	0.909	0.911	0.916	0.953	0.976	0.986	0.994
168	0.909	0.910	0.912	0.917	0.953	0.976	0.986	0.994

ATOMIC MASS (amu)	180°	170°	160°	145°	90°	60°	45°	30°
170	0.910	0.911	0.913	0.918	0.954	0.977	0.986	0.994
172	0.911	0.912	0.914	0.919	0.955	0.977	0.986	0.994
174	0.912	0.913	0.915	0.920	0.955	0.977	0.987	0.994
176	0.913	0.914	0.916	0.921	0.956	0.978	0.987	0.994
178	0.914	0.915	0.917	0.921	0.956	0.978	0.987	0.994
180	0.915	0.916	0.917	0.922	0.957	0.978	0.987	0.994
182	0.916	0.916	0.918	0.923	0.957	0.978	0.987	0.994
184	0.917	0.917	0.919	0.924	0.957	0.978	0.987	0.994
186	0.918	0.918	0.920	0.925	0.958	0.979	0.987	0.994
188	0.918	0.919	0.921	0.925	0.958	0.979	0.988	0.994
190	0.919	0.920	0.922	0.926	0.959	0.979	0.988	0.994
192	0.920	0.921	0.922	0.927	0.959	0.979	0.988	0.994
194	0.921	0.921	0.923	0.928	0.960	0.980	0.988	0.994
196	0.922	0.922	0.924	0.928	0.960	0.980	0.988	0.995
198	0.922	0.923	0.925	0.929	0.960	0.980	0.988	0.995
200	0.923	0.924	0.925	0.930	0.961	0.980	0.988	0.995

2

RUTHERFORD SCATTERING CROSS SECTION OF THE ELEMENTS FOR 1 MeV ^{4}He^a

ELE-MENT	AT. NO. (Z_2)	AVRG. MASS (amu)	$d\sigma/d\Omega$ in 10^{-24} cm^2/Steradian							
			179.5°	170°	160°	145°	90°	60°	45°	30°
BE	4	9.01	0.053	0.055	0.058	0.069	0.297	1.294	3.836	18.454
B	5	10.81	0.097	0.098	0.104	0.122	0.482	2.038	6.008	28.848
C	6	12.01	0.147	0.150	0.159	0.185	0.704	2.944	8.661	41.550
N	7	14.01	0.214	0.218	0.230	0.266	0.974	4.023	11.803	56.568
O	8	16.00	0.291	0.297	0.312	0.360	1.285	5.267	15.429	73.896
F	9	19.00	0.384	0.390	0.410	0.471	1.642	6.681	19.542	93.540
NE	10	20.18	0.478	0.486	0.511	0.586	2.032	8.254	24.131	115.486
NA	11	22.99	0.590	0.599	0.629	0.720	2.471	9.998	29.210	139.749
MG	12	24.31	0.707	0.718	0.754	0.862	2.945	11.903	34.767	166.318
AL	13	26.98	0.838	0.851	0.893	1.021	3.466	13.979	40.812	195.201
SI	14	28.09	0.974	0.991	1.039	1.187	4.023	16.216	47.335	226.391
P	15	30.97	1.129	1.146	1.201	1.371	4.626	18.623	54.348	259.896
S	16	32.06	1.285	1.306	1.370	1.563	5.267	21.192	61.838	295.707
CL	17	35.45	1.460	1.483	1.555	1.773	5.954	23.933	69.818	333.833
AR	18	39.95	1.647	1.672	1.752	1.997	6.685	26.840	78.283	374.272
K	19	39.10	1.835	1.861	1.951	2.223	7.446	29.904	87.221	417.011
CA	20	40.08	2.033	2.064	2.163	2.465	8.253	33.136	96.646	462.065
SC	21	44.96	2.249	2.285	2.394	2.727	9.108	36.542	106.561	509.435
TI	22	47.90	2.476	2.513	2.633	2.998	10.001	40.110	116.956	559.113
V	23	50.94	2.706	2.751	2.882	3.281	10.935	43.844	127.835	611.101
CR	24	52.00	2.947	2.997	3.139	3.574	11.909	47.740	139.194	665.397
MN	25	54.94	3.208	3.255	3.410	3.882	12.926	51.806	151.039	722.006
FE	26	55.85	3.469	3.522	3.690	4.200	13.982	56.034	163.365	780.923
CO	27	58.93	3.741	3.802	3.983	4.533	15.082	60.431	176.177	842.153
NI	28	58.71	4.024	4.089	4.283	4.875	16.219	64.990	189.469	905.690
CU	29	63.54	4.330	4.392	4.601	5.235	17.404	69.721	203.250	971.543
ZN	30	65.37	4.633	4.702	4.925	5.604	18.627	74.615	217.511	1039.703
GA	31	69.72	4.947	5.026	5.264	5.989	19.894	79.677	232.258	1110.176
GE	32	72.59	5.272	5.358	5.611	6.384	21.201	84.903	247.486	1182.958
AS	33	74.92	5.606	5.700	5.970	6.791	22.549	90.294	263.198	1258.051
SE	34	78.96	5.970	6.054	6.340	7.213	23.940	95.853	279.394	1335.465
BR	35	79.91	6.326	6.416	6.720	7.644	25.370	101.575	296.072	1415.167
KR	36	83.80	6.693	6.791	7.112	8.090	26.843	107.465	313.235	1497.192
RB	37	85.47	7.070	7.175	7.514	8.547	28.357	113.520	330.880	1581.526
SR	38	87.62	7.457	7.569	7.927	9.017	29.912	119.740	349.009	1668.171
Y	39	88.91	7.855	7.974	8.351	9.499	31.508	126.126	367.620	1757.125
ZR	40	91.22	8.263	8.390	8.786	9.994	33.146	132.679	386.716	1848.391
NB	41	92.91	8.681	8.816	9.232	10.501	34.825	139.397	406.294	1941.967
MO	42	95.94	9.110	9.253	9.690	11.021	36.547	146.282	426.358	2037.855

^{a} Differential scattering cross section, $d\sigma/d\Omega$ in 10^{-24} cm^2/Steradian as calculated from the Rutherford scattering cross section formula, Eq. (2.26). The mass M_1 of ^{4}He is taken as 4.0026 amu; the mass M_2 for the scattering element is an average over the isotopic masses and is listed in the third column.

ELE-MENT	AT. NO. (Z_2)	AVRG. MASS (amu)	$d\sigma/d\Omega$ in 10^{-24} cm²/Steradian							
			179.5°	170°	160°	145°	90°	60°	45°	30°
TC	43	99.00	9.549	9.701	10.159	11.554	38.310	153.333	446.904	2136.053
RU	44	101.10	9.998	10.159	10.639	12.099	40.114	160.549	467.933	2236.560
RH	45	102.90	10.458	10.627	11.129	12.657	41.959	167.930	489.446	2339.379
PD	46	106.40	10.928	11.107	11.631	13.228	43.846	175.479	511.443	2444.508
AG	47	107.90	11.408	11.596	12.143	13.810	45.774	183.192	533.922	2551.948
CD	48	112.40	11.899	12.097	12.668	14.406	47.746	191.073	556.886	2661.699
IN	49	114.80	12.400	12.608	13.203	15.014	49.757	199.119	580.333	2773.760
SN	50	118.70	12.951	13.130	13.749	15.635	51.811	207.331	604.264	2888.132
SB	51	121.80	13.475	13.662	14.306	16.269	53.905	215.709	628.678	3004.814
TE	52	127.60	14.008	14.205	14.875	16.916	56.043	224.254	653.576	3123.808
I	53	126.90	14.552	14.757	15.452	17.572	58.219	232.961	678.955	3245.109
XE	54	131.30	15.107	15.321	16.043	18.243	60.438	241.837	704.820	3368.723
CS	55	132.90	15.671	15.894	16.643	18.926	62.698	250.878	731.167	3494.647
BA	56	137.30	16.246	16.479	17.256	19.622	65.001	260.085	757.998	3622.882
LA	57	138.90	16.832	17.074	17.878	20.330	67.343	269.458	785.312	3753.426
CE	58	140.10	17.427	17.678	18.512	21.050	69.728	278.996	813.108	3886.281
PR	59	140.90	18.034	18.294	19.156	21.783	72.153	288.700	841.389	4021.446
ND	60	144.20	18.650	18.920	19.812	22.529	74.621	298.570	870.154	4158.923
PM	61	147.00	19.277	19.557	20.479	23.287	77.130	308.607	899.401	4298.709
SM	62	150.40	19.914	20.205	21.157	24.058	79.681	318.809	929.133	4440.808
EU	63	152.00	20.562	20.863	21.846	24.841	82.273	329.177	959.347	4585.216
GD	64	157.30	21.220	21.532	22.547	25.638	84.907	339.712	990.047	4731.935
TB	65	158.90	21.888	22.211	23.258	26.446	87.582	350.412	1021.228	4880.964
DY	66	162.50	22.567	22.901	23.980	27.267	90.299	361.278	1052.893	5032.304
HO	67	164.90	23.256	23.601	24.713	28.100	93.057	372.309	1085.042	5185.955
ER	68	167.30	23.955	24.312	25.457	28.946	95.856	383.507	1117.673	5341.915
TM	69	168.90	24.665	25.032	26.212	29.805	98.697	394.870	1150.788	5500.186
YB	70	173.00	25.385	25.765	26.979	30.676	101.579	406.400	1184.388	5660.769
LU	71	175.00	26.115	26.507	27.755	31.559	104.503	418.095	1218.470	5823.660
HF	72	178.50	26.856	27.260	28.544	32.456	107.469	429.956	1253.035	5988.864
TA	73	181.00	27.607	28.023	29.343	33.364	110.475	441.983	1288.084	6156.376
W	74	183.90	28.369	28.797	30.153	34.285	113.524	454.176	1323.617	6326.200
RE	75	186.20	29.141	29.581	30.975	35.219	116.613	466.534	1359.633	6498.335
OS	76	190.20	29.923	30.376	31.807	36.166	119.745	479.059	1396.133	6672.781
IR	77	192.20	30.716	31.182	32.650	37.124	122.917	491.750	1433.115	6849.534
PT	78	195.10	31.519	31.997	33.505	38.096	126.132	504.606	1470.581	7028.602
AU	79	197.00	32.332	32.824	34.370	39.079	129.387	517.628	1508.531	7209.979
HG	80	200.60	33.156	33.661	35.246	40.076	132.684	530.817	1546.964	7393.666
TL	81	204.40	34.096	34.509	36.134	41.085	136.023	544.171	1585.881	7579.665
PB	82	207.20	34.943	35.367	37.032	42.106	139.403	557.691	1625.281	7767.972
BI	83	209.00	35.801	36.235	37.942	43.140	142.824	571.377	1665.164	7958.590
PO	84	210.00	36.669	37.114	38.862	44.186	146.287	585.228	1705.531	8151.519
AT	85	210.00	37.547	38.002	39.792	45.244	149.791	599.245	1746.380	8346.758
RN	86	222.00	38.435	38.905	40.737	46.318	153.339	613.431	1787.716	8544.311
FR	87	223.00	39.335	39.815	41.690	47.402	156.926	627.780	1829.533	8744.169
RA	88	226.00	40.244	40.736	42.655	48.498	160.555	642.295	1871.833	8946.343
AC	89	227.00	41.164	41.668	43.630	49.607	164.225	656.976	1914.617	9150.824
TH	90	232.00	42.094	42.611	44.617	50.729	167.937	671.823	1957.885	9357.617
PA	91	231.00	43.035	43.562	45.614	51.863	171.689	686.836	2001.635	9566.719
U	92	238.00	43.986	44.527	46.623	53.010	175.485	702.015	2045.870	9778.132

3

⁴He STOPPING CROSS SECTIONS[a]

Atom	No.	400	600	800	1000	1200	1400	1600
H	1	14.02	14.11	13.5	12.49	11.34	10.19	9.154
He	2	16.72	17.88	18.03	17.52	16.63	15.56	14.46
Li	3	22.28	21.99	21.46	20.64	19.6	18.42	17.28
Be	4	27.09	26.76	25.89	24.71	23.4	22.06	20.8
B	5	32.6	33.49	32.67	31.27	29.48	27.59	25.74
C	6	33.32	36.58	37.21	36.19	34.27	31.99	29.72
N	7	46.23	48.45	48.12	46.24	43.54	40.58	37.71
O	8	44.34	47.72	48.39	47.34	45.29	42.81	40.27
F	9	40.07	43.99	45.66	45.73	44.76	43.15	41.24
Ne	10	39.32	43.59	45.54	45.86	45.1	43.68	41.92
Na	11	42.02	44.08	44.95	44.88	44.24	43.14	41.96
Mg	12	56.04	57.26	56.78	55.26	53.21	50.99	48.82
Al	13	55.39	54.86	53.81	52.43	50.85	49.18	47.5
Si	14	70.15	71.09	69.44	66.3	62.5	58.62	55.02
P	15	64.66	68.45	67.57	65.13	62.06	58.88	55.87
S	16	62.12	68.61	69.72	67.75	64.72	61.48	58.37
Cl	17	83.26	86.5	84.96	80.68	75.18	69.52	64.36
Ar	18	83.61	88.7	87.82	83.47	77.52	71.27	65.56
K	19	83.16	89.08	90.61	88.88	85.56	80.68	75.61
Ca	20	93.78	97.3	97.14	94.47	90.5	85.61	80.58
Sc	21	92.58	96.27	96.3	93.86	90.42	85.73	81.12
Ti	22	91.07	95.41	95.76	93.54	89.87	85.55	81.14
V	23	86.19	90.13	90.55	88.7	85.58	81.89	78.13
Cr	24	79.42	84.81	86.62	85.97	83.76	80.68	77.24
Mn	25	77.08	82.69	84.4	83.6	81.35	78.41	75.3
Fe	26	80.15	86.9	89.26	88.64	86.13	82.59	78.65
Co	27	72.11	79.07	82.04	82.29	80.82	78.38	75.5
Ni	28	68.29	74.6	77.74	78.66	78.07	76.56	74.54
Cu	29	62.41	68.2	71.77	73.58	74.05	73.5	72.24
Zn	30	65.53	70.47	72.85	73.47	72.98	71.71	69.97
Ga	31	74.23	78.12	79.41	79.12	77.79	75.86	73.6
Ge	32	77.76	81.9	82.76	82.1	80.26	77.76	75.18
As	33	81.41	87.03	87.98	87.02	84.84	82.01	79.14
Se	34	83.2	89.4	89.8	87.8	84.9	81.6	78.4
Br	35	95.55	101.1	101.1	97.91	93.04	87.7	82.65

[a] Semiempirical values of the stopping cross section, ϵ, in eV/(10^{15} atoms/cm²) for ⁴He at energies from 400 to 4000 keV in all elements.

  Data from J. F. Ziegler and W. K. Chu, At. and Nucl. Data Tables **13**, 481 (1974).

Atom	No.	400	600	800	1000	1200	1400	1600
Kr	36	102.2	108.2	108	104.2	98.67	92.74	87.26
Rb	37	98.18	108.3	110.1	107.4	102.6	97.34	92.4
Sr	38	109	117	117.4	114.2	109	103.1	97.75
Y	39	110	120.4	121.1	117.3	111.6	105.5	99.6
Zr	40	115.4	126	126.8	123.2	117.9	112	106.2
Nb	41	118.1	128.2	128.7	125.1	119.8	114	108.4
Mo	42	109.8	120.5	122.2	119.6	115.1	110	104.8
Tc	43	116	126.8	128.9	126.3	121.2	115.4	109.2
Ru	44	104.1	116.8	120.5	119.5	116	111.3	105.8
Rh	45	100.9	113.6	117.7	117.2	113.9	109.5	104.6
Pd	46	89.09	104.9	111.9	112.9	110.3	105.8	100.5
Ag	47	88.63	101.9	108.4	110.2	108.8	105.4	100.9
Cd	48	96.33	107	112	113	111.4	108.1	103.8
In	49	104.3	110.1	113.7	115.2	114.8	112.7	109.3
Sn	50	108.2	115.8	118.6	118.3	115.8	112.1	107.9
Sb	51	116.2	122.2	122.2	119.9	116.8	113.3	110
Te	52	121.3	127.2	126.5	123.4	119.4	115.3	111.2
I	53	135	141.7	141	135.8	128.5	120.6	113.2
Xe	54	144.7	149.7	148.2	143	136	128.7	122
Cs	55	129.7	141.5	143.1	139.7	134.4	128.7	123.2
Ba	56	141.2	150.7	151.4	147.4	141.3	134.9	128.7
La	57	144.7	156.5	156.9	152.3	145.7	138.6	131.7
Ce	58	136.4	146.1	147.7	144.5	139.1	133.2	127.4
Pr	59	134.1	143.8	145.7	142.9	137.8	132.2	126.5
Nd	60	131.9	141.6	143.5	141	136.5	131	125.5
Pm	61	129.7	139.4	141.4	139.2	135.1	129.8	124.4
Sm	62	127.7	137.7	139.4	137.4	133.2	128.4	123.3
Eu	63	125.8	135.6	137.4	135.6	131.7	127.1	122.1
Gd	64	130.1	139.9	141.7	139.9	135.8	131.1	125.8
Tb	65	122.2	131.7	133.6	132.2	128.7	124.5	119.7
Dy	66	111.5	123.9	128.1	127.9	125.4	121.8	117.7
Ho	67	107.5	118.4	122.4	122.3	120.1	117	113.2
Er	68	106.1	116.8	120.8	120.7	118.6	115.7	112
Tm	69	104.7	115.2	119.2	119.2	117.2	114.1	110.8
Yb	70	103.5	113.8	117.7	117.8	115.9	113	109.7
Lu	71	106.3	116.9	120.4	120.2	118.1	115	111.6
Hf	72	109.7	120.8	124.5	124.3	122.2	118.9	115.4
Ta	73	105.8	117.5	121.7	121.8	119.8	116.7	113.1
W	74	103.4	114.2	118	118.2	116.5	113.9	110.9
Re	75	114.4	125.8	129.8	129.8	127.3	124.1	120.4
Os	76	112.5	124.5	129	129.5	127.3	124.3	120.7
Ir	77	110.7	123.2	128.2	129.3	127.3	124.5	121.1
Pt	78	103.1	117.6	124.2	126.2	125.6	123.7	121.2
Au	79	109.9	122.7	128	129.1	127.9	125.3	122.3
Hg	80	103.5	116.9	122.7	124.2	123.5	121.6	119.2
Tl	81	113.4	125	129.5	130	128.4	125.8	122.9
Pb	82	126.4	138.1	141.9	141.6	139.1	135.9	132.4
Bi	83	124.6	136	139.2	138.3	135.4	131.9	128.3
Po	84	127.3	140	143	141.8	138.4	134.6	130.6
At	85	128.2	142.7	146.1	145.1	141.1	137	132.7
Rn	86	127.7	144.4	148.7	147.7	143.5	139.2	134.7
Fr	87	143.7	158.2	160.9	158.5	153	147.6	142.2
Ra	88	155.2	167.8	169.8	166.1	160.1	154	147.9
Ac	89	158.1	171.3	173.8	170	164.1	157.6	151
Th	90	159.4	173.4	176.6	173.2	167.2	160.5	153.7
Pa	91	153.1	166.8	170.4	168	162.8	156.8	150.8
U	92	150.7	164.4	168.4	166.6	161.9	156.1	150.3

Atom	No.	1800	2000	2400	2800	3200	3600	4000
H	1	8.289	7.606	6.75	6.081	5.534	5.108	4.683
He	2	13.44	12.52	11.12	10.02	9.117	8.416	7.714
Li	3	16.24	15.35	13.63	12.3	11.22	10.33	9.587
Be	4	19.65	18.64	16.55	14.93	13.61	12.61	11.68
B	5	24.1	22.7	20.09	18.09	16.51	15.22	14.14
C	6	27.68	25.97	23.1	20.8	18.99	17.59	16.36
N	7	35.15	32.98	29.39	26.68	24.45	22.6	21.05
O	8	37.91	35.84	32.39	29.54	27.21	25.18	23.55
F	9	39.24	37.31	34.44	31.79	29.46	27.45	25.75
Ne	10	40.06	38.24	35.79	33.56	31.55	29.66	27.99
Na	11	40.73	39.56	36.57	34.04	31.86	29.95	28.24
Mg	12	46.85	45.11	41.57	38.67	36.2	34.16	32.33
Al	13	45.85	44.25	40.38	37.38	34.96	32.92	31.08
Si	14	51.88	49.26	44.71	41.27	38.44	36.11	34.09
P	15	53.08	50.67	45.88	42.17	39.14	36.66	34.55
S	16	55.41	52.89	47.81	43.88	40.68	38.06	35.83
Cl	17	60.04	56.65	51.15	46.98	43.53	40.68	38.34
Ar	18	60.81	57.13	51.75	47.56	44.07	41.18	38.79
K	19	70.84	66.64	60.13	54.97	50.8	47.36	44.45
Ca	20	75.64	71.19	64.19	58.63	54.1	50.37	47.21
Sc	21	76.65	72.53	65.71	60.17	55.6	51.8	48.58
Ti	22	77.0	73.31	67.61	62.56	58.04	54.17	50.84
V	23	74.6	71.45	66.28	61.75	57.64	53.95	50.79
Cr	24	73.78	70.51	66.16	62.13	58.54	55.17	52.12
Mn	25	72.36	69.72	65.29	61.39	58.01	54.85	52.0
Fe	26	74.71	71.05	66.57	62.86	59.59	56.64	54.02
Co	27	72.55	69.75	65.5	62.01	58.96	56.23	53.73
Ni	28	72.3	70.04	66.0	61.0	56.55	52.64	49.94
Cu	29	70.51	68.48	64.9	62.0	59.34	56.91	54.72
Zn	30	68.12	66.04	62.25	59.06	56.35	53.91	51.77
Ga	31	71.25	68.82	64.46	60.89	57.91	55.29	53.0
Ge	32	72.46	69.8	65.08	61.25	58.1	55.38	53.02
As	33	75.97	73.05	67.85	63.66	60.24	57.32	54.8
Se	34	75.3	72.4	67.06	62.88	59.43	56.5	53.99
Br	35	78.29	74.75	69.06	64.64	50.96	57.9	55.17
Kr	36	82.66	79.04	63.04	68.14	64.21	60.94	58.0
Rb	37	87.67	83.47	76.51	71.01	66.59	62.85	59.7
Sr	38	92.56	87.93	80.3	74.20	69.51	65.46	62.04
Y	39	93.9	88.8	84.35	77.65	72.11	67.92	64.37
Zr	40	100.7	97.5	87.64	81.02	75.72	71.2	67.44
Nb	41	103.0	98.1	90.26	83.66	78.21	73.66	69.88
Mo	42	99.88	95.21	88.09	81.86	76.49	72.01	68.18
Tc	43	103.3	97.85	90.14	83.58	78.09	73.43	69.48
Ru	44	100.7	95.86	89.05	82.95	77.71	73.21	69.35
Rh	45	99.85	95.42	89.21	83.44	78.33	73.87	70.03
Pd	46	95.32	90.65	86.26	81.88	77.72	63.77	70.04

Atom	No.	1800	2000	2400	2800	3200	3600	4000
Ag	47	96.02	91.22	86.66	82.54	78.53	73.98	68.91
Cd	48	99.24	94.71	89.47	84.81	80.43	76.31	72.53
In	49	105.0	100.0	94.46	89.9	85.69	81.81	78.17
Sn	50	103.6	99.49	93.82	89.04	84.81	81.14	77.7
Sb	51	106.8	103.7	97.58	92.45	88.01	84.26	80.61
Te	52	107.4	103.9	97.5	92.23	87.74	83.82	80.34
I	53	106.7	101.5	95.02	89.84	85.31	81.53	78.07
Xe	54	116.2	111.5	104.1	98.28	93.26	89.05	85.31
Cs	55	117.9	113.1	105.0	98.36	92.91	88.36	84.29
Ba	56	122.9	117.6	108.7	101.6	95.71	90.86	86.53
La	57	125.1	119.0	109.9	102.6	96.71	91.72	87.5
Ce	58	121.8	116.7	108.0	101.0	95.22	90.56	86.27
Pr	59	121.0	116.0	107.5	100.5	94.84	90.27	86.02
Nd	60	120.2	115.3	106.9	99.99	94.4	89.8	85.71
Pm	61	119.2	114.5	106.2	99.41	93.91	89.35	85.35
Sm	62	118.3	113.6	105.5	98.8	93.37	88.86	84.96
Eu	63	117.3	112.7	104.7	98.15	92.81	88.34	84.53
Gd	64	120.8	116.0	107.7	100.8	95.23	90.57	86.59
Tb	65	115.2	110.8	103.1	96.8	91.62	87.24	83.63
Dy	66	113.5	109.5	102.3	96.4	91.36	87.11	83.41
Ho	67	109.5	105.7	98.56	92.71	87.83	83.66	80.21
Er	68	108.4	104.8	97.7	91.98	87.17	83.07	79.65
Tm	69	107.2	103.7	96.82	91.22	86.46	82.43	79.06
Yb	70	106.2	102.9	96.06	90.57	85.87	81.89	78.54
Lu	71	108.0	104.5	97.42	91.75	86.89	82.79	79.34
Hf	72	111.6	108.0	100.6	94.7	89.62	85.32	81.72
Ta	73	109.4	105.6	98.96	93.14	88.4	84.19	80.53
W	74	107.6	104.4	97.87	92.17	87.42	83.2	79.61
Re	75	116.5	112.7	105.2	98.97	93.58	88.98	85.12
Os	76	117.0	113.3	106.0	99.75	94.32	89.69	85.79
Ir	77	117.7	114.0	106.8	100.7	95.29	90.62	86.67
Pt	78	118.2	115.0	108.4	102.5	97.21	92.6	88.64
Au	79	118.9	115.5	110.0	104.9	99.92	95.42	91.39
Hg	80	116.3	113.3	107.1	101.5	96.45	91.9	87.95
Tl	81	119.6	116.4	109.8	104.0	98.82	94.11	89.99
Pb	82	128.7	125.0	117.7	111.5	106.0	100.9	96.47
Bi	83	124.5	120.8	113.6	107.6	102.2	97.4	93.07
Po	84	126.5	122.6	115.2	109.0	103.6	98.76	94.38
At	85	128.5	124.4	116.7	110.4	104.9	100.0	95.6
Rn	86	130.3	126.0	118.1	111.7	106.0	101.2	96.72
Fr	87	137.0	132.1	123.2	116.0	109.9	104.7	99.92
Ra	88	142.1	136.8	127.1	119.4	112.9	107.4	102.5
Ac	89	144.9	139.4	129.2	121.3	114.5	108.9	103.9
Th	90	147.4	141.7	131.2	123.0	116.0	110.2	105.2
Pa	91	145.0	139.5	129.6	121.8	115.1	109.5	104.7
U	92	144.6	139.3	129.5	121.7	115.2	109.7	105.0

4

ELECTRON CONFIGURATIONS AND IONIZATION POTENTIALS OF ATOMS[a]

Z	Element	1s	2s 2p	3s 3p 3d	4s 4p 4d 4f	5s 5p 5d	6s 6p 6d	7s	Ionization potential, eV[b]
1	H	1							13.595
2	He	2							24.580
3	Li	2	1						5.390
4	Be	2	2						9.320
5	B	2	2 1						8.296
6	C	2	2 2						11.260
7	N	2	2 3						14.532
8	O	2	2 4						13.614
9	F	2	2 5						17.422
10	Ne	2	2 6						21.564
11	Na	2	2 6	1					5.138
12	Mg	2	2 6	2					7.644
13	Al	2	2 6	2 1					5.984
14	Si	2	2 6	2 2					8.149
15	P	2	2 6	2 3					10.486
16	S	2	2 6	2 4					10.357
17	Cl	2	2 6	2 5					12.967
18	Ar	2	2 6	2 6					15.759
19	K	2	2 6	2 6	1				4.339
20	Ca	2	2 6	2 6	2				6.111
21	Sc	2	2 6	2 6 1	2				6.540
22	Ti	2	2 6	2 6 2	2				6.280
23	V	2	2 6	2 6 3	2				6.740
24	Cr	2	2 6	2 6 5	1				6.764
25	Mn	2	2 6	2 6 5	2				7.432
26	Fe	2	2 6	2 6 6	2				7.870
27	Co	2	2 6	2 6 7	2				7.864
28	Ni	2	2 6	2 6 8	2				7.633
29	Cu	2	2 6	2 6 10	1				7.724
30	Zn	2	2 6	2 6 10	2				9.391
31	Ga	2	2 6	2 6 10	2 1				6.00
32	Ge	2	2 6	2 6 10	2 2				7.88

(Continued)

[a] Occupied electron orbitals for elements.
[b] Energy required to singly ionize the atom.

321

Z	Element	1s	2s	2p	3s	3p	3d	4s	4p	4d	4f	5s	5p	5d	5f	6s	6p	6d	7s	Ionization potential, eV[b]
33	As	2	2	6	2	6	10	2	3											9.81
34	Se	2	2	6	2	6	10	2	4											9.75
35	Br	2	2	6	2	6	10	2	5											11.84
36	Kr	2	2	6	2	6	10	2	6											13.996
37	Rb	2	2	6	2	6	10	2	6			1								4.176
38	Sr	2	2	6	2	6	10	2	6			2								5.692
39	Y	2	2	6	2	6	10	2	6	1		2								6.377
40	Zr	2	2	6	2	6	10	2	6	2		2								6.835
41	Nb	2	2	6	2	6	10	2	6	4		1								6.881
42	Mo	2	2	6	2	6	10	2	6	5		1								7.131
43	Tc	2	2	6	2	6	10	2	6	(5)		(2)?								7.23
44	Ru	2	2	6	2	6	10	2	6	7		1								7.365
45	Rh	2	2	6	2	6	10	2	6	8		1								7.461
46	Pd	2	2	6	2	6	10	2	6	10										8.33
47	Ag	2	2	6	2	6	10	2	6	10		1								7.574
48	Cd	2	2	6	2	6	10	2	6	10		2								8.991
49	In	2	2	6	2	6	10	2	6	10		2	1							5.785
50	Sn	2	2	6	2	6	10	2	6	10		2	2							7.332
51	Sb	2	2	6	2	6	10	2	6	10		2	3							8.639
52	Te	2	2	6	2	6	10	2	6	10		2	4							9.01
53	I	2	2	6	2	6	10	2	6	10		2	5							10.44
54	Xe	2	2	6	2	6	10	2	6	10		2	6							12.127
55	Cs	2	2	6	2	6	10	2	6	10		2	6			1				3.893
56	Ba	2	2	6	2	6	10	2	6	10		2	6			2				5.210
57	La	2	2	6	2	6	10	2	6	10		2	6	1		2				5.61
58	Ce	2	2	6	2	6	10	2	6	10	1	2	6	1		2				6.91
59	Pr	2	2	6	2	6	10	2	6	10	3	2	6			2				5.70
60	Nd	2	2	6	2	6	10	2	6	10	4	2	6			2				6.31
61	Pm	2	2	6	2	6	10	2	6	10	5	2	6			2				
62	Sm	2	2	6	2	6	10	2	6	10	6	2	6			2				5.6
63	Eu	2	2	6	2	6	10	2	6	10	7	2	6			2				5.67
64	Gd	2	2	6	2	6	10	2	6	10	7	2	6	1		2				6.16
65	Tb	2	2	6	2	6	10	2	6	10	8	2	6	1		2				6.74
66	Dy	2	2	6	2	6	10	2	6	10	9	2	6	1		2		?		6.32
67	Ho	2	2	6	2	6	10	2	6	10	10	2	6	1		2		?		
68	Er	2	2	6	2	6	10	2	6	10	11	2	6			2		?		
69	Tm	2	2	6	2	6	10	2	6	10	13	2	6			2				
70	Yb	2	2	6	2	6	10	2	6	10	14	2	6			2				6.22
71	Lu	2	2	6	2	6	10	2	6	10	14	2	6	1		2				6.15
72	Hf	2	2	6	2	6	10	2	6	10	14	2	6	2		2				5.5
73	Ta	2	2	6	2	6	10	2	6	10	14	2	6	3		2				7.7
74	W	2	2	6	2	6	10	2	6	10	14	2	6	4		2				7.98
75	Re	2	2	6	2	6	10	2	6	10	14	2	6	5		2				7.87
76	Os	2	2	6	2	6	10	2	6	10	14	2	6	6		2				8.7
77	Ir	2	2	6	2	6	10	2	6	10	14	2	6	7		2				9.2
78	Pt	2	2	6	2	6	10	2	6	10	14	2	6	9		1		?		9.0
79	Au	2	2	6	2	6	10	2	6	10	14	2	6	10		1				9.22
80	Hg	2	2	6	2	6	10	2	6	10	14	2	6	10		2				10.434
81	Tl	2	2	6	2	6	10	2	6	10	14	2	6	10		2	1			6.106
82	Pb	2	2	6	2	6	10	2	6	10	14	2	6	10		2	2			7.415
83	Bi	2	2	6	2	6	10	2	6	10	14	2	6	10		2	3			7.287
84	Po	2	2	6	2	6	10	2	6	10	14	2	6	10		2	4			8.43
85	At	2	2	6	2	6	10	2	6	10	14	2	6	10		2	5			9.2
86	Rn	2	2	6	2	6	10	2	6	10	14	2	6	10		2	6			10.745
87	Fr	2	2	6	2	6	10	2	6	10	14	2	6	10		2	6		1	4.0
88	Ra	2	2	6	2	6	10	2	6	10	14	2	6	10		2	6		2	5.277
89	Ac	2	2	6	2	6	10	2	6	10	14	2	6	10		2	6	1	2?	6.9
90	Th	2	2	6	2	6	10	2	6	10	14	2	6	10		2	6	2	2?	
91	Pa	2	2	6	2	6	10	2	6	10	14	2	6	10	2	2	6	1	2	
92	U	2	2	6	2	6	10	2	6	10	14	2	6	10	3	2	6	1	2?	4
93	Np	2	2	6	2	6	10	2	6	10	14	2	6	10	4	2	6	1	2?	
94	Pu	2	2	6	2	6	10	2	6	10	14	2	6	10	5	2	6		2	
95	Am	2	2	6	2	6	10	2	6	10	14	2	6	10	6	2	6		2	
96	Cm	2	2	6	2	6	10	2	6	10	14	2	6	10	7	2	6	1	2?	
97	Bk	2	2	6	2	6	10	2	6	10	14	2	6	10	8	2	6	1	2?	
98	Cf	2	2	6	2	6	10	2	6	10	14	2	6	10	9	2	6	1	2?	

5

ELECTRON BINDING ENERGIES (eV)[a]

	$1s_{1/2}$ K	$2s_{1/2}$ L_I	$2p_{1/2}$ L_{II}	$2p_{3/2}$ L_{III}	$3s_{1/2}$ M_I	$3p_{1/2}$ M_{II}	$3p_{3/2}$ M_{III}	$3d_{3/2}$ M_{IV}	$3d_{3/2}$ M_V	$4s_{1/2}$ N_I	$4p_{1/2}$ N_{II}	$4p_{3/2}$ N_{III}	$4d_{3/2}$ N_{IV}	$4d_{5/2}$ N_V	$4f_{3/2}$ N_{VI}	$4f_{1/2}$ N_{VII}
1 H	14															
2 He	25															
3 Li	55															
4 Be	111															
5 B	188			5												
6 C	284			7												
7 N	399			9												
8 O	532	24		7												
9 F	686	31		9												
10 Ne	867	45		18												
11 Na	1 072	63		31	1											
12 Mg	1 305	89		52	2											
13 Al	1 560	118	74	73	1											
14 Si	1 839	149	100	99	8	3										
15 P	2 149	189	136	135	16	10										
16 S	2 472	229	165	164	16	8										
17 Cl	2 823	270	202	200	18	7										
18 A	3 203	320	247	245	25	12										
19 K	3 608	377	297	294	34	18										
20 Ca	4 038	438	350	347	44	26			5							
21 Sc	4 493	500	407	402	54	32			7							
22 Ti	4 965	564	461	455	59	34			3							
23 V	5 465	628	520	513	66	38			2							
24 Cr	5 989	695	584	575	74	43			2							
25 Mn	6 539	769	652	641	84	49			4							
26 Fe	7 114	846	723	710	95	56			6							
27 Co	7 709	926	794	779	101	60			3							
28 Ni	8 333	1 008	872	855	112	68			4							
29 Cu	8 979	1 096	951	931	120	74			2							
30 Zn	9 659	1 194	1 044	1 021	137	87			9							
31 Ga	10 367	1 298	1 143	1 116	158	107	103	18				1				
32 Ge	11 104	1 413	1 249	1 217	181	129	122	29				3				
33 As	11 867	1 527	1 359	1 323	204	147	141	41				3				
34 Se	12 658	1 654	1 476	1 436	232	168	162	57				6				
35 Br	13 474	1 782	1 596	1 550	257	189	182	70	69	27		5				
36 Kr	14 326	1 921	1 727	1 675	289	223	214	89		24		11				
37 Rb	15 200	2 065	1 864	1 805	322	248	239	112	111	30	15	14				
38 Sr	16 105	2 216	2 007	1 940	358	280	269	135	133	38		20				
39 Y	17 039	2 373	2 155	2 080	395	313	301	160	158	46		26		3		
40 Zr	17 998	2 532	2 307	2 223	431	345	331	183	180	52		29		3		
41 Nb	18 986	2 698	2 465	2 371	469	379	363	208	205	58		34		4		
42 Mo	20 000	2 866	2 625	2 520	505	410	393	230	227	62		35		2		
43 Tc	21 044	3 043	2 793	2 677	544	445	425	257	253	68		39		2		
44 Ru	22 117	3 224	2 967	2 838	585	483	461	284	279	75		43		2		
45 Rh	23 220	3 412	3 146	3 004	627	521	496	312	307	81		48		3		

[a] Binding energies of electrons in the various shells of the elements.
From K. Siegbahn et al., *ESCA, Atomic, Molecular and Solid State Structure Studied by Means of Electron Spectroscopy*, Almquist and Wiksells, Uppsala (1967).

	$1s_{1/2}$ K	$2s_{1/2}$ L_I	$2p_{1/2}$ L_{II}	$2p_{3/2}$ L_{III}	$3s_{1/2}$ M_I	$3p_{1/2}$ M_{II}	$3p_{3/2}$ M_{III}	$3d_{3/2}$ M_{IV}	$3d_{5/2}$ M_V	$4s_{1/2}$ N_I	$4p_{1/2}$ N_{II}	$4p_{3/2}$ N_{III}	$4d_{3/2}$ N_{IV}	$4d_{5/2}$ N_V	$4f_{3/2}$ N_{VI}	$4f_{1/2}$ N_{VII}
46 Pd	24 350	3 605	3 331	3 173	670	559	531	340	335	86	51			1		
47 Ag	25 514	3 806	3 524	3 351	717	602	571	373	367	95	62	56		3		
48 Cd	26 711	4 018	3 727	3 538	770	651	617	411	404	108	67			9		
49 In	27 940	4 238	3 938	3 730	826	702	664	451	443	122	77			16		
50 Sn	29 200	4 465	4 156	3 929	884	757	715	494	485	137	89			24		
51 Sb	30 491	4 699	4 381	4 132	944	812	766	537	528	152	99			32		
52 Te	31 814	4 939	4 612	4 341	1 006	870	819	582	572	168	110			40		
53 I	33 170	5 188	4 852	4 557	1 072	931	875	631	620	186	123			50		
54 Xe	34 561	5 453	5 104	4 782	1 145	999	937	685	672	208	147			63		
55 Cs	35 985	5 713	5 360	5 012	1 217	1 065	998	740	726	231	172	162	79	77		
56 Ba	37 441	5 987	5 624	5 247	1 293	1 137	1 063	796	781	253	192	180	93	90		
57 La	38 925	6 267	5 891	5 483	1 362	1 205	1 124	849	832	271	206	192	99			
58 Ce	40 444	6 549	6 165	5 724	1 435	1 273	1 186	902	884	290	224	208	111		1	
59 Pr	41 991	6 835	6 441	5 965	1 511	1 338	1 243	951	931	305	237	218	114		2	
60 Nd	43 569	7 126	6 722	6 208	1 576	1 403	1 298	1 000	978	316	244	225	118		2	
61 Pm	45 185	7 428	7 013	6 460	1 650	1 472	1 357	1 052	1 027	331	255	237	121		4	
62 Sm	46 835	7 737	7 312	6 717	1 724	1 542	1 421	1 107	1 081	347	267	249	130		7	
63 Eu	48 519	8 052	7 618	6 977	1 800	1 614	1 481	1 161	1 131	360	284	257	134		0	
64 Gd	50 239	8 376	7 931	7 243	1 881	1 689	1 544	1 218	1 186	376	289	271	141		0	
65 Tb	51 996	8 708	8 252	7 515	1 968	1 768	1 612	1 276	1 242	398	311	286	148		3	
66 Dy	53 788	9 047	8 581	7 790	2 047	1 842	1 676	1 332	1 295	416	332	293	154		4	
67 Ho	55 618	9 395	8 919	8 071	2 128	1 923	1 741	1 391	1 351	436	343	306	161		4	
68 Er	57 486	9 752	9 265	8 358	2 207	2 006	1 812	1 453	1 409	449	366	320	177	168	4	
69 Tm	59 390	10 116	9 618	8 648	2 307	2 090	1 885	1 515	1 468	472	386	337	180		5	
70 Yb	61 332	10 488	9 978	8 943	2 397	2 172	1 949	1 576	1 527	487	396	343	197	184	6	
71 Lu	63 314	10 870	10 349	9 244	2 491	2 264	2 024	1 640	1 589	506	410	359	205	195	7	
72 Hf	65 351	11 272	10 739	9 561	2 601	2 365	2 108	1 716	1 662	538	437	380	224	214	19	18
73 Ta	67 417	11 680	11 136	9 881	2 708	2 469	2 194	1 793	1 735	566	465	405	242	230	27	25
74 W	69 525	12 099	11 542	10 205	2 820	2 575	2 281	1 872	1 810	595	492	426	259	246	37	34
75 Re	71 677	12 527	11 957	10 535	2 932	2 682	2 367	1 949	1 883	625	518	445	274	260	47	45
76 Os	73 871	12 968	12 385	10 871	3 049	2 792	2 458	2 031	1 960	655	547	469	290	273	52	50
77 Ir	76 111	13 419	12 824	11 215	3 174	2 909	2 551	2 116	2 041	690	577	495	312	295	63	60
78 Pt	78 395	13 880	13 273	11 564	3 298	3 027	2 646	2 202	2 121	724	608	519	331	314	74	70
79 Au	80 725	14 353	13 733	11 918	3 425	3 150	2 743	2 291	2 206	759	644	546	352	334	87	83
80 Hg	83 103	14 839	14 209	12 284	3 562	3 279	2 847	2 385	2 295	800	677	571	379	360	103	99
81 Tl	85 531	15 347	14 698	12 657	3 704	3 416	2 957	2 485	2 390	846	722	609	407	386	122	118
82 Pb	88 005	15 861	15 200	13 035	3 851	3 554	3 067	2 586	2 484	894	764	645	435	413	143	138
83 Bi	90 526	16 388	15 709	13 418	3 999	3 697	3 177	2 688	2 580	939	806	679	464	440	163	158
84 Po	93 105	16 939	16 244	13 814	4 149	3 854	3 302	2 798	2 683	995	851	705	500	473	184	
85 At	95 730	17 493	16 785	14 214	4 317	4 008	3 426	2 909	2 787	1 042	886	740	533	507	210	
86 Rn	98 404	18 049	17 337	14 619	4 482	4 159	3 538	3 022	2 892	1 097	929	768	567	541	238	
87 Fr	101 137	18 639	17 906	15 031	4 652	4 327	3 663	3 136	3 000	1 153	980	810	603	577	268	
88 Ra	103 922	19 237	18 484	15 444	4 822	4 490	3 792	3 248	3 105	1 208	1 058	879	636	603	299	
89 Ac	106 755	19 840	19 083	15 871	5 002	4 656	3 909	3 370	3 219	1 269	1 080	890	675	639	319	
90 Th	109 651	20 472	19 693	16 300	5 182	4 831	4 046	3 491	3 332	1 330	1 168	968	714	677	344	335

X-RAY WAVELENGTHS
(IN Å* UNITS)^a

Element	E(Kα)^b (keV)	Kα2 strong	Kα1 very strong	Kβ1 weak	K edge	Lα1	L_III edge
1 H							
2 He							
3 Li	0.05		228		226.5		
4 Be	0.11		114		111		
5 B	0.18		67.6				
6 C	0.28		44.7		43.68		
7 N	0.39		31.6		30.99		
8 O	0.52		23.62		23.32		
9 F	0.68		18.32				
10 Ne	0.85		14.610	14.452	14.3018		
11 Na	1.04		11.9101	11.575	11.569		405
12 Mg	1.25		9.8900	9.521	9.5122		250.7
13 Al	1.49	8.34173	8.33934	7.960	7.94813		170.4
14 Si	1.74	7.12791	7.12542	6.753	6.738		123
15 P	2.01	6.160	6.157	5 796	5.784		94
16 S	2.31	5.37496	5.37216	5.0316	5.0185		
17 Cl	2.62	4.7307	4.7278	4.4034	4.3971		
18 A	2.96	4.19474	4.19180	3.8860	3.87090		
19 K	3.31	3.7445	3.7414	3.4539	3.4365		42.1
20 Ca	3.69	3.36166	3.35839	3.0897	3.0703	36.33	35.49
21 Sc	4.09	3.0342	3.0309	2.7796	2.762	31.35	
22 Ti	4.51	2.75216	2.74851	2.51391	2.49734	27.42	27.29
23 V	4.95	2.50738	2.50356	2.28440	2.2691	24.25	
24 Cr	5.41	2.293606	2.28970	2.08487	2.07020	21.64	20.7
25 Mn	5.90	2.10578	2.101820	1.91021	1.89643	19.45	
26 Fe	6.40	1.939980	1.936042	1.75661	1.74346	17.59	17.525
27 Co	6.93	1.792850	1.788965	1.62079	1.60815	15.972	15.915
28 Ni	7.47	1.661747	1.657910	1.500135	1.48807	14.561	14.525
29 Cu	8.04	1.544390	1.540562	1.392218	1.38059	13.336	13.288
30 Zn	8.63	1.439000	1.435155	1.29525	1.2834	12.254	12.131

^{a} Å* unit is defined by the wavelength of the WK_{α_1} line = 0.2090100 Å*. The Å* unit is believed to be equal to the angstrom Å to within 5 parts per million and involves a conversion factor of 1.002056 ± 0.000005 Å/kX. Because of the still remaining uncertainty in this conversion factor it was decided to introduce the Å* unit. The distinction between Å and Å* is negligible except in work of the very highest accuracy.
^{b} E(K_{α}) is the energy hv of the unresolved K_{α} line to the nearest 0.01 keV.

Data from J. A. Bearden, Rev. Mod. Phys. **39**, 78 (1967).

Taken from B. D. Cullity, *Elements of X-ray Diffraction*, Addison-Wesley, Reading, MA (1978).

Element	$E(K\alpha)^b$ (keV)	$K\alpha_2$ strong	$K\alpha_1$ very strong	$K\beta_1$ weak	K edge	$L\alpha_1$	L_{III} edge
31 Ga	9.24	1.34399	1.340083	1.20789	1.1958	11.292	11.100
32 Ge	9.88	1.258011	1.254054	1.12894	1.11658	10.4361	10.187
33 As	10.53	1.17987	1.17588	1.05730	1.0450	9.6709	9.367
34 Se	11.21	1.10882	1.10477	0.99218	0.97974	8.9900	8.646
35 Br	11.91	1.04382	1.03974	0.93279	0.9204	8.3746	7.984
36 Kr	12.63	0.9841	0.9801	0.8785	0.86552	7.817	7.392
37 Rb	13.38	0.92969	0.925553	0.82868	0.81554	7.3183	6.862
38 Sr	14.14	0.87943	0.87526	0.78292	0.76973	6.8628	6.387
39 Y	14.93	0.83305	0.82884	0.74072	0.72766	6.4488	5.962
40 Zr	15.75	0.79015	0.78593	0.70173	0.68883	6.0705	5.579
41 Nb	16.58	0.75044	0.74620	0.66576	0.65298	5.7243	5.230
42 Mo	17.44	0.713590	0.709300	0.632288	0.61978	5.40655	4.913
43 Tc	18.33	0.67932	0.67502	0.60130	0.58906	5.1148	4.630
44 Ru	19.24	0.647408	0.643083	0.572482	0.56051	4.84575	4.369
45 Rh	20.17	0.617630	0.613279	0.545605	0.53395	4.59743	4.1299
46 Pd	21.12	0.589821	0.585448	0.520520	0.5092	4.36767	3.9074
47 Ag	22.11	0.563798	0.5594075	0.497069	0.48589	4.15443	3.6999
48 Cd	23.11	0.539422	0.535010	0.475105	0.46407	3.95635	3.5047
49 In	24.14	0.516544	0.512113	0.454545	0.44371	3.77192	3.3237
50 Sn	25.20	0.495053	0.490599	0.435236	0.42467	3.59994	3.1557
51 Sb	26.28	0.474827	0.470354	0.417085	0.40668	3.43941	3.0003
52 Te	27.38	0.455784	0.451295	0.399995	0.38974	3.28920	2.8555
53 I	28.51	0.437829	0.433318	0.383905	0.37381	3.14860	2.7196
54 Xe	29.67	0.42087	0.41634	0.36872	0.3584	3.0166	2.5926
55 Cs	30.86	0.404835	0.400290	0.354364	0.34451	2.8924	2.4740
56 Ba	32.07	0.389668	0.385111	0.340811	0.33104	2.77595	2.3629
57 La	33.31	0.375313	0.370737	0.327983	0.31844	2.66570	2.261
58 Ce	34.57	0.361683	0.357092	0.315816	0.30648	2.5615	2.166
59 Pr	35.87	0.348749	0.344140	0.304261	0.29518	2.4630	2.0791
60 Nd	37.19	0.336472	0.331846	0.293299	0.28453	2.3704	1.9967
61 Pm	38.54	0.324803	0.320160	0.28290	0.27431	2.2822	1.9191
62 Sm	39.92	0.313698	0.309040	0.27301	0.26464	2.1998	1.8457
63 Eu	41.33	0.303118	0.298446	0.263577	0.25553	2.1209	1.7761
64 Gd	42.77	0.293038	0.288353	0.25460	0.24681	2.0468	1.7117
65 Tb	44.24	0.283423	0.278724	0.24608	0.23841	1.9765	1.6497
66 Dy	45.73	0.274247	0.269533	0.23788	0.23048	1.90881	1.5916
67 Ho	47.26	0.265486	0.260756	0.23012	0.22291	1.8450	1.5368
68 Er	48.83	0.257110	0.252365	0.22266	0.21567	1.78425	1.4835
69 Tm	50.42	0.249095	0.244338	0.21556	0.20880	1.7268	1.4334
70 Yb	52.04	0.241424	0.236655	0.20884	0.20224	1.67189	1.3862
71 Lu	53.70	0.234081	0.229298	0.20231	0.19585	1.61951	1.3405
72 Hf	55.40	0.227024	0.222227	0.19607	0.18982	1.56958	1.2972
73 Ta	57.11	0.220305	0.215497	0.190089	0.18394	1.52197	1.2553
74 W	58.87	0.213828	0.2090100	0.184374	0.17837	1.47639	1.2155
75 Re	60.67	0.207611	0.202781	0.178880	0.17302	1.43200	1.1773
76 Os	62.50	0.201639	0.196794	0.173611	0.16787	1.39121	1.1408
77 Ir	64.36	0.195904	0.191047	0.168542	0.16292	1.35128	1.1058
78 Pt	66.26	0.190381	0.185511	0.163675	0.15818	1.31304	1.0723
79 Au	68.20	0.185075	0.180195	0.158982	0.153593	1.27640	1.04000
80 Hg	70.18	0.179958	0.175068	0.154487	0.14918	1.24120	1.0091
81 Tl	72.19	0.175036	0.170136	0.150142	0.14495	1.20739	0.9793
82 Pb	74.25	0.170294	0.165376	0.145970	0.140880	1.17501	0.95073
83 Bi	76.34	0.165717	0.160789	0.141948	0.13694	1.14386	0.9234
84 Po	78.48	0.16130	0.15636	0.13807		1.11386	
85 At	80.66	0.15705	0.15210	0.13432		1.08500	
86 Rn	82.88	0.15294	0.14798	0.13069		1.05723	
87 Fr	85.14	0.14896	0.14399	0.12719		1.03049	
88 Ra	87.46	0.14512	0.14014	0.12382		1.00473	0.8028
89 Ac	89.81	0.14141	0.136417	0.12055		0.97993	
90 Th	92.22	0.137829	0.132813	0.117396	0.11307	0.95600	0.7607
91 Pa	94.67	0.134343	0.129325	0.114345		0.93284	
92 U	97.18	0.130968	0.125947	0.111394	0.10723	0.910639	0.7223

MASS ABSORPTION COEFFICIENTS AND DENSITIES^a

Absorber	Density (gm/cm^3)	Mo		Cu		Co		Cr	
		$K\alpha$ 0.711 Å	$K\beta$ 0.632 Å	$K\alpha$ 1.542 Å	$K\beta$ 1.392 Å	$K\alpha$ 1.790 Å	$K\beta$ 1.621 Å	$K\alpha$ 2.291 Å	$K\beta$ 2.085 Å
1 H	0.08375×10^{-3}	0.3727	0.3699	0.3912	0.3882	0.3966	0.3928	0.4116	0.4046
2 He	0.1664×10^{-3}	0.2019	0.1972	0.2835	0.2623	0.3288	0.2966	0.4648	0.4001
3 Li	0.533	0.1968	0.1866	0.4770	0.3939	0.6590	0.5283	1.243	0.9639
4 Be	1.85	0.2451	0.2216	1.007	0.7742	1.522	1.152	3.183	2.388
5 B	2.47	0.3451	0.2928	2.142	1.590	3.357	2.485	7.232	5.385
6 C	2.27 (graphite)	0.5348	0.4285	4.219	3.093	6.683	4.916	14.46	10.76
7 N	1.165×10^{-3}	0.7898	0.6054	7.142	5.215	11.33	8.330	24.42	18.23
8 O	1.332×10^{-3}	1.147	0.8545	11.03	8.062	17.44	12.85	37.19	27.88
9 F	1.696×10^{-3}	1.584	1.154	15.95	11.66	25.12	18.57	53.14	39.99
10 Ne	0.8387×10^{-3}	2.209	1.597	22.13	16.24	34.69	25.72	72.71	54.91
11 Na	0.966	2.939	2.098	30.30	22.23	47.34	35.18	98.48	74.66
12 Mg	1.74	3.979	2.825	40.88	30.08	63.54	47.38	130.8	99.62
13 Al	2.70	5.043	3.585	50.23	37.14	77.54	58.08	158.0	120.7
14 Si	2.33	6.533	4.624	65.32	48.37	100.4	75.44	202.7	155.6
15 P	1.82 (yellow)	7.870	5.569	77.28	57.44	118.0	89.05	235.5	181.6
16 S	2.09	9.625	6.835	92.53	68.90	141.2	106.6	281.9	217.2
17 Cl	3.214×10^{-3}	11.64	8.261	109.2	81.79	164.7	125.3	321.5	250.2
18 A	1.663×10^{-3}	12.62	8.949	119.5	89.34	180.9	137.3	355.5	275.8
19 K	0.862	16.20	11.51	148.4	111.7	222.0	169.9	426.8	334.2
20 Ca	1.53	19.00	13.56	171.4	129.0	257.4	196.4	499.6	389.3
21 Sc	2.99	21.04	15.00	186.0	140.8	275.5	212.2	520.9	410.7
22 Ti	4.51	23.25	16.65	202.4	153.2	300.5	231.0	571.4	449.0
23 V	6.09	25.24	18.07	222.6	168.0	332.7	254.7	75.06	501.0
24 Cr	7.19	29.25	20.99	252.3	191.1	375.0	288.1	85.71	65.79
25 Mn	7.47	31.86	22.89	272.5	206.7	405.1	311.2	96.08	73.75
26 Fe	7.87	37.74	27.21	304.4	233.6	56.25	345.5	113.1	86.77
27 Co	8.8	41.02	29.51	338.6	258.7	62.86	47.71	124.6	96.06
28 Ni	8.91	47.24	34.18	48.83	282.8	73.75	56.05	145.7	112.5
29 Cu	8.93	49.34	35.77	51.54	38.74	78.11	59.22	155.2	119.5
30 Zn	7.13	55.46	40.26	59.51	45.30	88.71	68.00	171.7	133.5
31 Ga	5.91	56.90	41.69	62.13	46.65	94.15	71.39	186.9	144.0
32 Ge	5.32	60.47	44.26	67.92	51.44	102.0	77.79	199.9	154.5
33 As	5.78	65.97	48.57	75.65	57.01	114.0	86.76	224.0	173.3
34 Se	4.81	68.82	51.20	82.89	62.32	125.1	95.11	246.1	190.4
35 Br	3.12 (liquid)	74.68	55.56	90.29	68.07	135.8	103.5	266.2	206.2

^{a} Although these coefficients are given to four significant figures, the actual accuracy is much less; the uncertainty ranges from <2% to >15%, depending on absorber and wavelength.

Data from *International Tables for X-ray Crystallography*, Kynoch Press, Birmingham. England (1952), Vol. 3, pp. 46–56, Vol. 4, pp. 61–66; *Metals Handbook*, American Society for Metals, Cleveland (1948).

Taken from B. D. Cullity, *Elements of X-ray Diffraction*, Addison-Wesley, Reading, MA (1978).

Absorber	Density (gm/cm³)	Mo Kα 0.711 Å	Mo Kβ 0.632 Å	Cu Kα 1.542 Å	Cu Kβ 1.392 Å	Co Kα 1.790 Å	Co Kβ 1.621 Å	Cr Kα 2.291 Å	Cr Kβ 2.085 Å
36 Kr	3.488 × 10⁻³	79.10	58.64	97.02	73.22	145.7	111.2	284.6	220.7
37 Rb	1.53	83.00	62.07	106.3	80.16	159.6	121.8	311.7	241.8
38 Sr	2.58	88.04	65.59	115.3	86.77	173.5	132.2	339.3	263.4
39 Y	4.48	97.56	72.57	127.1	96.19	190.2	145.4	368.9	286.9
40 Zr	6.51	16.10	75.20	136.8	103.3	204.9	156.6	398.6	309.7
41 Nb	8.58	16.96	81.22	148.8	112.3	222.9	170.4	431.9	336.4
42 Mo	10.22	18.44	13.29	158.3	119.7	236.6	181.0	457.4	356.5
43 Tc	11.50	19.78	14.30	167.7	126.9	250.8	191.9	485.5	378.0
44 Ru	12.36	21.33	15.40	180.8	137.0	269.4	206.6	517.9	404.4
45 Rh	12.42	23.05	16.65	194.1	147.1	289.0	221.8	555.2	433.7
46 Pd	12.00	24.42	17.63	205.0	155.6	304.3	234.0	580.9	455.1
47 Ag	10.50	26.38	19.10	218.1	165.8	323.5	248.9	617.4	483.5
48 Cd	8.65	27.73	20.13	229.3	174.0	341.8	262.1	658.8	513.5
49 In	7.29	29.13	21.18	242.1	183.3	362.7	277.1	705.8	548.0
50 Sn	7.29	31.18	22.62	253.3	193.1	374.1	288.7	708.8	556.6
51 Sb	6.69	33.01	23.91	266.5	203.6	391.3	303.1	733.4	578.8
52 Te	6.25	33.92	24 67	273.4	208.4	404.4	311.7	768.9	602.7
53 I	4.95	36.33	26.53	291.7	221.9	434.0	333.2	835.2	650.8
54 Xe	5.495 × 10⁻³	38.31	27.86	309.8	235.9	459.0	353.4	755.4	685.2
55 Cs	1.91 (−10°C)	40.44	29.51	325.4	247.5	483.8	371.5	802.7	725.1
56 Ba	3.59	42.37	31.00	336.1	256.0	499.0	383.6	587.3	644.6
57 La	6.17	45.34	33.10	353.5	270.8	519.0	401.9	222.9	661.5
58 Ce	6.77	48.56	35.54	378.8	289.6	559.1	431.5	240.4	509.4
59 Pr	6.78	50.78	37.09	402.2	306.8	596.2	458.8	260.5	205.3
60 Nd	7.00	53.28	38 88	417.9	319.8	531.7	475.9	271.3	213.4
61 Pm		55.52	40.52	441.1	336.4	401.4	503.2	284.7	223.8
62 Sm	7.54	57.96	42.40	453.5	346.6	411.8	446.3	295.0	231.5
63 Eu	5.25	61.18	44.74	417.9	369.0	165.2	476.9	312.7	244.9
64 Gd	7.87	62.79	45.95	426.7	377.2	169.5	346.7	318.9	250.3
65 Tb	8.27	66.77	48.88	321.9	399.9	178.7	367.1	330.2	265.2
66 Dy	8.53	68.89	50.38	336.6	360.2	184.9	142.9	351.7	275.0
67 Ho	8.80	72.14	52.76	128.4	272.4	189.8	146.3	363.3	283.4
68 Er	9.04	75.61	55.07	134.3	291.7	198.4	153.0	379.7	296.1
69 Tm	9.33	78.98	57.94	140.2	288.5	207.4	159.8	397.0	309.6
70 Yb	6.97	80.23	59.22	144.7	110.9	214.0	164.9	409.6	319.4
71 Lu	9.84	84.18	62.04	152.0	116.5	224.6	173.2	429.5	335.1
72 Hf	13.28	86.33	64.15	157.7	121.0	232.9	179.6	445.3	347.2
73 Ta	16.67	89.51	66.07	161.5	123.9	238.3	183.9	454.7	355.0
74 W	19.25	95.76	70.57	170.5	131.5	249.7	193.7	470.4	369.1
75 Re	21.02	98.74	72.47	178.3	137.3	261.8	202.7	495.5	388.0
76 Os	22.58	100.2	74.13	183.8	141.3	270.3	209.0	512.4	401.1
77 Ir	22.55	103.4	77.20	192.2	147.6	283.4	218.8	539.6	421.6
78 Pt	21.44	108.6	80.23	198.2	151.2	295.2	226.4	571.6	443.9
79 Au	19.28	111.3	82.33	207.8	160.6	303.3	235.7	568.0	446.7
80 Hg	13.55	114.7	85.30	216.2	166.6	317.0	245.7	597.9	468.9
81 Tl	11.87	119.4	88.25	222.2	171.1	326.3	252.7	616.9	483.3
82 Pb	11.34	122.8	90.55	232.1	178.6	340.8	263.8	644.5	504.9
83 Bi	9.80	125.9	93.50	242.9	187.5	355.3	275.8	667.2	524.2
86 Rn	4.40 (liq., −62°C)	117.2	100.7	263.7	203.0	387.1	299.7	731.4	573.2
90 Th	11.72	99.46	73.34	306.8	236.6	449.0	348.4	844.1	663.0
92 U	19.05	96.67	72.63	305.7	236.2	446.3	346.9	774.0	657.1
94 Pu	19.81	48.84	78.99	352.9	271.2	519.6	401.6	803.2	771.6

7

KLL AUGER ENERGIES (eV)[a]

		$2s^0 2p^6$	$2s^1 2p^5$				$2s^2 2p^4$			
		$^1S_0$	$^1P_1$	$^3P_0$	$^3P_1$	$^3P_2$	$^1S_0$	$^1D_2$	$^3P_0$	$^3P_2$
		KL_1L_1	KL_1L_2	KL_1L_2	KL_1L_3	KL_1L_3	KL_2L_2	KL_2L_3	KL_3L_3	KL_3L_3
C	6	0.243	0.252	0.258	0.258	0.258	0.265	0.266	0.267	0.267
N	7	0.356	0.362	0.369	0.369	0.369	0.373	0.375	0.377	0.377
O	8	0.474	0.486	0.495	0.495	0.495	0.504	0.507	0.509	0.509
F	9	0.610	0.627	0.638	0.638	0.638	0.650	0.654	0.657	0.657
Ne	10	0.761	0.781	0.794	0.794	0.794	0.808	0.813	0.816	0.816
Na	11	0.928	0.952	0.967	0.967	0.967	0.984	0.989	0.993	0.993
Mg	12	1.105	1.135	1.151	1.151	1.151	1.172	1.179	1.183	1.183
Al	13	1.301	1.336	1.354	1.354	1.354	1.379	1.387	1.392	1.392
Si	14	1.516	1.554	1.574	1.574	1.575	1.602	1.611	1.616	1.617
P	15	1.742	1.784	1.805	1.806	1.806	1.835	1.845	1.851	1.852
S	16	1.982	2.034	2.057	2.058	2.059	2.096	2.107	2.114	2.115
Cl	17	2.249	2.305	2.329	2.330	2.331	2.370	2.382	2.389	2.391
A	18	2.527	2.586	2.612	2.613	2.614	2.656	2.669	2.677	2.679
K	19	2.815	2.881	2.909	2.910	2.912	2.959	2.973	2.981	2.984
Ca	20	3.122	3.195	3.224	3.225	3.227	3.279	3.294	3.303	3.306
Sc	21	3.456	3.533	3.563	3.564	3.567	3.622	3.638	3.647	3.651
Ti	22	3.799	3.886	3.916	3.919	3.922	3.985	4.002	4.011	4.016
V	23	4.168	4.259	4.290	4.293	4.298	4.362	4.381	4.391	4.397
Cr	24	4.557	4.651	4.683	4.687	4.692	4.757	4.778	4.788	4.795
Mn	25	4.956	5.056	5.089	5.094	5.100	5.169	5.191	5.202	5.211
Fe	26	5.374	5.480	5.514	5.519	5.527	5.598	5.622	5.634	5.644
Co	27	5.808	5.923	5.957	5.964	5.972	6.049	6.075	6.088	6.099
Ni	28	6.264	6.384	6.419	6.426	6.436	6.514	6.542	6.556	6.568
Cu	29	6.732	6.861	6.896	6.905	6.916	7.000	7.030	7.045	7.059
Zn	30	7.214	7.348	7.384	7.394	7.407	7.493	7.526	7.543	7.558

[a] Values are given for all the KLL-type transitions. The notation $2s^0 2p^6$, $2s^1 2p^5$, and $2s^2 2p^4$ indicates the final electron configuration.

From K. Siegbahn et al., *ESCA, Atomic, Molecular and Solid State Structure Studied by Means of Electron Spectroscopy*, Almquist and Wiksells, Uppsala, (1967).

		$2s^02p^6$		$2s^12p^5$				$2s^22p^4$		
		$^1S_0$ KL_1L_1	$^1P_1$ KL_1L_2	$^3P_0$ KL_1L_2	$^3P_1$ KL_1L_3	$^3P_2$ KL_1L_3	$^1S_0$ KL_2L_2	$^1D_2$ KL_2L_3	$^3P_0$ KL_3L_3	$^3P_2$ KL_3L_3
Ga	31	7.712	7.852	7.888	7.900	7.915	8.000	8.037	8.057	8.073
Ge	32	8.216	8.365	8.401	8.416	8.433	8.523	8.563	8.586	8.603
As	33	8.749	8.903	8.939	8.957	8.975	9.063	9.107	9.133	9.152
Se	34	9.283	9.447	9.483	9.504	9.524	9.616	9.665	9.695	9.715
Br	35	9.840	10.014	10.049	10.074	10.096	10.189	10.244	10.279	10.300
Kr	36	10.412	10.594	10.630	10.658	10.682	10.777	10.837	10.877	10.899
Rb	37	10.995	11.186	11.221	11.255	11.280	11.376	11.442	11.487	11.511
Sr	38	11.595	11.795	11.830	11.870	11.897	11.992	12.066	12.118	12.143
Y	39	12.213	12.422	12.457	12.503	12.532	12.626	12.708	12.767	12.793
Zr	40	12.851	13.069	13.104	13.157	13.188	13.279	13.370	13.437	13.464
Nb	41	13.505	13.731	13.766	13.827	13.860	13.948	14.049	14.125	14.153
Mo	42	14.179	14.414	14.449	14.519	14.554	14.639	14.750	14.836	14.865
Te	43	14.867	15.111	15.146	15.226	15.263	15.343	15.466	15.563	15.593
Ru	44	15.574	15.827	15.862	15.952	15.991	16.066	16.202	16.310	16.341
Rh	45	16.298	16.560	16.595	16.697	16.738	16.806	16.956	17.077	17.109
Pd	46	17.040	17.312	17.347	17.462	17.504	17.565	17.729	17.864	17.897
Ag	47	17.797	18.078	18.113	18.242	18.286	18.339	18.519	18.668	18.702
Cd	48	18.568	18.857	18.892	19.037	19.082	19.125	19.322	19.488	19.523
In	49	19.354	19.653	19.688	19.849	19.896	19.930	20.144	20.327	20.364
Sn	50	20.157	20.465	20.501	20.680	20.728	20.750	20.984	21.185	21.223
Sb	51	20.977	21.295	21.331	21.529	21.579	21.588	21.844	22.065	22.104
Te	52	21.814	22.142	22.179	22.398	22.449	22.444	22.722	22.965	23.005
I	53	22.668	23.006	23.043	23.284	23.338	23.316	23.618	23.884	23.925
Xe	54	23.527	23.879	23.916	24.182	24.237	24.201	24.530	24.822	24.863
Cs	55	24.426	24.783	24.820	25.111	25.167	25.109	25.463	25.781	25.823
Ba	56	25.330	25.697	25.735	26.053	26.111	26.033	26.416	26.762	26.805

		$2s^02p^6$		$2s^12p^5$				$2s^22p^4$		
		$^1S_0$ KL_1L_1	$^1P_1$ KL_1L_2	$^3P_0$ KL_1L_2	$^3P_1$ KL_1L_3	$^3P_2$ KL_1L_3	$^1S_0$ KL_2L_2	$^1D_2$ KL_2L_3	$^3P_0$ KL_3L_3	$^3P_2$ KL_3L_3
La	57	26.251	26.631	26.669	27.018	27.077	26.978	27.393	27.769	27.813
Ce	58	27.201	27.590	27.628	28.009	28.069	27.945	28.393	28.802	28.847
Pr	59	28.171	28.572	28.610	29.024	29.086	28.936	29.420	29.863	29.909
Nd	60	29.163	29.574	29.612	30.063	30.126	29.947	30.468	30.948	30.995
Pm	61	30.170	30.592	30.631	31.120	31.184	30.976	31.537	32.056	32.104
Sm	62	31.199	31.631	31.671	32.200	32.266	32.024	32.627	33.186	33.235
Eu	63	32.247	32.690	32.730	33.303	33.370	33.092	33.740	34.345	34.395
Gd	64	33.315	33.769	33.809	34.429	34.497	34.182	34.877	35.528	35.579
Tb	65	34.402	34.868	34.909	35.576	35.646	35.291	36.036	36.736	36.788
Dy	66	35.512	35.988	36.029	36.749	36.820	36.421	37.220	37.972	38.025
Ho	67	36.640	37.127	37.169	37.944	38.016	37.570	38.425	39.234	39.287
Er	68	37.788	38.287	38.329	39.162	39.236	38.740	39.655	40.522	40.576
Tm	69	38.958	39.469	39.512	40.406	40.481	39.934	40.911	41.840	41.895
Yb	70	40.151	40.674	40.716	41.675	41.752	42.192	41.149	43.186	43.242
Lu	71	41.361	41.897	41.940	42.967	43.045	42.383	43.496	44.559	44.617
Hf	72	42.589	43.137	43.181	44.280	44.359	43.635	44.821	45.957	46.015
Ta	73	43.831	44.391	44.436	45.611	45.691	44.900	46.164	47.377	47.436
W	74	45.097	45.671	45.715	46.971	47.053	46.193	47.538	48.831	48.891
Re	75	46.385	46.972	47.018	48.357	48.440	47.507	48.938	50.315	50.376
Os	76	47.690	48.291	48.337	49.767	49.851	48.839	50.361	51.830	51.892
Fr	77	49.022	49.636	49.682	51.205	51.291	50.195	51.812	53.375	53.437
Pt	78	50.375	51.003	51.050	52.672	52.759	51.575	53.292	54.954	55.017

		$2s^02p^6$	$2s^12p^5$					$2s^22p^4$		
		$^1S_0$ KL_1L_1	$^1P_1$ KL_1L_2	$^3P_0$ KL_1L_2	$^3P_1$ KL_1L_3	$^3P_2$ KL_1L_3	$^1S_0$ KL_2L_2	$^1D_2$ KL_2L_3	$^3P_0$ KL_3L_3	$^3P_2$ KL_3L_3
Au	79	51.752	52.393	52.440	54.167	54.255	52.978	54.801	56.568	56.633
Hg	80	53.149	53.802	53.849	55.685	55.774	54.397	56.330	58.206	58.272
Tl	81	54.554	55.227	55.275	57.225	57.316	55.840	57.890	59.882	59.948
Pb	82	55.992	56.677	56.726	58.799	58.891	57.302	59.476	61.591	61.658
Bi	83	57.451	58.155	58.205	60.402	60.495	58.799	61.098	63.338	63.406
Po	84	58.918	59.640	59.690	62.026	62.120	60.299	62.739	65.118	65.187
At	85	60.427	61.163	61.213	63.689	63.784	61.836	64.416	66.935	67.005
Rn	86	61.980	62.720	62.771	65.392	65.489	63.397	66.124	68.789	68.860
Fr	87	63.523	64.286	64.337	67.114	67.212	64.983	67.868	70.690	70.762
Ra	88	65.103	65.887	65.939	68.879	68.978	66.604	69.654	72.640	72.712
Ac	89	66.720	67.509	67.562	70.673	70.774	68.232	71.453	74.611	74.684
Th	90	68.341	69.153	69.207	72.498	72.600	69.898	73.302	76.640	76.714
Pu	91	70.016	70.842	70.896	74.373	74.476	71.599	75.190	78.714	78.789
U	92	71.704	72.550	72.604	76.280	76.384	73.327	77.116	80.839	80.916
Np	93	73.437	74.297	74.351	78.236	78.342	75.085	79.086	83.019	83.096
Pa	94	75.204	76.080	76.135	80.237	80.344	76.884	81.103	85.254	85.332
Am	95	77.060	77.930	77.985	82.317	82.425	78.727	83.177	87.558	87.637
Cm	96	78.867	79.590	79.646	84.386	84.495	80.240	85.099	89.888	89.968
Bk	97	80.594	81.528	81.585	86.408	86.518	82.388	87.331	92.204	92.284
Cf	98	83.286	84.187	84.245	89.453	89.565	85.017	90.348	95.607	95.688
Es	99	85.219	86.146	86.204	91.701	91.814	86.997	92.617	98.165	98.248
Fm	100	87.205	88.144	88.203	93.998	94.113	89.006	94.926	100.774	100.857
Md	101	89.221	90.192	90.251	96.356	96.471	91.085	97.315	103.472	103.556
No	102	91.267	92.260	92.320	98.763	98.880	93.173	99.744	106.240	106.325
Lr	103	93.373	94.388	94.448	101.250	101.368	95.322	102.252	109.108	109.194
Ku	104	95.518	96.555	96.615	103.796	103.915	97.510	104.820	112.055	112.142

TABLE OF THE ELEMENTS

ELE-MENT	AT. NO. (Z)	ISOTOPIC MASS (amu)	RELATIVE ABUN-DANCE	ATOMIC WEIGHT (amu)	ATOMIC DENSITY (atom/cm^3)	SPECIFIC GRAVITY
H	1	1.007825	0.9999	1.008		
He	2	4.002603	1.0000	4.003		
LI	3	6.015125	0.0756	6.940	4.700E+22	0.542
		7.016004	0.9244			
Be	4	9.012186	1.0000	9.012	1.21E+23	1.82
B	5	10.012939	0.1961	10.814	1.30E+23	2.47
		11.009305	0.8039			
C	6	12.000000	0.9889	12.011	1.76E+23	3.516
		13.003354	0.0111			
N	7	14.003074	0.9963	14.007		
		15.000108	0.0037			
C	8	15.994915	0.9976	15.999		
		16.999133	0.0004			
		17.999160	0.0020			
F	9	18.998405	1.0000	18.998		
Ne	10	19.992441	0.9092	20.171	4.36E+22	1.51
		20.993849	0.0026			
		21.991385	0.0882			
Na	11	22.989771	1.0000	22.990	2.652E+22	1.013
Mg	12	23.985042	0.7870	24.310	4.30E+22	1.74
		24.985839	0.1013			
		25.982593	0.1117			
Al	13	26.981539	1.0000	26.982	6.02E+22	2.70

From W. K. Chu, J. W. Mayer, and M. A. Nicolet, *Backscattering Spectrometry*, Academic Press, New York (1978).

333

ELE-MENT	AT. NO. (Z)	ISOTOPIC MASS (amu)	RELATIVE ABUN-DANCE	ATOMIC WEIGHT (amu)	ATOMIC DENSITY (atom/cm^3)	SPECIFIC GRAVITY
Si	14	27.976929	0.9221	28.086	5.00E+22	2.33
		28.976496	0.0470			
		29.973763	0.0309			
P	15	30.973765	1.0000	30.974		
S	16	31.972074	0.9500	32.061		
		32.971462	0.0076			
		33.967865	0.0422			
		35.967090	0.0001			
Cl	17	34.968851	0.7577	35.453		
		36.965899	0.2423			
Ar	18	35.967545	0.0034	39.948	2.66E+22	1.77
		37.962728	0.0006			
		39.962384	0.9960			
K	19	38.963710	0.9310	39.097	1.402E+22	0.910
		39.964000	0.0001			
		40.961832	0.0688			
Ca	20	39.962589	0.9697	40.081	2.30E+22	1.53
		41.958625	0.0064			
		42.958780	0.0015			
		43.955491	0.0206			
		47.952531	0.0019			
Sc	21	44.955919	1.0000	44.956	4.27E+22	2.99
Ti	22	45.952632	0.0793	47.879	5.66E+22	4.51
		46.951769	0.0728			
		47.947950	0.7394			
		48.947870	0.0551			
		49.944786	0.0534			
Y	23	49.947164	0.0024	50.942	7.22E+22	6.09
		50.943961	0.9976			
Cr	24	49.946055	0.0435	51.996	8.33E+22	7.19
		51.940513	0.8376			
		52.940653	0.0951			
		53.983882	0.0238			
Mn	25	54.938050	1.0000	54.938	8.18E+22	7.47
Fe	26	53.939617	0.0582	55.847	8.50E+22	7.87
		55.934936	0.9166			
		56.935398	0.0219			
		47.933282	0.0033			
Co	27	58.933189	1.0000	58.933	8.97E+22	8.9
Ni	28	57.935342	0.6788	58.728	9.14E+22	8.91
		59.930787	0.2623			
		60.931056	0.0119			
		61.928342	0.0366			
		63.927958	0.0108			
Cu	29	62.929592	0.6917	63.546	8.45E+22	8.93
		64.927786	0.3083			

ELE-MENT	AT. NO. (Z)	ISOTOPIC MASS (amu)	RELATIVE ABUN-DANCE	ATOMIC WEIGHT (amu)	ATOMIC DENSITY (atom/cm³)	SPECIFIC GRAVITY
Zn	30	63.929145	0.4889	65.387	6.55E+22	7.13
		65.926052	0.2781			
		66.927145	0.0411			
		67.924857	0.1857			
		69.925334	0.0062			
Ga	31	68.925574	0.6040	69.717	5.10E+22	5.91
		70.924706	0.3960			
Ge	32	69.924252	0.2052	72.638	4.42E+22	5.32
		71.922082	0.2743			
		72.923463	0.0776			
		73.921181	0.3654			
		75.921405	0.0776			
As	33	74.921596	1.0000	74.922	4.65E+22	5.77
Se	34	73.922476	0.0087	78.990	3.67E+22	4.81
		75.919207	0.0902			
		76.919911	0.0758			
		77.917314	0.2352			
		79.916527	0.4982			
		81.916707	0.0919			
Br	35	78.918329	0.5069	79.904	2.36E+22	4.05
		80.916292	0.4931			
Kr	36	77.920403	0.0035	83.801	2.17E+22	3.09
		79.916380	0.0227			
		81.913482	0.1156			
		82.914131	0.1155			
		83.911503	0.5690			
		85.910616	0.1737			
Rb	37	84.911800	0.7215	85.468	1.148E+22	1.629
		86.909187	0.2785			
Sr	38	83.913430	0.0056	87.616	1.78E+22	2.58
		85.909285	0.0986			
		86.908892	0.0702			
		87.905641	0.8256			
Y	39	88.905872	1.0000	88.906	3.02E+22	4.48
Zr	40	89.904700	0.5146	91.224	4.29E+22	6.51
		90.905642	0.1123			
		91.905031	0.1711			
		93.906313	0.1740			
		95.908286	0.0280			
Nb	41	92.906382	1.0000	9.2906	5.56E+22	8.58
Mo	42	91.906810	0.1584	95.890	6.42E+22	10.22
		93.905090	0.0904			
		94.905839	0.1572			
		95.904674	0.1653			
		96.906022	0.0946			
		97.905409	0.2378			
		99.907475	0.0963			

ELE-MENT	AT. NO. (Z)	ISOTOPIC MASS (amu)	RELATIVE ABUN-DANCE	ATOMIC WEIGHT (amu)	ATOMIC DENSITY (atom/cm^3)	SPECIFIC GRAVITY
Tc	43	0.0	0.0		7.04E+22	11.50
Ru	44	95.907598	0.0551	101.046	736E+22	12.36
		97.905289	0.0187			
		98.905636	0.1272			
		99.904218	0.1262			
		100.905577	0.1707			
		101.904348	0.3161			
		103.905430	0.1858			
Rh	45	102.905511	1.0000	102.906	7.26E+22	12.42
Pd	46	101.905609	0.0096	106.441	6.80E+22	12.00
		103.904011	0.1097			
		104.905064	0.2223			
		105.903479	0.2733			
		107.903891	0.2671			
		109.905164	0.1181			
Ag	47	106.905094	0.5183	107.868	5.85E+22	10.50
		108.904756	0.4817			
Cd	48	105.906463	0.0122	112.434	4.64E+22	8.65
		107.904187	0.0088			
		109.903012	0.1239			
		110.904188	0.1275			
		111.902763	0.2407			
		112.904409	0.1226			
		113.903360	0.2886			
		115.904762	0.0758			
In	49	112.904089	0.0428	114.818	3.83E+22	7.29
		114.903871	0.9572			
Sn	50	111.904835	0.0086	118.734	3.62E+22	5.76
		113.902773	0.0066			
		114.903346	0.0035			
		115.901745	0.1430			
		116.902958	0.0761			
		117.901606	0.2403			
		118.903313	0.0858			
		119.902198	0.3285			
		121.903441	0.0472			
		123.905272	0.0594			
Sb	51	120.903816	0.5725	121.759	3.31E+22	6.69
		122.904213	0.4275			
Te	52	119.904023	0.0009	127.628	2.94E+22	6.25
		121.903066	0.0246			
		122.904277	0.0087			
		123.902842	0.0461			
		124.904418	0.0699			
		125.903322	0.1871			
		127.904476	0.3179			
		129.906238	0.3448			

ELE-MENT	AT. NO. (Z)	ISOTOPIC MASS (amu)	RELATIVE ABUN-DANCE	ATOMIC WEIGHT (amu)	ATOMIC DENSITY (atom/cm^3)	SPECIFIC GRAVITY
I	53	126.904470	1.0000	126.904	2.36E+22	4.95
Xe	54	123.906120	0.0010	131.305	1.64E+22	3.78
		125.904288	0.0009			
		127.903540	0.0192			
		128.904784	0.2644			
		129.903509	0.0408			
		130.905085	0.2118			
		131.904161	0.2689			
		133.905367	0.1044			
		135.907221	0.0887			
Cs	55	132.905355	1.0000	132.905	9.05E+21	1.997
Ba	56	129.906245	0.0010	137.327	1.60E+22	3.59
		131.905120	0.0010			
		133.904612	0.0242			
		134.905550	0.0659			
		135.904300	0.0781			
		136.905500	0.1132			
		137.905000	0.7166			
La	57	137.906910	0.0009	138.905	2.70E+22	6.17
		138.906140	0.9991			
Ce	58	135.907100	0.0019	140.101	2.91E+22	6.77
		137.905830	0.0025			
		139.905392	0.8848			
		141.909140	0.1107			
Pr	59	140.907596	1.0000	140.908	2.92E+22	6.78
Nd	60	141.907663	0.2711	144.241	2.93E+22	7.00
		142.909779	0.1217			
		143.910039	0.2385			
		144.912538	0.0830			
		145.913086	0.1722			
		147.916869	0.0573			
		149.920915	0.0562			
Pm	61	0.0	0.0			
Sm	62	143.911989	0.0309	150.363	3.030E+22	7.54
		146.914867	0.1497			
		147.914791	0.1124			
		148.917180	0.1383			
		149.917276	0.0744			
		151.919756	0.2672			
		153.922282	0.2271			
Eu	63	150.919838	0.4782	151.964	2.04E+22	5.25
		152.921242	0.5218			
Gd	64	151.919794	0.0020	157.256	3.02E+22	7.89
		153.920929	0.0215			
		154.922664	0.1473			
		155.922175	0.2047			
		156.924025	0.1568			

ELE-MENT	AT. NO. (Z)	ISOTOPIC MASS (amu)	RELATIVE ABUN-DANCE	ATOMIC WEIGHT (amu)	ATOMIC DENSITY (atom/cm³)	SPECIFIC GRAVITY
		157.924178	0.2487			
		159.927115	0.2190			
Tb	65	158.925351	1.0000	158.925	3.22E+22	8.27
Dy	66	155.923930	0.0005	162.484	3.17E+22	8.53
		157.924449	0.0009			
		159.925202	0.0229			
		160.926945	0.1888			
		161.926803	0.2553			
		162.928755	0.2497			
		163.929200	0.2818			
Ho	67	164.930421	1.0000	164.930	3.22E+22	8.80
Er	68	161.928740	0.0014	167.261	3.26E+22	9.04
		163.929287	0.0156			
		165.930307	0.3341			
		166.932060	0.2294			
		167.932383	0.2707			
		169.935560	0.1488			
Tm	69	168.934245	1.0000	168.934	3.32E+22	9.32
Yb	70	167.934160	0.0014	173.036	3.02E+22	6.97
		169.935020	0.0303			
		170.936430	0.1431			
		171.936360	0.2182			
		172.938060	0.1613			
		173.938740	0.3184			
		175.942680	0.1273			
Lu	71	174.940640	0.9741	174.967	3.39E+22	9.84
		175.942660	0.0255			
Hf	82	173.940360	0.0018	178.509	4.52E+22	13.20
		175.941570	0.0520			
		176.943400	0.1850			
		177.943880	0.2714			
		178.946030	0.1375			
		179.946820	0.3524			
Ta	73	179.947544	0.0001	180.948	5.55E+22	16.66
		180.948007	0.9999			
W	74	179.947000	0.0014	183.842	6.30E+22	19.25
		181.948301	0.2641			
		182.950324	0.1440			
		183.951025	0.3064			
		185.954440	0.2841			
Re	75	184.953059	0.3707	186.213	6.80E+22	21.03
		186.955833	0.6293			
Cs	76	183.952750	0.0002	190.333	7.14E+22	22.58
		185.953870	0.0159			
		186.955832	0.0164			
		187.956081	0.1330			

ELE-MENT	AT. NO. (Z)	ISOTOPIC MASS (amu)	RELATIVE ABUN-DANCE	ATOMIC WEIGHT (amu)	ATOMIC DENSITY (atom/cm^3)	SPECIFIC GRAVITY
		188.958300	0.1610			
		189.958630	0.2640			
		191.961450	0.4100			
Ir	77	190.960640	0.3730	192.216	7.06E+22	22.55
		192.963012	0.6270			
Pt	78	189.959950	0.0001	195.081	6.62E+22	21.47
		191.961150	0.0078			
		193.962725	0.3290			
		194.964813	0.3380			
		195.964967	0.2530			
		197.967895	0.0721			
Au	79	196.966541	1.0000	196.967	5.90E+22	19.28
Hg	80	195.965820	0.0015	200.617	4.26E+22	14.26
		197.966756	0.1002			
		198.968279	0.1684			
		199.968327	0.2313			
		200.970308	0.1322			
		201.970642	0.2980			
		203.973495	0.0685			
Tl	81	202.972353	0.2950	204.384	3.50E+22	11.87
		204.974442	0.7050			
Pb	82	203.973044	0.0148	207.177	3.30E+22	11.34
		205.974468	0.2360			
		206.975903	0.2260			
		207.976650	0.5230			
Bi	83	208.980394	1.0000	208.980	2.82E+22	9.80

TABLE OF FLUORESCENCE
YIELDS FOR K, L, AND M SHELLS[a]

Z Element	ω_K		$\overline{\omega}_L$		$\overline{\omega}_M$		Z Element	ω_K		$\overline{\omega}_L$		$\overline{\omega}_M$
4 Be	4.5	-04					45 Rh	8.07	-01			
5 B	10.1	-04					46 Pd	8.19	-01			
6 C	2.0	-03					47 Ag	8.30	-01	5.6	-02	
7 N	3.5	-03					48 Cd	8.40	-01			
8 O	5.8	-03					49 In	8.50	-01			
9 F	9.0	-03					50 Sn	8.59	-01			
10 Ne	1.34	-02					51 Sb	8.67	-01	1.2	-01	
11 Na	1.92	-02					52 Te	8.75	-01	1.2	-01	
12 Mg	2.65	-02					53 I	8.82	-01			
13 Al	3.57	-02					54 Xe	8.89	-01	1.1	-01	
14 Si	4.70	-02					55 Cs	8.95	-01	8.9	-02	
15 P	6.04	-02					56 Ba	9.01	-01	9.3	-02	
16 S	7.61	-02					57 La	9.06	-01	1.0	-01	
17 Cl	9.42	-02					58 Ce	9.11	-01	1.6	-01	
18 Ar	1.15	-01					59 Pr	9.15	-01	1.7	-01	
19 K	1.38	-01					60 Nd	9.20	-01	1.7	-01	
20 Ca	1.63	-01					61 Pm	9.24	-01			
21 Sc	1.90	-01					62 Sm	9.28	-01	1.9	-01	
22 Ti	2.19	-01					63 Eu	9.31	-01	1.7	-01	
23 V	2.50	-01	2.4	-03			64 Gd	9.34	-01	2.0	-01	
24 Cr	2.82	-01	3.0	-03			65 Tb	9.37	-01	2.0	-01	
25 Mn	3.14	-01					66 Dy	9.40	-01	1.4	-01	
26 Fe	3.47	-01					67 Ho	9.43	-01			

[a] Fluorescence yields, ω, for the elements and various subshells. The values ω_L and ω_M indicate average values for the L subshells and M subshells; the multipliers in each column are in negative powers of 10 (e.g., ω_L for Cr is 3×10^{-3}).

From R. Woldseth, *X-ray Energy Spectrometry*, Kevex Corporation, Burlingame, CA (1973).

Z Element	ω_K		$\overline{\omega}_L$		$\overline{\omega}_M$		Z Element	ω_K		$\overline{\omega}_L$		$\overline{\omega}_M$	
27 Co	3.81	-01					68 Er	9.45	-01				
28 Ni	4.14	-01					69 Tm	9.48	-01				
29 Cu	4.45	-01	5.6	-03			70 Yb	9.50	-01				
30 Zn	4.79	-01					71 Lu	9.52	-01	2.9	-01		
31 Ga	5.10	-01	6.4	-03			72 Hf	9.54	-01	2.6	-01		
32 Ge	5.40	-01					73 Ta	9.56	-01	2.3	-01		
33 As	5.67	-01					74 W	9.57	-01	3.0	-01		
34 Se	5.96	-01					75 Re	9.59	-01				
35 Br	6.22	-01					76 Os	9.61	-01	3.5	-01		
36 Kr	6.46	-01	1.0	-02			77 Ir	9.62	-01	3.0	-01		
37 Rb	6.69	-01	1.0	-02			78 Pt	9.63	-01	3.3	-01		
38 Sr	6.91	-01					79 Au	9.64	-01	3.9	-01		
39 Y	7.11	-01	3.2	-02			80 Hg	9.66	-01	3.9	-01		
40 Zr	7.30	-01					81 Tl			4.6	-01		
41 Nb	7.48	-01					82 Pb	9.68	-01	3.8	-01	2.9	-02
42 Mo	7.64	-01	6.7	-02			83 Bi			4.1	-01	3.6	-02
43 Tc	7.79	-01					92 U	9.76	-01	5.2	-01	6	-02
44 Ru	7.93	-01											

10

PHYSICAL CONSTANTS, CONVERSIONS, AND USEFUL COMBINATIONS

Physical Constants

Avogadro constant $N_A = 6.022 \times 10^{23}$ particles/mole

Boltzmann constant $k = 8.617 \times 10^{-5}$ eV/K

Elementary charge $e = 1.602 \times 10^{-19}$ Coulombs

Planck constant $h = 6.626 \times 10^{-34}$ J s

$h = 4.136 \times 10^{-15}$ eV s

$\hbar = 6.582 \times 10^{-16}$ eV s

Speed of light $c = 2.998 \times 10^{10}$ cm/s

Useful Combinations

Electron rest mass $mc^2 = 0.511$ MeV

Bohr radius $a_0 = \hbar^2/me^2 = 0.529$ Å

Bohr velocity $v_0 = e^2/\hbar = 2.188 \times 10^8$ cm/s

Fine structure constant $\alpha = e^2/\hbar c = 7.297 \times 10^{-3} = 1/137$

Hydrogen binding energy $e^2/2a_0 = 13.606$ eV

Classical electron radius $r_e = e^2/mc^2 = 2.818 \times 10^{-13}$ cm

Electron charge $e^2 = 14.395$ eV Å

Electron Compton wavelength $\lambdabar = \hbar/mc = 3.861 \times 10^{-11}$ cm

Photon energy-wavelength $hc = 12398.5$ eV Å

Conversions

1 Å $= 10^{-8}$ cm $= 10^{-1}$ nm

1 eV $= 1.602 \times 10^{-19}$ J

1 eV/particle $= 23.06$ kcal/mol

m_p (proton mass) $= 1.6726 \times 10^{-27}$ kg

m (electron mass) $= 9.1095 \times 10^{-31}$ kg

$m_p/m = 1836.1$

343

ACRONYMS

AEM	Analytical Electron Microscopy	PES	Photoelectron Spectroscopy
AES	Auger Electron Spectroscopy	PIXE	Proton-Induced X-ray Emission
EDS	Energy Dispersive Spectroscopy	PRA	Prompt Reaction Analysis
EELS	Electron Energy-loss Spectroscopy	RBS	Rutherford Backscattering Spectrometry
EMA	Electron Microprobe Analysis	RHEED	Reflection High Energy Electron Diffraction
ESCA	Electron Spectroscopy for Chemical Analysis	SEM	Scanning Electron Microscopy
EXAFS	Extended X-ray Absorption Fine Structure	SEXAFS	Surface Extended X-ray Absorption Fine Structure
FRS	Forward Recoil Spectroscopy	SIMS	Secondary Ion Mass Spectroscopy
HEED	High Energy Electron Diffraction	SNMS	Secondary Neutral Mass Spectroscopy
HEIS	High Energy Ion Scattering	TEM	Transmission Electron Microscopy
IIXS	Ion-Induced X-ray Spectroscopy	UPS	Ultraviolet Photoelectron Spectroscopy
IMMA	Ion Microprobe Mass Analysis	WDS	Wavelength Dispersive Spectroscopy
LEED	Low Energy Electron Diffraction	XES	X-ray Energy Spectroscopy
LEIS	Low Energy Ion Scattering	XPS	X-ray Photoelectron Spectroscopy
NAA	Neutron Activation Analysis	XRD	X-ray Diffraction
NRA	Nuclear Reaction Analysis	XRF	X-ray Fluorescence

INDEX

345

SURFACE ANALYSIS TECHNIQUES

Method	Particle Measured	Process	Information	Monolayer Sensitivity	Effective Probing Depth	Depth Profiling
Auger Emission Spectroscopy (A.E.S.)	Auger Electrons	e → e	—Elements Li-U	10^{-1}	20 Å	In Combination with Sputtering
Scanning Electron Microscopy-Energy Dispersive X-Ray (SEM-EDX)	X-Rays	e → x	—Elements Na-U	None	10000 Å	None
Transmission Electron Microscopy (TEM)	Diffracted Electrons	e → e e e e	—Compound Identification	None	500 Å	None
X-Ray Photo Electron Spectroscopy (XPS or ESCA)	Photo Electrons	x → e	—Elements Li-U, —Chemical Bonding	10^{-1}	30 Å	None
Secondary Ion Mass Spectrometry (SIMS)	Sputtered Ions	i → i	—Elements H-U	10^{-5}	10 Å	Via Sputtering
Rutherford Backscattering Spectrometry (RBS)	Scattered Ions (H^+, He^+)	He^+ → He^+	—Elements Be-U	$10 \cdot 10^{-4}$	100 Å	Via Energy Loss
Particle Induced X-Ray Emission (PIXE)	X-Rays	i → x	—Elements Na-U	None	1000 Å - 10000 Å	None
Nuclear Reaction Analysis (NRA)	Light Ion Reaction Product	i → i	—Light Elements (H, Be, B, C O, F...)	10^{-1}	100 Å - 10000 Å	Via Energy Loss

PERIODIC TABLE OF THE ELEMENTS

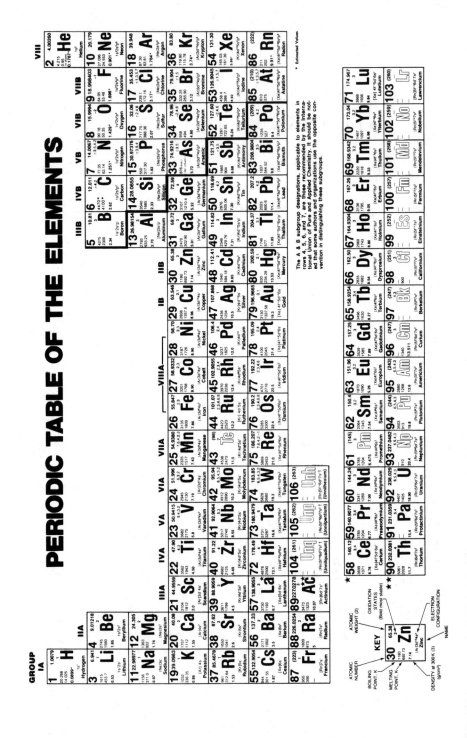